Neutrinos play a decisive part in nuclear and elementary particle physics, as well as in astrophysics and cosmology. Because they interact so weakly with matter, some of their most basic properties, such as their mass and charge conjugation symmetry, are largely unknown. This book, while describing all aspects of neutrino physics, focuses on what we know and may hope to know about the mass of the neutrino and its particle-antiparticle symmetry. Topics include neutrino mixing, neutrino decay, neutrino oscillations, double beta decay, solar neutrinos, supernova neutrinos and related issues. The authors stress the physical concepts, and discuss both theoretical and experimental techniques.

This second edition is completely up to date and differs from the first in that it contains an expanded coverage of new experimental results and recent theoretical advances. Since publication of the first edition, many issues that were at that time unresolved, such as tritium beta decay and reactor neutrino oscillations, have been clarified and these are discussed in the new edition. Also included is an expanded coverage of solar and supernova neutrinos.

This book deals with one of the most intriguing issues in modern physics, and will be of value to researchers, graduate students and advanced undergraduates specializing in experimental and theoretical particle physics and nuclear physics.

Physics of Massive Neutrinos

Physics of Massive Neutrinos

Felix Boehm
California Institute of Technology

Petr Vogel
California Institute of Technology

Second edition

Published by the Press Syndicate of the University of Cambridge
The Pitt Building, Trumpington Street, Cambridge CB2 1RP
40 West 20th Street, New York, NY 10011-4211, USA
10 Stamford Road, Oakleigh, Victoria 3166, Australia

First published 1987
Second edition 1992

A catalogue record of this book is available from the British Library

Library of Congress cataloguing in publication data available

ISBN 0 521 41824 0 hardback
ISBN 0 521 42849 1 paperback

Transferred to digital printing 2003

Contents

Preface to the First Edition

In preparing this book, we have set ourselves the goal of presenting a unified picture of the physics of neutrinos as it now emerges from studies of nuclear physics, particle physics, astrophysics, and cosmology. While describing all aspects of neutrino physics, we focus strongly on what we know and hope to learn about the neutrino's mass and particle-antiparticle symmetry, as we consider these to be among the most burning questions.

As our readers we have in mind students in nuclear and particle physics at graduate level as well as researchers in these fields. The book thus may serve as a text in a specialized course, or as a supplement to a standard text in nuclear or particle physics. We have written each chapter sufficiently selfcontained that it can be read independently. To keep the length of the text within limits, we have provided each chapter with references to specialized review articles. Also, for the same reason, we have elected only to touch upon several topics that are not directly related to the problem of neutrino mass and mixing, such as neutral current interactions of neutrinos, electron neutrino scattering, and deep inelastic neutrino nucleon scattering.

Throughout, we have attempted to stress physical concepts, treating theoretical developments and experimental techniques on an equal footing, while including material up to Fall 1986. The prospect of finding massive neutrinos and its repercussions on our fundamental understanding of particles is exciting, as is the ingenuity of the experimental approaches in exploring neutrino mass.

We are grateful to Peter Rosen and Garry Steigman for reading the manuscript and for offering useful suggestions.

Pasadena, January 1987

Felix Boehm
Petr Vogel

Preface to the Second Edition

In this second edition we have followed the framework of the original edition, changing or expanding the text only when warranted by new developments. The field of neutrino physics itself has been expanding at a rapid pace (as witnessed by our bibliography which has increased by 50% in just four years). Among the notable advances in the intervening years let us mention the opening of neutrino astronomy with the observation of the neutrino signal of SN1987A. The next supernova in, or near, our galaxy, is eagerly awaited by a number of "supernova watch" detectors. The detection of solar neutrinos has become an important subject of neutrino physics. Four detectors are running now, and several more are being built. The suppression of the solar neutrino flux (at least at higher energies) has been confirmed, and many physicists believe that solar neutrinos will represent the first observational manifestation of the neutrino mass and mixing. Double beta decay is another topic that has expanded considerably. The two-neutrino decay has been seen in the laboratory in three nuclei, thus it has become well established. The lifetime limits and the corresponding neutrino mass limits from the neutrinoless decay are continually improving. Better understanding of the nuclear matrix elements also puts double beta decay on a firmer footing.

In the broader context of particle physics as a whole the standard electroweak model appears to describe all observable phenomena. Yet, despite its success, it must eventually be replaced by a broader theory, with fewer parameters and deeper insight. The prospect of finding massive neutrinos remains one of the main possible openings to the world "beyond the standard model".

Pasadena, September 1991

Felix Boehm
Petr Vogel

1

Neutrino Properties

Neutrino physics is playing a unique role in elucidating the properties of weak interactions. But even more important, neutrino studies are capable of providing new ingredients for future theoretical descriptions of the physics of elementary particles. Although impressive progress has been made in our understanding of neutrino interactions, many important issues remain to be settled, foremost among these are the questions: have neutrinos a nonvanishing rest mass, and are neutrinos, participating in weak interaction, pure states in a quantum mechanical sense? The role of neutrino mass and its various consequences, therefore, will form the central issue throughout this text.

The study of neutrino properties, unlike the study of other elementary particles, has always crossed the traditional disciplinary boundaries. While we think of particle properties as being the domain of high energy physics, much information on neutrino properties has come from low energy nuclear physics, as well as from astrophysics and cosmology. For that reason it will be necessary to explore a number of different disciplines and techniques.

Among the many excellent reviews on weak interaction, we mention the books by Bailin (82), Commins & Bucksbaum (83), Georgi (84), Okun (82), Pietschmann (83), and Taylor (78). Since the first edition of this book several related books have appeared, Grotz & Klapdor (90), Holstein (89), and Kayser, Gibrat-Debu & Perrier (89). More specialized reviews on various aspects of neutrino physics are mentioned at the beginning of each Chapter.

This Chapter provides an introduction to the formal description of neutrinos and their properties. To begin, we briefly recapitulate in Section 1.1 the major milestones in the discovery of neutrinos, and outline the goal and purpose of this book. We review the fundamental properties of neutrinos with regards to charge conjugation in Section 1.2, elaborating on the important distinction between Dirac and Majorana neutrinos. In Section 1.3 we discuss the neutrino mass matrix and show how diagonalization of this matrix leads to the concept of neutrino oscillations. Finally, we explain in Section 1.4 how various fundamental particle theories may lead to finite neutrino mass.

1.1 Introduction

The history of neutrino physics began with Wolfgang Pauli's often quoted letter to the Physical Society of Tubingen (Pauli 30), in which he postulates the existence of a new particle, the neutrino, in order to explain the observed continuous electron spectrum accompanying nuclear beta decay. If, as it was then believed, the electron decay were a two-body decay, the laws of energy and momentum conservation would have predicted a monochromatic electron peak instead. Pauli required his hypothetical particle to be neutral and have spin 1/2, to ensure conservation of electric charge and angular momentum. Its rest mass was expected to be small, but not necessarily vanishing. Learning of Pauli's idea, Fermi (34) proposed his famous theory of beta decay, based on which Bethe & Peierls (34) predicted the cross section for the interaction of the neutrino with matter to be extremely small. The first experimental evidence of a neutrino induced interaction was brought by Reines & Cowan (53). In addition to the electron neutrino emitted in nuclear beta decay, two other neutrinos were discovered, the muon neutrino, demonstratively distinct from the electron neutrino (Danby et al. 62), and the the tau neutrino (Perl et al. 75), even though the latter's existence has only been inferred so far. The concept that these neutrinos could mix was proposed by Maki et al. (62) and by Pontecorvo (67). Earlier, Pontecorvo (58) discussed the possibility of neutrino-antineutrino oscillations.

As more and more experimental evidence became available, our understanding of weak interaction (and neutrino physics) was greatly revolutionized by the concept of parity-nonconservation (Lee & Yang 56). Experimental and theoretical developments led the way to modern weak interaction theories (Glashow 61; Salam 68; Weinberg 67) which became part of the more encompassing framework of the "standard model". This model is capable of describing all the known physics of weak and electromagnetic interactions, incorporating all the experimental results at energies available at present accelerators. The observation of the neutral weak current (Hasert et al. 73) and the discovery of the intermediate vector bosons W and Z (Arnison et al. 83; Bagnaia et al. 83) contributed spectacularly to the success of the model.

Yet, despite its success, the standard model appears in need of extension and generalization. In its present form it is not capable of predicting the masses of the fermions, nor can it explain why there are several fermion families (electron, muon, tau, and their neutrinos, and the analogous quark families). The study of neutrino properties is one of the few avenues which could lead to new physics beyond the standard model, and this is the chief reason why the neutrino is such an interesting particle. Earlier reports, indicating that neutrinos may have nonvanishing mass (see, e.g., Lubimov 86), although rejected by more recent work, have stimulated these endeavors.

One of the most important aspects of neutrino physics, therefore, has to do with its possible rest mass. We have no idea why neutrinos are so much lighter than the charged leptons, and we may only speculate that this difference may be a reflection of some fundamental symmetry. One possibility is that neutrinos, unlike any other fermions, are Majorana particles, that is they cannot be distinguished from their antiparticles.

If the neutrino is a massive Majorana particle, then neutrinoless double beta decay can take place. A search for this process is one of the most challenging tasks today. Also, if the neutrino has mass, then the question arises, is it a pure eigenstate of weak interaction, as described in the standard model, or is it a superposition of other neutrino states, the eigenstates of the mass matrix, as suggested in theoretical models of grand unification? If the neutrino is a mixed state in this sense, then a number of phenomena may take place, such as neutrino oscillations and neutrino decay.

Another aspect of neutrino physics has to do with the problem of lepton families (electrons, muons, and taus) and the related concept of lepton number conservation. With the help of neutrinos it is possible to test the lepton conservation laws to great accuracy.

These fundamental issues briefly sketched here can be addressed by experiment, and it is our aim to describe in this book both experiment and theory on an equal basis. From a theoretical point of view, the crucial issue is the question of whether the world of physics can be described by the standard model of electroweak interaction. If that were the case, the neutrino is predicted to be massless. The experiments on neutrino oscillations, neutrinoless double beta decay, neutrino decay, or searches for heavy neutrinos should give a null result, as they presently do at the limit of today's sensitivity. (In a few cases possible indications of effects associated with the neutrino mass have been reported, see, e.g., Section 4.1) If in the future one or several of the mentioned processes are unequivocally found, it would point the way to some more general theory.

1.2 Dirac and Majorana Neutrinos

It was Dirac's equation that first led to the concept of particles and antiparticles, the positive electron being the earliest candidate for an antiparticle. While positive electrons are clearly distinct from negative electrons by their electromagnetic properties, it is not obvious in what way neutral particles should differ from their antiparticles. The neutral pion, for example, was found to be identical to its antiparticle. The neutral kaon, on the other hand, is clearly different from its antiparticle. The pion and kaon, both bosons, are not truly elementary particles, however, as they are composed of two charged fermions, the quarks and the antiquarks.

The concept of a particle which is identical to its antiparticle was formally introduced by Majorana as early as 1937 (Majorana 37). Thus, we refer to such a particle as a Majorana particle. In contrast, a Dirac particle is one which is distinct from its antiparticle. In this Section we shall describe the properties of Majorana particles in general, and neutrinos in particular, and how they can be distinguished from Dirac particles.

The difference between a Majorana and Dirac particle has to do with its transformation properties under charge conjugation. However, as far as we know, neutrinos interact only weakly, and weak interactions are not invariant with respect to charge conjugation. Consequently, an interacting Majorana neutrino cannot be an eigenstate of charge conjugation C alone. (If a Majorana neutrino had a definite value of C at one instant, it would no longer have this value at a later time, because weak interaction would mix in other eigenstates of C.) Therefore, we have to generalize the definition of a Majorana particle and also consider transformation properties with respect to other discrete symmetries, such as parity P and time reversal T (or their combinations CP and CPT). Moreover, weak interactions involve neutrino states of a selected chirality (left-handed or right-handed) and thus the formalism should be based on the chiral projections of the corresponding states.

To visualize the difference between Majorana and Dirac neutrinos, let us follow the arguments of Kayser (85) and assume the existence of a massive left-handed neutrino ν_L (see Figure 1.1). (We shall show later that there is an intimate relation between helicity, i.e., handedness, and chirality, i.e., the transformation property with respect to the application of γ_5.) By virtue of CPT invariance which we assume to be valid, the existence of ν_L implies the existence of the CPT mirror image, a right-handed antineutrino $\bar{\nu}_R$. As a massive ν_L travels slower than light, a frame of reference moving faster than ν_L exists. In this frame the neutrino is going the other way, but its spin is unchanged. Thus, the Lorentz transformation to the faster moving frame of reference turns ν_L into a right-handed ν_R, shown at the extreme left of Figure 1.1(a). Now, this ν_R may or may not be the same particle as the *CPT* mirror image $\bar{\nu}_R$ of ν_L. If it is *not* the same, as indicated in Figure 1.1(a), ν_R has its own *CPT* mirror image $\bar{\nu}_L$. Altogether, there are four states with the same mass. This quadruplet of states is called a Dirac neutrino ν^D. It has distinct particle and antiparticle states, may have a magnetic dipole moment, and, if *CP* is not conserved, even an electric dipole moment. Thus, for a Dirac neutrino, ν_L can be converted into the opposite helicity state both by a Lorentz transformation and by the torque exerted by an external $\vec{B}$ or $\vec{E}$ field.

If, on the other hand, the right-handed particle obtained by the Lorentz transformation to a moving reference frame is the *same* particle as the *CPT* image of the original ν_L (Figure 1.1(b)), there are only two states with a common mass. This pair of states represents a Majorana neutrino ν^M. It is easy to see that the dipole magnetic and electric moments of ν^M must vanish. Indeed, at rest in an external static field the neutrino would have the interaction energy $-\mu<\vec{s}\cdot\vec{B}>-d<\vec{s}\cdot\vec{E}>$, where $\vec{s}$ is the neutrino spin operator and μ and d are the corresponding dipole moments. Under *CPT* the fields $\vec{E}$ and $\vec{B}$ are unchanged, while the spin vector of a Majorana neutrino reverses its direction. Therefore, under *CPT* the dipole interaction

Figure 1.1. (a) The four distinct states of a Dirac neutrino ν^D. (b) The two distinct states of a Majorana neutrino ν^M.

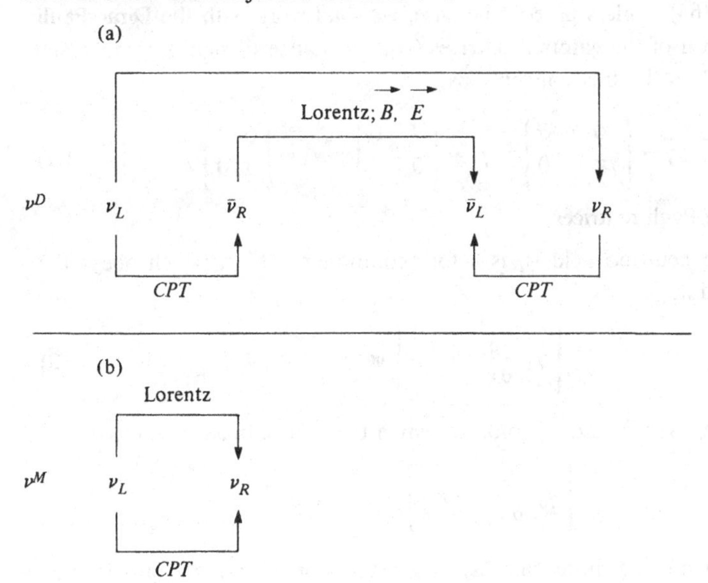

energy changes sign. Consequently, if *CPT* invariance holds, the dipole moments μ and d must vanish.

It should be stressed that the existence of the neutrino rest mass was essential for the above discussion of the differences between Dirac and Majorana neutrinos. A massless neutrino travels with the speed of light and we can no longer reverse its helicity by going to the faster moving Lorentz frame. Moreover, let us assume that the weak interaction involves only left-handed currents. (This agrees with experimental evidence, interactions involving right-handed currents have not been detected so far.) For neutrinos with left-handed interaction, the dipole moment is proportional to mass and vanishes for a massless particle. Thus, the helicity of a massless neutrino cannot be reversed by an external $\vec{E}$ or $\vec{B}$ field, either. Indeed, in the massless case the states $|\nu_L^D\rangle$ and $|\bar{\nu}_R^D\rangle$ are completely disconnected from the remaining two states $|\nu_R^D\rangle$ and $|\bar{\nu}_L^D\rangle$; the latter states need not even exist. As there are only two relevant states, the Dirac-Majorana distinction has vanished. Actually, the disappearance of the distinction between the Dirac and Majorana neutrinos is not abrupt. As the mass gets smaller (or as the mass/energy gets smaller), the ability to decide whether the observed neutrino states are the two spin states of a Majorana neutrino or half of the four states of a Dirac neutrino, gradually vanishes.

1.2.1 *Definitions and effect of discrete symmetries*

Throughout this Chapter we shall use the conventions and notation of Sakurai (64). Unless noted otherwise, we shall work with the Dirac-Pauli representation of the gamma matrices (with a change of sign of γ_5 to make it compatible with other conventions):

$$\vec{\gamma} = \begin{pmatrix} 0 & -i\vec{\sigma} \\ i\vec{\sigma} & 0 \end{pmatrix} ; \; \gamma_4 = \begin{pmatrix} I & 0 \\ 0 & -I \end{pmatrix} ; \; \gamma_5 = \begin{pmatrix} 0 & I \\ I & 0 \end{pmatrix} , \tag{1.1}$$

where σ are Pauli matrices.

A free neutrino field ψ_i is a four-component object which obeys the Dirac equation

$$\left(\gamma_\mu \frac{\partial}{\partial x_\mu} + m_i \right) \psi_i = 0 . \tag{1.2}$$

The field $\gamma_5\psi_i$ satisfies an equation in which the sign of mass is reversed,

$$\left(\gamma_\mu \frac{\partial}{\partial x_\mu} - m_i \right) \gamma_5\psi_i = 0 . \tag{1.3}$$

The fields with a definite chirality (eigenstates of γ_5) are the projections,

$\frac{(1 \pm \gamma_5)}{2}\psi$. It is obvious from (1.2) and (1.3) that a free massive particle cannot have a definite chirality at all times. On the other hand, in the ultrarelativistic limit $E/m \to \infty$ (or for massless particles) the chirality becomes a good quantum number. There is an intimate relation between the chirality and helicity (i.e., spin projection on the direction of motion). The helicity operator is $\vec{\Sigma}\cdot\hat{p}$, where $\vec{\Sigma} = -i\gamma_4\gamma_5\vec{\gamma}$. Then $\gamma_5 \to \vec{\Sigma}\cdot\hat{p}$ in the ultrarelativistic limit for positive energy states. We shall follow the often used, but not very precise custom, and refer to the corresponding chiral projections $\frac{(1 \pm \gamma_5)}{2}\psi$ as right- and left-handed, respectively.

The charge conjugate field ψ_i^c is defined as

$$\psi_i^c = \eta_C C \bar{\psi}^{\,T} , \tag{1.4}$$

where η_C is a phase factor, C is a 4×4 matrix, (the symbol C is customarily used for two meanings, as the symbol of the charge conjugation operation and as the 4×4 matrix) and the superscript T signifies a transposed matrix (as usual $\bar{\psi} = \psi^+\gamma_4$). In the Dirac-Pauli representation $C = \gamma_4\gamma_2$ so that $\psi^c = -\eta_C\gamma_2\psi^*$. Note that the following relations are independent of representation:

$$C^{-1}\gamma_\mu C = -\gamma_\mu^T ; \quad C^{-1}\gamma_5 C = \gamma_5^T .$$

If ψ is a chirality eigenstate ($\gamma_5\psi = \lambda\psi$), then ψ^c is also a chirality eigenstate with eigenvalue $\lambda^c = -\lambda$.

Next, we shall consider the effect of the parity transformation ($\vec{x}' = -\vec{x}$, $x_4' = x_4$) on the field ψ. Requiring that the Dirac equation in both systems describes the same situation leads to the rule

$$\psi(\vec{x}) \to \eta_P\gamma_4\psi(-\vec{x}) , \tag{1.5}$$

where η_P is a phase factor. The charge conjugate field ψ^c transforms under the parity transformation as

$$\psi^c \to \eta_C\eta_P^* C((\gamma_4\psi)^+\gamma_4)^T = -\eta_P^*\gamma_4\psi^c . \tag{1.6}$$

Thus, for real η_P, we obtain the well-known result that the intrinsic parity of the antifermion is opposite to that of a fermion. (This can be experimentally verified, for example, by a study of the positronium decay.) If, however, ψ should describe a Majorana particle (charge conjugation eigenstate), one can satisfy Eq. (1.6) only with pure imaginary η_P. Hence we come to important conclusion: *Majorana particles have imaginary intrinsic parity.*

The Majorana field can be defined as

$$\chi(x) = \frac{1}{\sqrt{2}} [\psi(x) + \eta_C\psi^c(x)] . \tag{1.7}$$

By the appropriate choice of phase (if $\eta_C = e^{2i\Phi}\lambda_C$, where $\lambda_C = \pm 1$, choosing $\chi' = e^{-i\Phi}\chi$) we obtain a field which is an eigenstate of charge conjugation with the eigenvalue $\lambda_C = \pm 1$. The Majorana field (1.7) has the CP phase $\pm i$; fields with $+i$ will be called even CP states, and those with $-i$ will be called odd CP states. (Until now we have considered only free fields. Once interactions, for example weak interactions of the neutrinos, are included, the requirement of C invariance alone is insufficient, as we mentioned above. Thus, we have to require that the Majorana field is an eigenstate of CPT, or at least of CP. For our purpose, however, it is sufficient to use the definition (1.7).) If the Majorana state χ, or state $\chi' = e^{-i\Phi}\chi$, has the charge conjugation eigenvalue λ_C, the state $\gamma_5\chi'$ has the opposite eigenvalue $-\lambda_C$.

As mentioned, the chiral projections, $\frac{(1 \pm \gamma_5)}{2}\psi$, do not obey the Dirac equation unless $m = 0$. We use the following notation for the chiral projections of the fields ψ and the charge conjugate field ψ^c

$$\psi_L \equiv \frac{(1-\gamma_5)}{2}\psi \; ; \; (\psi_L)^c = \frac{(1+\gamma_5)}{2}\psi^c = (\psi^c)_R \, , \tag{1.8a}$$

$$\psi_R \equiv \frac{(1+\gamma_5)}{2}\psi \; ; \; (\psi_R)^c = \frac{(1-\gamma_5)}{2}\psi^c = (\psi^c)_L \, . \tag{1.8b}$$

It is obvious that the charge conjugation eigenstates (1.7) cannot simultaneously be eigenstates of chirality.

1.2.2 Mass term for a single field

In the field theory of neutrinos the mass is determined by the mass term of the neutrino Lagrangian. This mass term must be Lorentz invariant and hermitian. This requirement restricts the possible mass terms to two groups: $\bar{\psi}\psi$ and $\bar{\psi^c}\psi^c$, as well as $\bar{\psi}\psi^c$ and its hermitian conjugate $\bar{\psi^c}\psi$. The Lagrangian of a single Dirac field has the form

$$\mathrm{L}_D = -\frac{1}{2}\int d^4x \, (\bar{\psi}\gamma_\mu\partial_\mu\psi + \bar{\psi} m_D \psi) \, , \tag{1.9}$$

and we see that it contains the mass term of the first kind.* (The parameter m_D is real to preserve hermicity of the Lagrangian.) This Dirac mass term is invariant under the global phase transformation

$$\psi \rightarrow e^{i\alpha}\psi \; ; \; \psi^c \rightarrow e^{-i\alpha}\psi^c \, .$$

The second possible Lorentz invariant mass term containing $\bar{\psi}\psi^c$ and $\bar{\psi^c}\psi$ is

* The first part of (1.9) is the kinetic energy term. We shall not discuss it further, but one has to make sure that any transformation of the mass part leaves the kinetic energy part invariant.

not invariant under this phase transformation. Thus, the Dirac mass term can be associated with a conserved quantum number (usually called the "lepton number") while the second Lorentz invariant mass term violates conservation of the lepton number by two units. It will become apparent shortly why this second mass term is known as the Majorana mass term.

Let us consider the most general mass Lagrangian

$$-2\mathrm{L}_M = \frac{1}{2}\left(\bar{\psi} m_D \psi + \bar{\psi}^c m_D \psi^c + \bar{\psi} m_M \psi^c + \bar{\psi}^c m_M^* \psi \right). \tag{1.10}$$

This Lagrangian is hermitian and Lorentz invariant; it depends on three real parameters m_D, m_1, m_2 ($m_M = m_1 + im_2$). It is instructive to rewrite this Lagrangian in the matrix form

$$-2\mathrm{L}_M = \frac{1}{2}\begin{pmatrix}\bar{\psi}, \bar{\psi}^c\end{pmatrix}\begin{pmatrix} m_D & m_M \\ m_M^* & m_D \end{pmatrix}\begin{pmatrix}\psi \\ \psi^c\end{pmatrix}. \tag{1.10a}$$

To find fields of a definite mass, we have to diagonalize (1.10a). When this is done, we find that the mass term (1.10a) has two real eigenvalues $m_D \pm |m_M|$, and the corresponding eigenvectors are

$$\begin{pmatrix}\varphi_+ \\ \varphi_-\end{pmatrix} = \frac{1}{\sqrt{2}}\begin{pmatrix} e^{-i\theta}\psi + e^{i\theta}\psi^c \\ -e^{-i\theta}\psi + e^{i\theta}\psi^c \end{pmatrix}, \tag{1.11}$$

where $\tan 2\theta = m_2/m_1$. Both eigenvectors are Majorana fields, i.e., they are eigenstates of charge conjugation with opposite eigenvalues. (If m_M=0 in (1.10), the mass term is already diagonal with a single eigenvalue m_D; the eigenvectors ψ and ψ^c in that case are not charge conjugation eigenstates and describe a Dirac particle.)

There is a difficulty associated with the diagonalization of (1.10). This mass term, being hermitian, has real eigenvalues. But these eigenvalues are not always positive, while it is customary to consider the particle rest mass a positive quantity. As long as $m_D > |m_M|$, this does not represent a problem, because both mass eigenvalues are positive. If, on the other hand, $|m_M| > m_D$, one of the eigenvalues is negative. When that happens, we can use the γ_5 trick (see (1.3)) and change the sign of m by using the field $\gamma_5\varphi_-$ instead of φ_-. In that case, however, both solutions have a positive C eigenvalue. In the special case of pure Majorana mass (m_D=0), the two mass eigenvalues are degenerate and we can use any combination of φ_+ and $\gamma_5\varphi_-$ as the appropriate eigenvector.

If, instead, we work with chiral projections of the field ψ, we can use the fact that $\bar{\psi}_L\varphi_L = \bar{\psi}_R\varphi_R = 0$ for any fields ψ and φ (since $(1+\gamma_5)(1-\gamma_5)=0$). Thus, the Dirac mass term, expressed in terms of chiral projections is either $\bar{\psi}_L\psi_R$ or $\bar{\psi}_R\psi_L$, while the Majorana mass term has the form $\bar{\psi}_L(\psi^c)_R$, etc.

The mass terms of both kinds "violate chirality" because, as we have remarked already, chirality is not a conserved quantity for massive particles.

When the mass term is expressed in terms of the chiral projections, the number of parameters, and the dimension of the mass matrix, doubles. The most general hermitian mass term, written in a form analogous to (1.10a) is now

$$-2L_M = \frac{1}{2}\left[\bar{\psi}_R, (\bar{\psi}^c)_R, \bar{\psi}_L, (\bar{\psi}^c)_L\right] \begin{bmatrix} 0 & M \\ M^+ & 0 \end{bmatrix} \begin{bmatrix} \psi_R \\ (\psi^c)_R \\ \psi_L \\ (\psi^c)_L \end{bmatrix} , \tag{1.12}$$

where the 2×2 submatrix M depends on three complex numbers, and can be expanded in terms of the Pauli matrices

$$M \equiv m_D I + m_1\sigma_x + m_2\sigma_y \, . \tag{1.13}$$

The mass term (1.12) is automatically *CPT* invariant. If, further, *CP* invariance is required, the parameters m_D and m_1 must be real and m_2 must be pure imaginary.

Diagonalization of the mass term (1.12) is relatively simple in the *CP* invariant situation. (The treatment of the general *CP* noninvariant case is discussed by Rosen (83); throughout this Section we use many results from that reference.) The eigenvalue equation is of the form

$$\lambda_\pm = \frac{1}{2}\left\{(m_R + m_L) \pm [(m_R - m_L)^2 + 4m_D^2]^{\frac{1}{2}}\right\} , \tag{1.14}$$

where

$$m_R = m_1 + |m_2| \, , \quad m_L = m_1 - |m_2|$$

for $m_2 = +i|m_2|$. (For $m_2 = -i|m_2|$ we must interchange m_L and m_R.) The eigenvectors, as in the nonchiral case, are C or *CP* eigenstates (but, naturally, not chirality eigenstates). Again, we may encounter negative mass eigenvalues, which could be removed by the "γ_5 trick". Eq. (1.12) is formally four-dimensional; in reality, eigenvalues and eigenvectors appear in pairs and only two are independent.

1.2.3 Concluding remarks

We have shown that the general mass term (1.10) or (1.12) has eigenvectors which are charge conjugation eigenstates, i.e., they describe Majorana particles. Typically, these states are also *CP* eigenstates. The eigenvalues of the general mass term are not necessarily positive. We can restore the required positive value of the mass eigenvalue by redefining the corresponding eigenvector, i.e., by using a γ_5 multiple of it. That, as we have seen, means change of sign of the C or CP eigenvalue. In Chapter 6 we shall see how this feature of the theory leads to possible cancellations in the neutrinoless double beta decay.

For a single field, we end up with two mass eigenvalues and correspondingly two eigenstates, whether we use chiral projections or not. In the case of Dirac particles, we had just one mass value m_D and two states which were not C or CP eigenstates, the Dirac particle and its charge conjugate. The standard Dirac particle is described by the more general theory as well. We are free to describe a Dirac particle as a linear combination of two Majorana particles with opposite C eigenvalues. The choice of description will depend upon the coupling of the field to other fields in the interactions in which it takes part.

Apart from exploring neutrinoless double beta decay, one could, in principle, distinguish the Majorana and Dirac neutrinos by their electromagnetic properties. Neutrinos are, naturally, electrically neutral. Dirac neutrinos, however, can have a nonvanishing magnetic dipole moment, and perhaps even an electric dipole moment. Majorana neutrinos, on the other hand, must have vanishing dipole moments. This distinction may not have much practical relevance, because, for example, the standard Weinberg-Salam-Glashow theory predicts the magnetic moment of a Dirac neutrino of mass m_ν to be (Marciano & Sanda 77)

$$\mu = \frac{3eG_F}{8\pi^2\sqrt{2}} m_\nu \approx 3\times10^{-19} \frac{m_\nu}{1 \text{ eV}} \mu_{Bohr} \,,$$

a very small quantity for light neutrinos. (Neutrino magnetic moments will be further discussed in Chapter 5 in the context of the solar neutrino problem.) This is a general feature of most experiments designed to distinguish Dirac and Majorana neutrinos: as the mass to energy ratio $m_\nu/E \to 0$, the ability to tell whether the particle is Dirac or Majorana gradually vanishes (Kayser 85). Other possible ways to separate Dirac and Majorana neutrinos depend on the phenomenon of neutrino decay, again implying neutrino mass. We shall mention some of these features in Chapter 4. Finally, if unusual weak interaction couplings are present, such as tensor coupling, it will be possible to observe the difference between Majorana and Dirac neutrinos by measuring the neutrino-electron scattering cross section (Rosen 82; Dass 85). However, reliance on exotic weak couplings makes this approach less general.

1.3 General Neutrino Mass Term

In this Section we shall discuss a more general case of n neutrino flavors. (We know that there exist at least three flavors, electron neutrino, muon neutrino, and tau neutrino.) From the results of the previous Section we expect that the corresponding mass term will be a $2n \times 2n$ matrix, and its eigenstates will be $2n$ Majorana neutrinos. We further expect that the most general theory will violate the conservation of the total lepton number, as well as (or in addition to) the violation of the conservation of individual flavor lepton numbers.

In the previous Subsection we actually did not use the fact that neutrinos participate in weak interactions, although, in anticipation, we formulated the theory in terms of chiral projections of the neutrino field. Now, we must remedy this, because the very concept of flavor is related to the weak interaction. In the standard theory of weak interactions, only the left-handed currents participate through their coupling to the corresponding intermediate vector boson W_L, i.e.,

$$-\mathbf{L}_W = \frac{g_L}{\sqrt{2}} J_L^\mu W_{L\mu}^+ + h.c. \tag{1.15}$$

is the corresponding part of the charged current weak interaction Lagrangian. (We shall consider the generalization of the charged current weak interaction Lagrangian (1.15) later, in connection with neutrino oscillations in Chapter 5, and with neutrinoless double beta decay in Chapter 6.) The left-handed lepton current J_L is of the form

$$J_L^\mu = \sum_l \bar{\nu}_{lL} \gamma^\mu l_L \,, \tag{1.16}$$

where the index l characterizes the lepton flavor and L signifies, as before, the corresponding chiral projection. In (1.16) the summation is over all n lepton flavors. We have assumed "universality", i.e., we have assumed that the coupling to the intermediate vector boson W_L is the same for all lepton flavors.

The Lagrangian (1.15) has two features which are of crucial importance to us: a) The neutrino field ν_l is not necessarily the eigenstate of the mass term. b) The right-handed neutrino does not appear at all; it does not participate in weak interactions governed by (1.15).

The interaction (1.15) conserves separately the electron, N_e, muon, N_μ, and tau, N_τ lepton numbers as defined in Table 1.1. (Other formulations of the lepton number conservation rules are possible. For example, in the Konopinski-Mahmoud (53) (see also Zel'dovich 52) scheme there is only

Table 1.1 *Lepton numbers of various particles. (Antiparticles have lepton numbers of opposite sign.)*

	ν_e	e^-	ν_μ	μ^-	ν_τ	τ^-	Hadrons,Photon
N_e	1	1	0	0	0	0	0
N_μ	0	0	1	1	0	0	0
N_τ	0	0	0	0	1	1	0

one conserved lepton charge, and e^- and μ^- have lepton numbers of opposite sign. In the multiplicative muon conservation scheme (Feinberg & Weinberg 61) the decay $\mu^+ \rightarrow e^+ + \nu_\mu + \overline{\nu}_e$ is possible. This decay has a branching ratio of less than 0.098 at 90% CL (Willis et al. 80) thus supporting the standard additive conservation law.)

The conservation of individual lepton numbers, $N_{e,\mu,\tau}$, has been extensively tested by searching for decays of the type $\mu \rightarrow e\gamma$, $\mu^+ \rightarrow e^+e^-e^+$, etc. Decays or reactions violating the conservation of individual lepton numbers have not so far been observed (we shall discuss some of these processes in Chapter 4).

Flavor neutrino oscillations. We shall now consider the possibility that lepton number conservation is violated in a subtle way, because the neutrino mass term $\mathbf{L}_M$ is not diagonal when expressed in terms of flavor neutrinos. For simplicity, we shall confine ourselves to the first two lepton flavors e and μ and we shall consider the neutrino mass term of the Dirac type and assume CP invariance (we shall also use the fields themselves, rather than the chiral projections). The most general mass term is then (Gribov & Pontecorvo 69)

$$-\mathbf{L}_M = m_{\nu_e\nu_e}\,\overline{\nu}_e\nu_e + m_{\nu_\mu\nu_\mu}\,\overline{\nu}_\mu\nu_\mu + m_{\nu_e\nu_\mu}\,(\overline{\nu}_e\nu_\mu + \overline{\nu}_\mu\nu_e)\,. \tag{1.17}$$

Since $\mathbf{L}_M$ is symmetric, we can diagonalize it by the substitution

$$\begin{aligned} \nu_e &= \cos\theta\ \nu_1 + \sin\theta\ \nu_2 \\ \nu_\mu &= -\sin\theta\ \nu_1 + \cos\theta\ \nu_2 \end{aligned} \tag{1.18}$$

The eigenvalues of $\mathbf{L}_M$ are

$$m_{1,2} = \frac{1}{2}\left\{ m_{\nu_e\nu_e} + m_{\nu_\mu\nu_\mu} \pm [(m_{\nu_e\nu_e} - m_{\nu_\mu\nu_\mu})^2 + 4m^2_{\nu_e\nu_\mu}]^{\frac{1}{2}} \right\},$$

and the angle θ is determined by the formula

$$\tan 2\theta = 2m_{\nu_e\nu_\mu} \,/\, (m_{\nu_\mu\nu_\mu} - m_{\nu_e\nu_e})\,.$$

The particles ν_1 and ν_2 have definite masses m_1 and m_2, respectively. They are Dirac particles, and, consequently, are not charge conjugation eigenstates. Depending on the sign of the determinant of $\mathbf{L}_M$, one of the eigenvalues could be negative; as stated before, the field $\gamma_5\nu_2$ should be used in such a case. The particles ν_1 and ν_2 then evolve in time as

$$|\nu_1(t)> = e^{-iE_1t}\,|\nu_1(0)>; \quad |\nu_2(t)> = e^{-iE_2t}\,|\nu_2(0)> , \qquad (1.19)$$

where $E_i = (p^2 + m_i^2)^{\frac{1}{2}}$ is the neutrino energy. We assume that these particles propagate without interactions; the time and distance are then interchangeable. A particle which has been created at t=0 as an electron neutrino ν_e evolves in time as

$$\begin{aligned} |\nu_e(t)> &= \cos\theta\, e^{-iE_1t}\,|\nu_1(0)> + \sin\theta\, e^{-iE_2t}\,|\nu_2(0)> \qquad (1.20) \\ &= (\, e^{-iE_1t}\cos^2\theta + e^{-iE_2t}\sin^2\theta\,)\,|\nu_e(0)> \\ &+ \cos\theta\sin\theta\,(\, e^{-iE_2t} - e^{-iE_1t}\,)\,|\nu_\mu(0)> . \end{aligned}$$

The probability that the neutrino, created as ν_e at time t=0, is in the state $|\nu_\mu>$ at time t, is given by the absolute square of the amplitude of $|\nu_\mu(0)>$ in (1.20), i.e.,

$$P(\nu_e \to \nu_\mu) = |<\nu_\mu|\nu_e(t)>|^2 = \frac{1}{2}\sin^2 2\theta \left[1 - \cos\frac{m_2^2 - m_1^2}{2p}t\right] , \qquad (1.21)$$

where we have performed a little bit of straightforward algebra and assumed that $p >> m_{1,2}$. Expression (1.21) describes the simplest type of neutrino flavor oscillation. The angle θ, the so called mixing angle, describes the amount of mixing. The probability $P(\nu_e \to \nu_\mu)$ varies periodically with time or distance, this periodicity is characterized by the "oscillation length", $L_{osc} = 2\pi\frac{2p}{|m_2^2 - m_1^2|}$. The oscillation is a typical interference effect caused by the nondiagonal form of the mass term (1.17). (For diagonal $\mathbf{L}_M$ one would require $m_{\nu_e\nu_\mu} = 0$ and, consequently, $\theta = 0$ or $\pi/2$, and there are no oscillations.)

The probability $P(\nu_e \to \nu_e)$ that the neutrino created as ν_e at t=0 is still ν_e at t, is given by the absolute square of the amplitude of $|\nu_e(0)>$ in (1.20). Because the state (1.20) is normalized, and the normalization is conserved, we obtain

$$P(\nu_e \to \nu_e) + P(\nu_e \to \nu_\mu) = 1 . \qquad (1.22)$$

Thus, we see that flavor oscillations, caused by the nondiagonal neutrino mass term, violate conservation of individual lepton numbers (N_e and N_μ in our simplest example). The total lepton number (N_e+N_μ in our case), however, remains conserved, as we see in Eq. (1.22). It is easy to generalize the preceding analysis to the case of more than two generations (see Chapter 5). Again, we encounter periodic behavior of the probability of finding different flavors; however, there are now more mixing angles and oscillation lengths.

The generalization of Eq. (1.18) is of the form

$$\nu_l = \sum_l U_{l,i}\, \nu_i \,, \tag{1.23}$$

i.e., the neutrino participating in weak interactions (1.16) is a linear combination of mass eigenstates; the corresponding coefficients $U_{l,i}$ form a unitary matrix. In addition, if CP is conserved, this matrix must be real. The weak lepton current (1.16) can be rewritten now in terms of the neutrino mass eigenstates as

$$J_L^\mu = \sum_{l,i} \nu_{iL} U_{l,i}^+ \gamma^\mu l_L \,. \tag{1.24}$$

Thus, we see that the weak interactions are nondiagonal when expressed in terms of the mass eigenstates.

In the previous Subsection we have discussed the transformation properties of the neutrino field under C and P. Now, we see that neutrino dynamics enters in a subtle way. If we combine (1.23) with the weak interaction Lagrangian (1.15), (1.16) to form (1.24), we see that the requirement of CP invariance of weak interaction implies that the phase factors η_P and η_C of the individual fields are no longer independent. We shall return to this problem in Chapter 6.

Oscillations of Majorana neutrinos. Again, we shall restrict our discussion to the case of two flavors and CP conserving theory. For simplicity, we shall still work with fields which are not chiral projections. As explained in the previous Subsection, the dimension of the problem doubles. So, the general hermitian mass term for two neutrino flavors is of the form (Bilenky & Pontecorvo 76)

$$-\mathbf{L}_M = \frac{1}{2}\left(\overline{\Psi}, \overline{\Psi^c}\right) \begin{pmatrix} M_D & M_M \\ M_M^+ & M_D^T \end{pmatrix} \begin{pmatrix} \Psi \\ \Psi^c \end{pmatrix} , \tag{1.25}$$

where M_D is a 2×2 hermitian matrix of Dirac masses, and M_M is a 2×2 symmetric matrix of Majorana masses. Furthermore, the requirement of CP invariance makes both of these matrices real. The symbol Ψ is a column made of two fields, ψ_e and ψ_μ, for example. The matrix (1.25) can be diagonalized in two steps. First, the transformation W is applied:

$$\begin{pmatrix}\Psi\\ \Psi^c\end{pmatrix} \rightarrow W\begin{pmatrix}\Psi\\ \Psi^c\end{pmatrix} = \frac{1}{\sqrt{2}}\begin{pmatrix}\Psi+\Psi^c\\ -\Psi+\Psi^c\end{pmatrix}; \quad W = \frac{1}{\sqrt{2}}\begin{pmatrix}I & I\\ -I & I\end{pmatrix}; \tag{1.26}$$

$$W\begin{pmatrix}M_D & M_M\\ M_M & M_D\end{pmatrix}W^+ = \begin{pmatrix}M_D+M_M & 0\\ 0 & M_D-M_M\end{pmatrix}. \tag{1.27}$$

Thus, there are two eigenstates, even under charge conjugation; their eigenvalues are the eigenvalues of M_D+M_M. Besides, there are two eigenstates, odd under charge conjugation; their eigenvalues are the eigenvalues of M_D-M_M. The two 2×2 matrices, generally, cannot be diagonalized simultaneously, because M_D and M_M do not commute. Therefore, there are altogether four Majorana mass eigenstates of the mass term (1.25) and, correspondingly, the mixing matrix is a 4×4 unitary matrix. The flavor eigenstates ν_e and ν_μ, and the corresponding antiparticles ν_e^c and ν_μ^c are expressed in terms of the *same* set of four Majorana mass eigenstates. As we will see in Chapter 5, the "neutrino-antineutrino" oscillations (as proposed earlier by Pontecorvo 58) are possible now, and both the individual lepton numbers, N_e, N_μ, and the total lepton number, N_e+N_μ, are no longer conserved.

1.4 Neutrino Mass and Grand Unification

Before discussing grand unified theories, let us review the problem of neutrino mass in the standard Glashow-Weinberg-Salam theory. In this theory, the first family of quarks and leptons are assigned as

$$\begin{pmatrix}u_\alpha\\ d_\alpha\end{pmatrix}_L, u_{\alpha R}, d_{\alpha R}, \begin{pmatrix}\nu_e\\ e^-\end{pmatrix}_L, e_R^-, \tag{1.28}$$

where α=1,2,3 is a color index and $L(R)$ denotes the chiral projections discussed above. (For simplicity we have neglected the Cabibbo mixing among quarks.) In the minimal electroweak theory, the only Higgs scalar forms a doublet (Φ^+,Φ^0). This assignment for the Higgs scalar leads, for example, to the successful prediction of the Z boson mass.

With these assignments, the neutrino must be massless. The Dirac mass term cannot be present, because ν_R is absent by fiat. On the other hand, the Majorana mass term cannot be present either, because the theory implies lepton number conservation, and, as mentioned earlier, a Majorana mass term would violate lepton number conservation.

There are several ways in which the minimal model can be extended and neutrino mass introduced at the electroweak level. None of these possibilities, however, is compelling or at least really attractive. One possibility is an extension of (1.28) by the introduction of the right-handed neutrino ν_R, which is a singlet under $SU(2)\times U(1)$, as is its charged partner e_R^-. In this case, the neutrino Dirac mass term can be constructed in complete analogy to the charged lepton or quark mass terms (which are Dirac mass terms). It is difficult to understand, however, why m_ν is so much smaller than the other masses. Also, the new field ν_R would not interact directly with other fermions; it would be practically sterile. Another possibility is an extension of the Higgs sector, for example, by postulating a new Higgs triplet. In such a case, the neutrino can acquire Majorana mass without the necessity of introducing a new field ν_R. (The Majorana mass term would couple $\bar{\nu}_L$ with $(\nu^c)_R$.) Due to the different mechanism of mass generation, it is, perhaps, not surprising that the neutrino is so light. The model of the Higgs triplet, in addition, predicts the existence of a massless boson, the so called majoron (Gelmini & Rondacelli 81; Georgi et al. 81). These majorons could accompany neutrinoless double beta decay (see Chapter 6), although recent measurement of the decay width of the Z^0 boson essentially excludes this possibility. Another scheme, the singlet majoron mechanism of neutrino mass generation (Chikashige, Mohapatra & Peccei 81), not excluded by the Z^0 width measurement, belongs to the see-saw class of models discussed below.

As we stressed before, the standard model is extremely successful, as it describes essentially all known physics up to the highest energies probed at existing accelerators. On the other hand, the standard model leaves too many unanswered questions to be the ultimate theory of physics. Possible extensions of the standard model are grand unified theories (GUTs), which help with some, but not all, of the shortcomings of the standard model. For example, many GUTs are successful in predicting the weak angle $\sin^2\theta_W$, but shed little light on the fermion masses and family structure. Furthermore, the simplest GUT, the minimal $SU(5)$ (Georgi & Glashow 74), predicts a proton lifetime which is shorter than the experimental limit (although extensions with longer lifetimes are possible). While it seems unlikely that any GUT is the complete theory, it is likely that the ultimate theory will incorporate various features of grand unification. Below, we give a brief overview of the neutrino mass problem in GUTs.

The simplest GUT is the minimal $SU(5)$ theory of Georgi & Glashow (74). In this theory the baryon number, N_{baryon}, and the lepton number, N_{lepton}, are not conserved separately, but their difference is a conserved quantity. As in the standard model, the right-handed neutrino, ν_R, is absent, and the neutrino is massless. The Dirac mass is absent, because ν_R is absent by fiat, and the Majorana mass is absent, because the

$N_{baryon} - N_{lepton}$ is conserved. This theory, attractive because of its simplicity, is somewhat discredited by the mentioned prediction of too short a proton lifetime. As in the standard model, extensions which allow finite neutrino mass are possible (explicit ν_R, extension of the Higgs sector) but unattractive.

GUTs based on larger groups typically contain massive neutrinos. The neutrino mass in this class of theories could be generated by the see-saw mechanism of Gell-Mann, Ramond, & Slansky (79) and Yanagida (79) (see also Stech 80). This mechanism is based on the Majorana mass and is therefore unique to neutrinos. Thus, it explains naturally why neutrinos have a very different mass than the charged leptons.

To see how this situation arises, we will describe in greater detail the steps necessary in order to diagonalize the general mass term (1.12) in the basis with chiral projections. The diagonalization can be performed in two steps. First, the basis is rearranged by combining the upper and lower components,

$$\Psi \rightarrow \Psi' \equiv VU\Psi \; ; \quad U = \begin{pmatrix} I & 0 \\ 0 & \sigma_x \end{pmatrix} \; , \quad V = \frac{1}{\sqrt{2}} \begin{pmatrix} I & I \\ I & -I \end{pmatrix} \; , \qquad (1.29)$$

where I is the unit 2×2 matrix and σ_x is the usual Pauli matrix. The mass matrix in (1.12) is transformed to the block diagonal form,

$$VU \begin{pmatrix} 0 & M \\ M^+ & 0 \end{pmatrix} U^+ V^+ = \begin{pmatrix} M' & 0 \\ 0 & -M' \end{pmatrix} \; ; \quad M' = \begin{pmatrix} m_R & m_D \\ m_D & m_L \end{pmatrix} . \qquad (1.30)$$

The two upper components of Ψ' are charge conjugation eigenstates (Majorana neutrinos) $\psi_R + (\psi_R)^c$ and $\psi_L + (\psi_L)^c$, while the two lower ones are the differences of the same fields $\psi_R - (\psi_R)^c$ and $\psi_L - (\psi_L)^c$.

Now, we can consider the underlying physics. Let us assume that, besides the familiar ν_L, which couples to electrons in weak charged current, there exists a weak interaction singlet neutrino ν_R of the Majorana mass m_R. This right-handed neutrino must be heavy, otherwise it would have been observed already. The GUTs will make the following assignments

$$m_L \approx 0 \; , \quad m_R \approx M_{GUT} \approx 10^{14} - 10^{16} \text{ GeV} \; , \quad m_D \approx m_{fermion} \; , \qquad (1.31)$$

where $m_{fermion}$ represents either charged lepton or quark masses. The value M_{GUT} is the GUT mass scale. (This is the energy value where the three running coupling constants, corresponding to the groups, $SU(3)$, $SU(2)$, and $U(1)$, come together as a function of the momentum transfer.) The diagonalization of M' is trivial under these circumstances. There are two eigenvalues, a very heavy Majorana neutrino with $m_R \approx M_{GUT}$, and a light one $m_L \approx m^2_{fermion}/M_{GUT}$. (Naturally, the same result is obtained from Eq. (1.14), the negative mass can be avoided by the γ_5 transformation described earlier.) Substituting the corresponding fermion masses, we see that the light neutrino, dominantly the Majorana partner of left-handed charged leptons, will have mass as low as 10^{-13} eV. The light and heavy neutrinos will mix with a very small mixing angle $\tan^2\theta \approx m_D^2/M_{GUT}^2$.

The see-saw mechanism predicts that the neutrino mass is proportional to the square of the mass of a fermion (lepton or quark) from the same family. Thus, one expects that

$$m_{\nu_e} : m_{\nu_\mu} : m_{\nu_\tau} = m_e^2 : m_\mu^2 : m_\tau^2 \ , \tag{1.32a}$$

or, more likely,

$$m_{\nu_e} : m_{\nu_\mu} : m_{\nu_\tau} = m_u^2 : m_c^2 : m_t^2 \ , \tag{1.32b}$$

where u, c, t denote the up, charm, and top quarks, respectively. According to this model, the tau neutrino would be the heaviest, and the electron neutrino the lightest. Many modifications of this idea have been proposed, often replacing M_{GUT} by a smaller mass scale and thus making the light neutrinos correspondingly heavier. It is also possible that the heavy neutrino mass m_R follows the family hierarchy. In that case the quadratic dependence in Eqs. (1.32) will be replaced by the linear dependence on $m_{fermion}$.

It is, perhaps, significant that the mass range obtained in the see-saw mechanism for $m_R \approx M_{GUT}$ is quite similar to the corresponding range obtained if one interprets the result of the solar neutrino experiment as an indication of the resonant neutrino matter oscillation, as described in Chapter 5. On the other hand, none of the popular theoretical models leads naturally to the electron neutrino mass as large as 10 eV, the present experimental upper limit (Chapter 2). Fine tuning and modifications are possible to obtain this result, as for example Witten (80) has shown in the framework of the $O(10)$ GUT, where he obtained a neutrino mass $(\alpha/\pi)^{-2}$ times larger than our estimate above. In Table 1.2 we summarize the predictions of various models (Langacker 88, 91), including representative examples of models with $m_R << M_{GUT}$. We also include a column denoted by $<m_{\nu_e}>$, a quantity which appears in nuclear double beta decay (Chapter 6); the corresponding entry should be understood as an order of magnitude estimate.

Table 1.2 *Major classes of neutrino mass models, with their most natural scales.*

Model	m_{ν_e}	$<m_{\nu_e}>$
Extended $SU(2)\times U(1)$ ν_R added,Dirac mass	1 - 10 MeV	0
Extended $SU(2)\times U(1)$ Higgs triplet added Majorana mass	Arbitrary	m_{ν_e}
GUTs beyond $SU(5)$ See-saw mechanism	$10^{-9} - 10^{-13}$ eV	m_{ν_e}
Intermediate see-saw ($m_R \approx 10^9$GeV)	$10^{-4} - 10^{-7}$ eV	m_{ν_e}
TeV see-saw ($m_R \approx 10^3$GeV)	10^{-1} eV	m_{ν_e}

2

Kinematic Tests for Neutrino Mass

In this Chapter we shall discuss how neutrino mass can be determined from purely kinematic considerations of weak interaction processes. For simplicity, we assume here that only one neutrino is involved in a reaction or decay, leaving the more general situation of neutrino mixing until Chapter 4. We shall explore how finite neutrino mass will affect the momenta and energies of the charged particles emitted in a weak decay, as well as the overall decay rate. A complete review of all aspects of lepton spectroscopy can be found in Bullock & Devenish (83) and in the review article by Robertson & Knapp (88).

In treating the kinematics of weak decays, we shall first focus our attention on nuclear beta decay. We shall show that beta decay is a very sensitive tool for studying the electron antineutrino mass. Near the endpoint of the beta spectrum, a massive neutrino has so little kinetic energy that it becomes nonrelativistic, with the result that the decay probability depends linearly on mass. Of particular interest here is the low energy beta decay of tritium which, as we shall see, has the potential of determining neutrino mass down to a few eV. (It is customary to give the mass in units eV as well as eV/c^2.) As this is by far the most sensitive kinematic test for neutrino mass, we present in Section 2.1 a comprehensive discussion of several tritium experiments. We then turn to the muon neutrino and discuss the two-body decays of the pion into charged particles and neutrinos (Section 2.2), followed by the brief Section 2.3 on the tau neutrino mass. In studying the decays into muon and tau neutrinos, we find that the neutrino mass enters quadratically. For this reason, combined with the fact that much larger kinetic energies are involved in these decays, the mass limits for the muon and tau neutrino are much less restrictive than those for the electron neutrino.

2.1 Electron Neutrino Mass: The Beta Spectrum of Tritium

In an allowed nuclear beta decay,

$$(Z, A) \rightarrow (Z+1, A) + e^- + \bar{\nu}_e \,, \tag{2.1}$$

(or corresponding positron decay), the neutrino mass m_ν must be included in the total decay energy E_0, leading to the relation between the maximum energy of the electron, $E_e^{Max}(m_\nu)$, and the masses of the initial atom, M_i, and of the final ion, M_f,

$$E_e^{Max}(m_\nu) = E_0 - m_\nu = M_i - M_f - m_\nu \,. \tag{2.2}$$

(As the electron energy is much smaller than the mass of the nucleus, we have neglected here the recoil energy of the nucleus, which has a maximum value of $(E_0 - m_e^2)/2M_f$.)

The electron spectrum (in dimensional units) is given by

$$\frac{dN}{dE} = G_F^2 \frac{m_e^5 c^4}{2\pi^3 \hbar^7} \cos^2\theta_C |M|^2 F(Z,E) pE(E_0-E)[(E_0-E)^2 - m_\nu^2]^{\frac{1}{2}}, \tag{2.3}$$

where G_F is the Fermi coupling constant, M contains the nuclear matrix elements, $F(Z,E)$ is the Coulomb function and θ_C is the Cabbibo angle.

From expression (2.3), it can be seen that appreciable sensitivity to neutrino mass exists only near the endpoint of the spectrum. The spectrum terminates abruptly (with infinite slope) at an energy $E_0 - m_\nu$, manifestly distiguishing it from the spectrum for the case of a massless neutrino. In an experiment, the spectrum needs to be studied up to an electron energy E near E_0, so that $E_0 - E \equiv \Delta E$ is of the order of m_ν. However, at this energy E, the number of decays, dN/dE, becomes extremely small. It is easy to see that for a low energy electron decay with maximum kinetic energy $Q = E_0 - m_e << E_0$, the fraction of decays in the interval ΔE is given by $(\Delta E/Q)^3$. In the case of the tritium decay (Q = 18.6 keV), the number of electrons in an energy interval of, say, ΔE = 9 eV (the present neutrino mass limit) is only 10^{-10} of all decays in the spectrum. Also, with higher resolution, the spectrometer acceptance and thus the focal area of the source becomes smaller (proportional to ΔE^2), further reducing the number of electrons reaching the detector. Clearly, the fraction $(\Delta E/Q)^3$ becomes rapidly smaller for beta decays with higher Q values, such as ^{63}Ni (Q = 67 keV) or ^{35}S (Q = 167 keV). It is, therefore, imperative to select a candidate with low Q. In addition, another important criterion in selecting a decay, is that its lifetime should be relatively short. The longer the lifetime, the larger the mass thickness required for equal decay rates and, as a consequence, the larger the line broadening, as a result of electron attenuation in the source.

The influence of neutrino mass on the beta spectrum near the endpoint can best be seen in a linearized spectrum obtained by transforming (2.3) in the following way,

$$K(E) \equiv \left[\frac{dN/dE}{pEF(Z,E)}\right]^{\frac{1}{2}} \sim \left\{(E_0-E)\left[(E_0-E)^2-m_\nu^2\right]^{\frac{1}{2}}\right\}^{\frac{1}{2}}. \qquad (2.4)$$

An example of this linearized spectrum (also referred to as a Kurie plot) is shown in Figure 2.1, visibly demonstrating the difference for the cases $m_\nu = 0$ and $m_\nu \neq 0$.

Our single branch decay formula (2.3) describes the decay of a bare nucleus. In reality, the decaying nucleus is surrounded by electrons in various energy states. All decay branches from one or more initial states into several final states, varying in energy by the difference in electron binding energies, must be taken into account.

In the case of the free tritium atom, one can calculate the excitation energies and branchings with high accuracy. Calculations have also been

Figure 2.1. Illustration of linearized beta spectrum near the endpoint for neutrino mass $m_\nu = 0$ and $m_\nu \neq 0$.

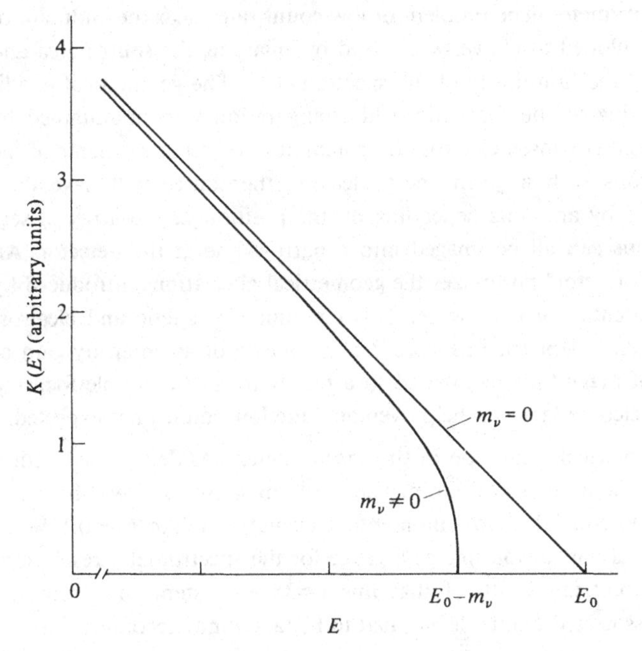

performed for the $^3H_2$ molecule, as well as for more complicated hydrocarbon molecules, and we shall return to the complications caused by the atomic final states later in this Chapter. Suffice it here to note that the presence of a multitude of decay branches modifies the compound spectrum near the endpoint in such a way as to reduce the slope of the linearized spectrum and thus to counteract the effect of the neutrino mass.

Another important aspect in a measurement of the beta spectrum is the resolution function of the spectrometer. In the neighborhood of the endpoint, the finite resolution causes distortions of the spectrum which again tend to counteract the effect of the neutrino mass. The idealized situation with infinite slope depicted in Figure 2.1, therefore, can never be realized in an experiment. As to the spectrometer resolution required, it must be of the order of the neutrino mass or better. With a current upper limit of 9.4 eV for the mass, the resolution in energy in a tritium experiment thus must be at least 10^{-3}.

In the following Subsections we shall briefly describe several tritium experiments and their results.

2.1.1 *The tritium decay experiments*

Bergkvist's experiment. In a classic experiment, Bergkvist (72) has studied the tritium beta spectrum with the help of a double focusing ($\sqrt{2}\pi$) magnetic spectrometer. The problem of low count rates near the endpoint of the spectrum, alluded to above, was solved by enlarging the source area and thus increasing the luminosity of the spectrometer. The geometrical conditions for focusing in the magnetic field configuration were maintained by applying a properly chosen electrostatic potential to different segments of the source. Electrons with a given energy leaving the source will have their energy changed by amounts depending on the position in the array of segments, and thus can all be imaged into a narrow line at the detector. An electrostatic "corrector" minimizes the geometrical aberrations introduced by the electric potential of the source. This combined magnetic and electrostatic spectrometer, depicted in Figure 2.2, is capable of an intensity gain of 2—3 orders of magnitude, compared to a purely magnetic double focusing geometry, a welcome factor to help overcome the low count rates expected.

The geometrical resolution of the spectrometer was determined with a K-conversion line at 22.9 keV emitted by a ^{170}Tm source and was found to be $\Delta E = 40$ eV, equivalent to a momentum resolution of $\Delta p/p = 0.11\%$. In using an internal conversion line as a gauge for the spectrometer resolution, the sizable Lorentzian width of that line ($\approx$35 eV), stemming from the width of the associated atomic levels, had to be taken into account.

The tritium source was prepared by implanting tritium molecules ($^3H_2$) into a 0.2 mm thick aluminum foil. At an implantation voltage of 1 kV the tritium was deposited at a depth of 1.4 μg cm^{-2} into the aluminum oxide surface layer giving rise to an average energy loss in the source of 20 eV. The actual experimental resolution thus was the composite of the geometrical resolution and the energy loss contribution and amounted to 55 eV. The entire segmented source foil represented a source strength of 20 millicuries.

There is another contribution to the resolution which must be considered, having to do with the atomic states in the final $^3He^+$ ion populated by the beta decay. In a free He^+ ion the 1s ground state is reached in about 70% of all beta decays, while the 2s level which lies 40.5 eV above the ground state is populated in about 25% of decays. The line broadening associated with this final state multiplicity is illustrated in Figure 2.3. To complicate matters further, there are also transitions to higher states and to the continuum (about 5% of all decays). Also, as the tritium is not in the form of a free atom, but rather is bound in a molecule (involving the aluminum oxide backing), the situation is quite a bit more complicated than the two level assumption used by Bergkvist, as we shall discuss in a Subsection below.

In analyzing the spectrum, the statistical errors of the data points (which must be corrected for background) grow rapidly in the neighborhood of the endpoint where the spectrum is most sensitive to neutrino mass. While the shape of the spectrum will be altered by the finite mass, it will

Figure 2.2. Electrostatic-magnetic beta spectrometer built by Bergkvist (72) to explore the tritium beta spectrum near the endpoint.

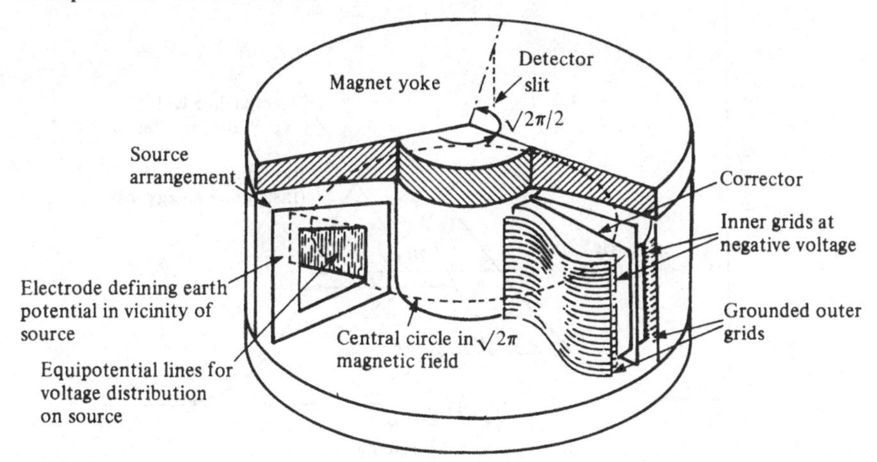

also be affected by the finite resolution and, most importantly, by back-scattering of electrons from the source material as well as scattering in the spectrometer. Distortions from scattering, however, are expected to be continuous over a large energy region and should be distinct from the distortion near the endpoint. Bergkvist's data agree well with the expected spectrum for zero mass. In comparison, the spectrum for a mass of 67 eV gives a much larger χ^2 and thus represents a poorer fit. The final result, which includes an uncertainty in the value of the endpoint, is quoted as $m_\nu < 60$ eV (90% CL).

The ITEP experiment. A long series of experiments performed at the Institute for Theoretical and Experimental Physics in Moscow (Lubimov et al. 80, 81; Boris et al. 85, 87; Lubimov 86) has provided positive evidence of neutrino mass, with mass values reported to be in the range of 17—40 eV. A toroidal spectrometer (Tretyakov et al. 75) was built for these experiments and the techniques were continuously perfected over the duration of the work. In the most recent experiment (Boris et al. 85, 87; Lubimov 86) the electron spectrum was scanned by varying an applied electrostatic potential of about 4 kV while keeping the magnetic field constant. Accelerating the electrons to an energy beyond the endpoint energy helps to reduce the source related background in the detector. A geometrical resolution of 20 eV was achieved as measured with the internal conversion lines from a ^{169}Yb source whose Lorentzian width was taken into account. In an inter-

Figure 2.3. Overall line shape resulting from the combined effect of finite experimental resolution and distribution in endpoint energy due to atomic final state effects in tritium beta decay (Bergkvist 72).

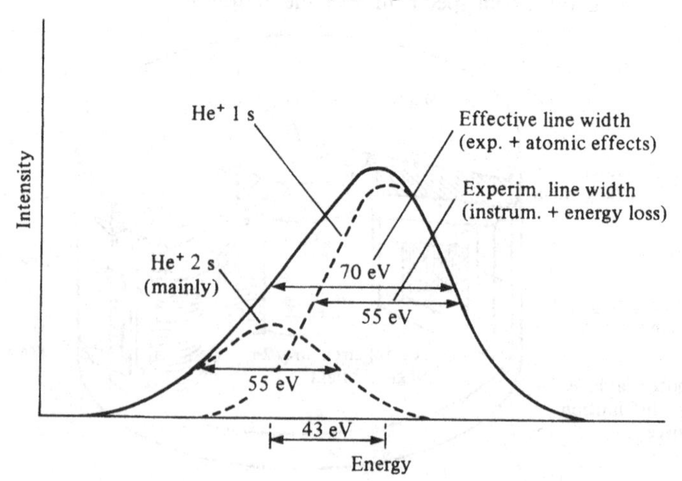

nal conversion process, owing to the sudden change of charge in the inner atomic shells, outer electrons may be shaken off, giving rise to a low energy tail of the conversion line. In deriving the geometrical resolution function, this effect was taken into account. The tritium source was a thin foil (2 μg cm^{-2}) of tritiated valine ($C_5H_{11}NO_2$) with a specific activity of 1 millicurie cm^{-2} and a total area of 5 cm^2 deposited on both sides of an aluminum foil. Applying Bergkvist's technique the source was cut into nine strips, each connected to the proper focusing potential.

One of the most difficult problems in this experiment had to do with the line profile from the valine source. What is the correct resolution function for deconvoluting the measured spectrum? By placing the ^{169}Yb calibration source behind tritiated valine foils of various thicknesses, the electron attenuation and the line profile could be explored. As Figure 2.4 shows, the low energy tail of the resolution function is quite significant compared to the area under the geometrical peak and may lead to sizable uncertainties in the spectrum shape.

The measured tritium spectrum then can be expressed by a folding integral over the total resolution function and the shape function (2.3). The latter, however, needed to be modified by the addition of a correction term,

Figure 2.4. Resolution function in the ITEP experiment. Curve A represents the geometrical resolution function, curve B shows the contribution from electron attenuation in the source and curve C gives the backscattering contribution.

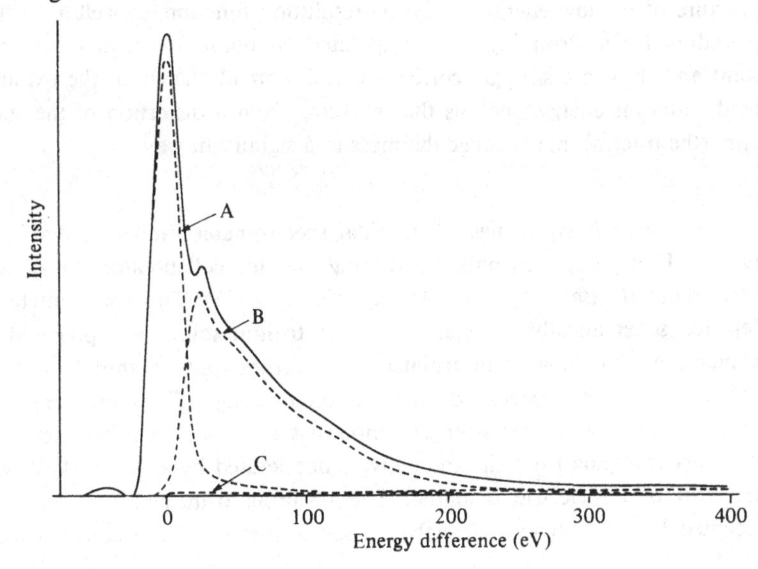

α, proportional to $(E_0 - E)^2$, which was required to account for an apparent nonlinearity in the spectrum (possibly caused by electrons scattered in the spectrometer, or by variations with energy in the spectrometer acceptance). To account for the multiplicity of the spectrum as a result of the large number of levels in the final state of the valine molecule (a total of 51 levels were used (Kaplan et al. 83); see also the discussion below), the required folding was included in the analysis. From a fit to the parameters m_ν, E_0, and α, Boris et al. (85, 87) have derived a neutrino mass, based on the valine final state, of

$$m_\nu = (30.3\,^{+2}_{-8})\ \text{eV}, \tag{2.5}$$

as well as a value for the total decay energy of $E_0 = (18{,}600.3 \pm 0.4)$ eV. For the case (probably unjustified) in which the final state configurations are allowed to vary over a wide range, a conservative range of mass values, $17 < m_\nu < 40$ eV, is obtained. The observed linearized spectrum near the endpoint, as well as the difference between the experimental spectrum and the calculated spectrum for $m_\nu = 0$ are shown in Figure 2.5.

The ITEP measurement, while not contradicting Bergkvist's results, is the only experiment to date that has provided positive evidence for a finite neutrino mass. Over the course of this series of experiments, the values reported for the mass have changed only little and appear to have been unaffected by several drastic changes in procedures such as improvement of the resolution or inclusion of the Lorentzian width (Simpson 83). While this seems at first reassuring, it is, on closer inspection, rather difficult to understand. As mentioned above, the resulting spectrum depends strongly on the structure of the low energy tail in the resolution function, as well as on the procedure for determining the extrapolated endpoint. In an analysis, endpoint and mass are strongly correlated, and a small change in the extrapolated endpoint energy, such as that resulting from a distortion of the spectrum (the α term), may change the mass in a significant way.

The Zurich experiment. A toroidal spectrometer similar to that used by the ITEP group was built by Kündig and his collaborators in Zurich, Switzerland (Fritschi et al. 86, 91; Kündig et al. 86). This spectrometer is depicted schematically in Figure 2.6. The tritium source was prepared by implanting a monolayer of tritiated hydrocarbon onto a thin SiO_2 layer which, in turn, was deposited on a carbon backing. The source area was 157 cm^2 and the source strength amounted to about 50 millicuries. The electrons emerging from the source were decelerated by about 15 keV with the help of an electric potential. The resulting truncated spectrum was accepted by the momentum analyzing spectrometer set at a constant mag-

Figure 2.5. (a) Observed tritium beta spectrum (Kurie plot) near the end-point for three runs (R_1,R_2,R_3) from the ITEP experiment of Boris et al. (85). The solid curves are overall fits based on the valine final state. The dashed curves are calculated spectra for $m_\nu = 0$. (b) Difference between "best fit" to the experimental data and calculated spectrum for $m_\nu = 0$. The best fit was achieved for a set of E_0, α and $m_\nu = 34.8$ eV. The $m_\nu = 0$ horizontal line is based on the same parameters E_0 and α.

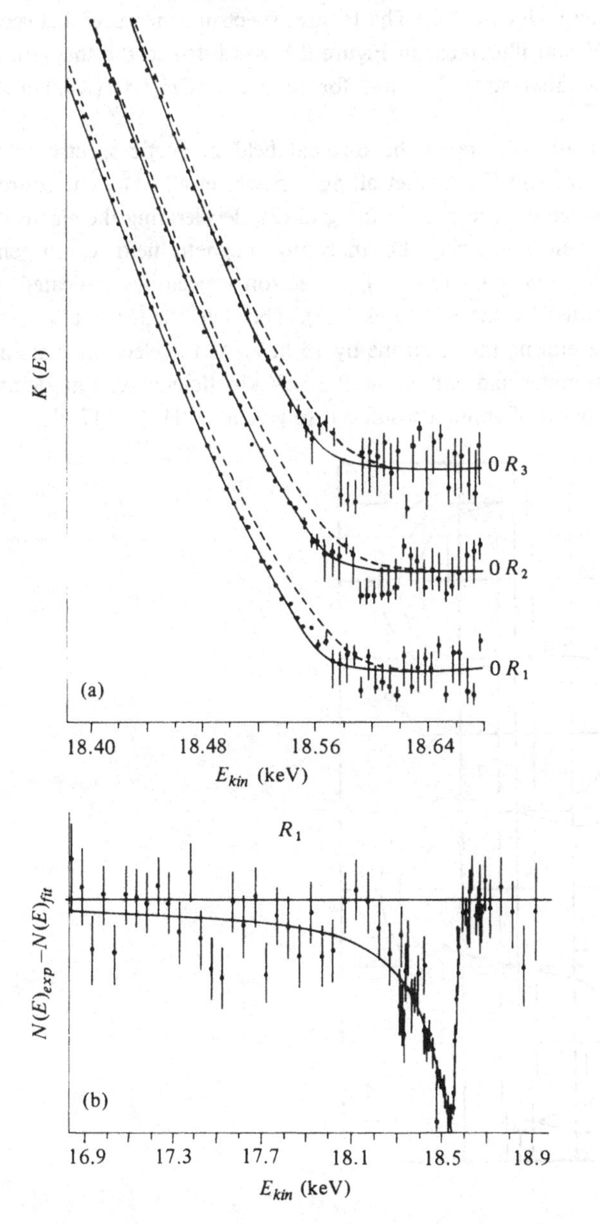

netic field permitting only 2.2 keV electrons to be transmitted. The scanning was accomplished by varying the retardation voltage. At the exit slit, the transmitted electrons were reaccelerated by 15 keV to facilitate detection in a proportional counter. The spectrometer resolution profile was studied with the help of an L_I internal conversion line at 19.4 keV in ^{119}Sn. The resolution achieved was 17 eV. The resolution function used for the analysis shown in Figure 2.7 exhibits a considerably lesser tail from energy loss than the profile in Figure 2.4. The tritium spectrum measured between 17.7 and 19.3 keV and illustrated in Figure 2.8 was fitted to the theoretical spectrum with the final state function for methane ($CH_3{}^3H$) (Kaplan &

Figure 2.6. Schematic picture of the toroidal field magnetic spectrometer for the Zurich experiment (Kündig et al. 86; Fritschi et al. 91). The source (1) is at a high voltage with respect to the grid (2), decelerating the electrons by about 15 keV before entering the analyzing magnetic field region generated by the rectangular coils (3,4). The electron trajectories indicated in the Figure are limited by the apertures (6,7). The detector (5) is also at a high voltage reaccelerating the electrons by 15 keV. As the electrons passing through the spectrometer had only about 2.2 keV kinetic energy, a moderate spectrometer resolution of about 1% sufficed to give a FWHM of 17 eV.

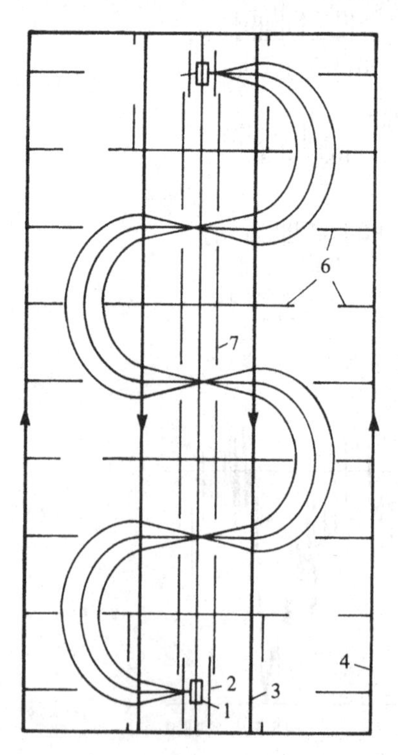

Smelov 86). In analyzing the data between 17.65 and 18.75 keV, free parameters for each of the following effects were introduced: backscattering, endpoint energy to the atomic ground state in ^{3}He, over-all normalization, constant background, and mass-parameter m_ν^2. A χ^2 for 270 degrees of freedom of 317 was found and the m_ν^2 was ($-$ 158 $\pm$ 150) eV^2. In addition to the quoted statistical value, there was a systematic error of $\pm$ 103 eV^2. In-as-much as this result is negative, it is, strictly speaking, non-physical. To restore its meaning one has to resort to a "prescription" (Review of Particle Properties 90) which leads, at 95% CL, to the following resulting limit

$$m_\nu^2 < 237 \text{ eV}^2, \ \textit{or} \ \ m_\nu < 15.4 \text{ eV}, \tag{2.6}$$

a result in strong disagreement with the ITEP result.

The Los Alamos experiment. We mention next the work by the Los Alamos group (Robertson et al. 86, 91; Wilkerson et al. 87, 91). The Los Alamos team, also using a toroidal spectrometer, has developed a novel gaseous source of both atomic and molecular tritium. In the work quoted, molecular tritium gas is injected into a 4 m long cylinder, mounted on the

Figure 2.7. Spectrometer resolution function in the Zurich experiment by Fritschi et al. (91). The measured values are given by the solid curve and the calculated results by the points.

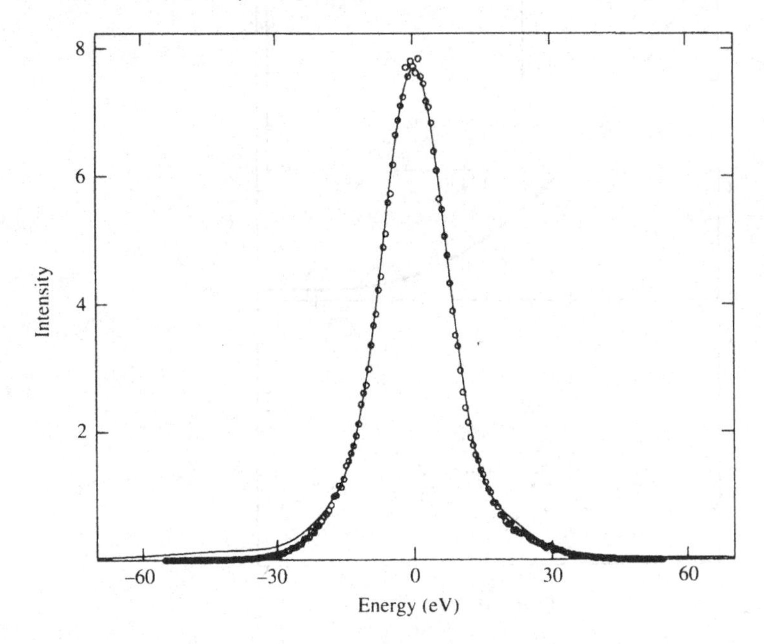

axis of the spectrometer. A schematic figure of the Los Alamos spectrometer is shown in Figure 2.9. This source, whose surface density is 10^{15} atoms cm^{-2}, resides in an axial magnetic field confining the electrons to motion along the axis, until they exit into the spectrometer field region. The detector was a 96-pad Si microstrip array. The spectrum is taken for a fixed electron momentum equivalent to 23 keV, with the decelerating voltage scanned. Before and after tritium runs, the 17.82 keV internal conversion line of $^{83}Kr^{m}$ is measured to determine the instrumental resolution and energy calibration. Owing to the inhomogenous field spectrometer focusing conditions each detector pad receives counts from a slightly different momentum region of the spectrum, the total range from one end of the detector to the other being 100 eV.

The atomic corrections to be applied for $^{3}H_2$ are much smaller than in methane and amount to about 10 eV^2. The uncertainty arising from the fit with a Gaussian was 12 eV^2. The effect from a tail on m_ν^2 was 15 eV^2 and the energy loss in the source was estimated to be 25 eV^2. It was found that

Figure 2.8. Tritium spectrum obtained by Fritschi et al. (91). The insert shows the endpoint region with the fit curve. The bottom part of the figure shows the difference of data minus fit divided by 1σ for $m_\nu = 0$.

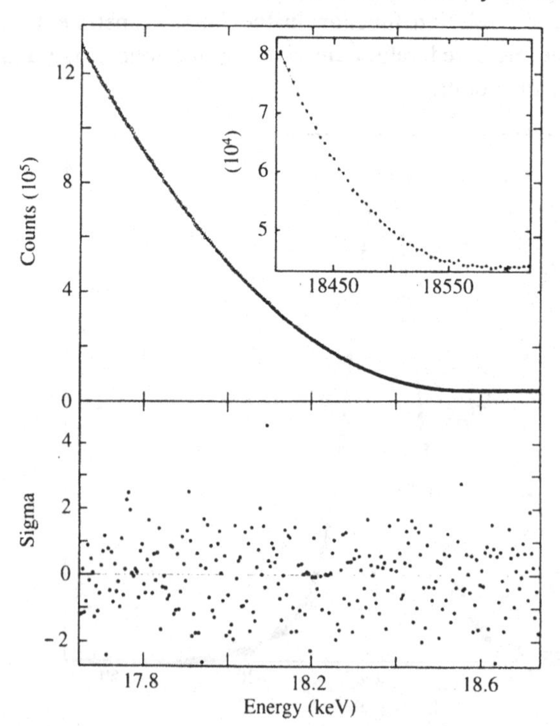

the spectrometer efficiency varied slightly with the accelerator voltage applied, introducing a distortion in the spectrum. This distortion was described by a linear and quadratic parameter, α_1 or α_2, to be treated as free parameters in the analysis. As the extrapolated slope differed slightly as the spectra were analyzed with α_1 or α_2, different values of m_ν^2 were obtained. Averages of the α_1 and α_2 fits were then taken resulting in a total uncertainty of 50 eV^2 for m_ν^2.

Four different runs were taken between 1988 and 1989. The ^3H spectra were scanned from 16.44 to 18.94 keV in 10 eV steps. The $^{83}Kr^m$ internal conversion line corrected for satellite lines caused by electron shake-up and shake-off was used for calibration. Its intrinsic Lorentzian line width was 2.26 eV. The total spectrometer resolution was 23 eV. The endpoint energy, the shape parameters α_1 or α_2, and the neutrino square mass m_ν^2 were determined with a maximum likelihood test. The best fit to m_ν^2 was, $m_\nu^2 = (-\ 147 \pm 68 \pm 41)\ eV^2$. Again, the principal value is negative, and the proper analysis (Review of Particle Properties 90) gives the following result,

$$m_\nu < 9.4 \text{ eV} \quad (95\% \text{ CL}).$$

If the value of m_ν^2 were shifted to 0, a neutrino mass limit of 12.5 eV would result. In either case the ITEP result is clearly ruled out by this experiment.

For illustration, Figure 2.10, taken from Robertson et al. (91), shows the residuals for fits to neutrino masses 0 (top) and 30 eV (bottom) showing clearly that a mass of 30 eV must be rejected.

Figure 2.9. Sketch of the Los Alamos spectrometer for the tritium beta spectrum measurement by Wilkerson et al. (91).

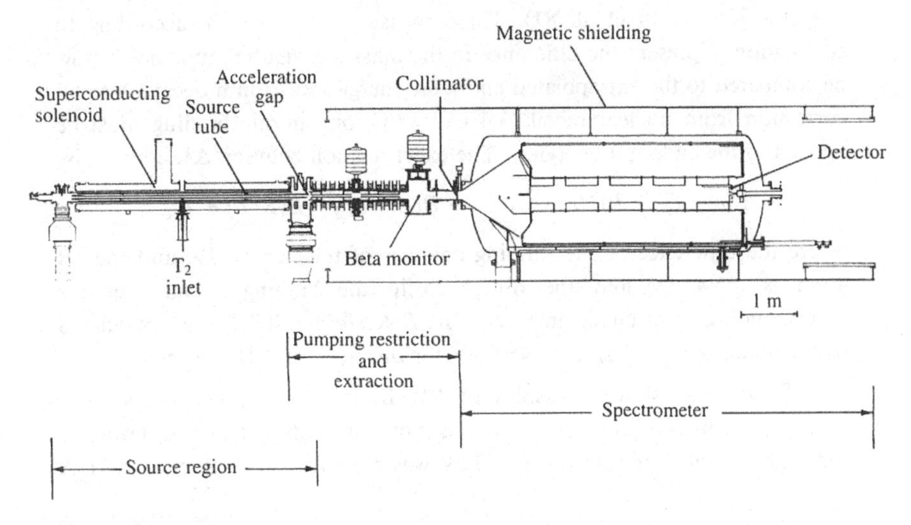

The Institute for *Nuclear Structure Tokyo* (*INS*) *experiment*. This experiment (Kawakami et al. 87, 91) makes use of an ironfree double focusing spectrometer (radius of the focal plane = 75 cm) at the INS in Tokyo. It is conceptually similar to that used in Berkvist's experiment. To increase acceptance, use is made of a large source as well as a large detector, both segmented. The detector is a position sensitive proportional chamber consisting of six independent cells. The overall spectrometer energy resolution was 15 eV. ^{109}Cd served as a reference source and its KLL Auger electrons were used for calibration. Final-state corrections for ^3H in methane were applied following the calculations of Kaplan and Smelov (86).

In the analysis, an empirical shape factor is required to fit the data. Other free parameters were: an energy dependent linear background, a constant background, an energy dependent fitting parameter (about 1%/keV), the endpoint energy, and the square mass m_ν^2. There were 354 data points between 17.88 and 18.95 keV. The weighted average of the value of the square mass over the six cells was $m_\nu^2 < (-65 \pm 85 \pm 65)\ \mathrm{eV}^2$. According to the procedure in the Review of Particle Properties (90), this leads to

$$m_\nu^2 < 166\ \mathrm{eV}^2; \quad m_\nu < 13\ \mathrm{eV} \quad (95\%\ \mathrm{CL}).$$

Again, this result is at variance with the ITEP value.

The ^3H–^3He *mass difference*. The mass difference ΔM between ^3H and ^3He provides an important check on the spectrum endpoint, and, owing to the correlation between endpoint and mass, on the neutrino mass itself. Very precise mass differences have been obtained from ion cyclotron resonance experiments (Lippmaa et al. 85). Mass differences were also obtained from the study of the radiofrequency mass spectrometer doublets (Smith & Wapstra 75; Smith et al. 81). These measurements (which, according to convention, represent the difference in the masses of neutral atoms) can now be compared to the extrapolated endpoint energies in tritium decay. A small correction from nuclear recoil (3.4 eV) and from atomic binding must be applied to the endpoint energies. The exact relation between ΔM and E_0 is

$$\Delta M = E_0 + B(He) - B(T) - B(R{:}He^+) + B(R{:}T) + E_{rec},$$

where the difference of the binding energies of the neutral He and the ^3H atom is 65.4 eV and the difference in the binding energies of the corresponding molecular fragment R, $B(R{:}He^+) - B(R{:}T)$ is calculated (Kaplan and Smelov 86) to be 48.6 eV for the $^3H_2$ or ^3H H molecule.

Figure 2.11 shows a display of ΔM from various experiments. The agreement between different results is not very satisfactory. A critically summarized value of 18599.4 ± 3.0 eV was presented by Audi et al. (85). A more recent measurement with the Los Alamos spectrometer by Staggs et al. (89) gives $\Delta M = 18589.0 \pm 2.6$ eV.

The final-state effect. As stated earlier, the shape of the beta spectrum is affected by the surrounding atomic electrons ("spectator electrons"). Their effect is twofold: a) Their charges modify the potential near the nucleus

Figure 2.10. Residual for the fit to the neutrino mass 0 (top) and 35 eV (bottom) demonstrating that a neutrino mass of 35 eV can clearly be rejected. (Taken from the Los Alamos experiment of Robertson et al. 91.)

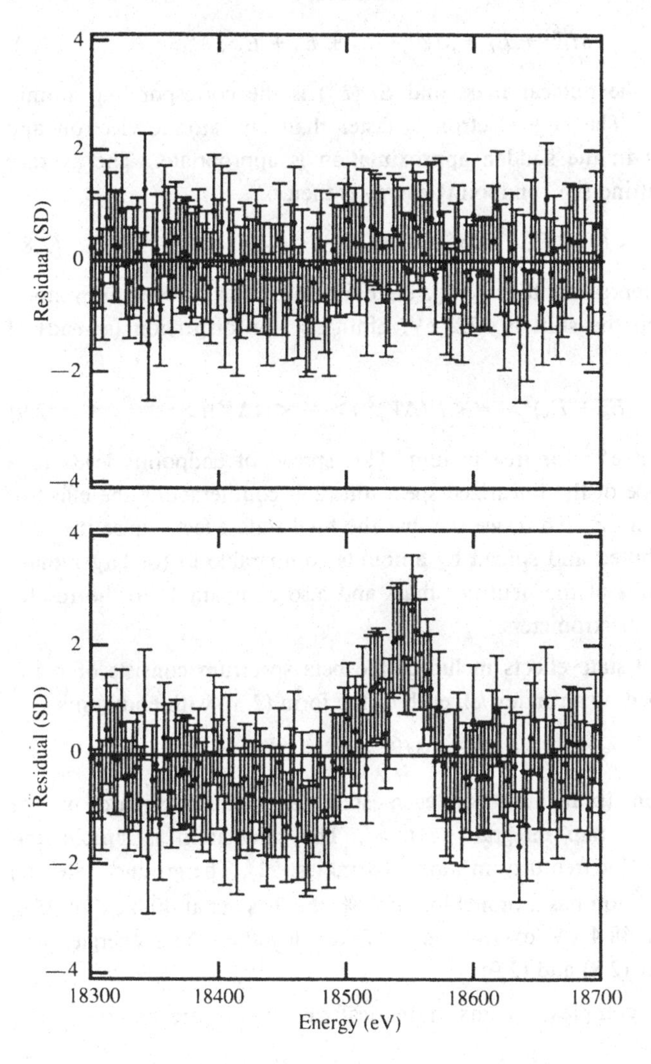

where beta decay takes place. This rather small and slightly energy dependent effect is taken into account by the appropriate screening correction in the function $F(Z,E)$ in (2.3). b) The sudden change of the nuclear charge Z causes a change ΔV in the atomic potential, which in turn shifts the energies of the surrounding electrons. Moreover, the final ion (or molecule) is typically not in an energy eigenstate but in a superposition of eigenstates. In order to extract the neutrino mass from the experimental beta spectrum, one must calculate the branching ratios and energies of all final states populated in the decay.

The importance of this effect can be seen in the following way. The energy balance, with atomic electrons included, is

$$M_i^{(n)} + E_i = M_f^{(n)} + E_f + E_e + E_\nu \,, \tag{2.7}$$

where $M^{(n)}$ is the nuclear mass and E_i (E_f) is the corresponding atomic binding energy. The beta electron is faster than any atomic electron and thus treatment in the sudden approximation is appropriate. The average electron + neutrino (i.e., endpoint) energy is then

$$<E_e + E_\nu> = M_i^{(n)} - M_f^{(n)} - <i|\Delta V|i> \,. \tag{2.8}$$

The last term represents the average endpoint shift. For a free tritium atom, $\Delta V = -e^2/r$, and the shift is 27.2 eV. Similarly, the dispersion (spread) of endpoints is

$$<(E_e + E_\nu)^2> = <i|(\Delta V)^2|i> - <i|\Delta V|i>^2 \,, \tag{2.9}$$

which is $(27.2)^2$ eV^2 for free tritium. This spread of endpoints leads to a decrease of slope of the linearized spectrum (2.4) counteracting the effect of finite neutrino mass. Thus, we see that the final state effects cause the endpoints to be shifted and spread by amounts comparable to (or larger than) the present limits of the neutrino mass and also comparable to the resolution of present spectrometers.

With final state effects included, the beta spectrum consists of many branches (labeled by an index k), each of the form (2.3), with endpoints

$$E_0^k = M_i^{(n)} - M_f^{(n)} + E_i - E_f^k \,.$$

The population (branching) of each state is just the square of the corresponding overlap integral $<i|f^k>^2$. The calculation is simple and essentially exact for free tritium atoms (Bergkvist 71). The ground state (1s) of the final $^3He^+$ ion has a branching of 70%, the 2s state at 40.8 eV of 25%, the 3s state at 48.4 eV of 1.4%, etc. Taken together these energies and branchings obey (2.8) and (2.9).

For more complex systems, numerical calculations are necessary. For the molecule $^3H_2$, we refer to the work by Fackler et al. (85), for valine and a number of other tritium containing compounds, and to the work by Kaplan et al. (82, 85). In Figure 2.12 we show the branching vs. energy for the ^{3}H atom, the $^3H_2$ molecule, the $CH_3{}^3H$ molecule (methane), and for valine.

Conclusion. At the present time there is no confirmed evidence for finite neutrino mass from tritium beta decay experiments, with an upper limit for the mass of 9.4 eV at 95% CL. Several experiments in progress should be capable of providing comparable mass sensitivity.

2.1.2 *Note on electron capture experiments*

The neutrinos emitted in nuclear beta decay (2.1) are of the type $\bar{\nu}_e$. The mass limits obtained for the $\bar{\nu}_e$ should also hold for the ν_e. This is a consequence of the *CPT* theorem which implies that the mass of a particle is equal to the mass of its antiparticle. (The validity of this theorem, expected to hold on theoretical grounds, has been demonstrated with great accuracy for charged leptons, mesons, and baryons. The most restrictive limit has been established from the $K_0 - \bar{K}_0$ mass difference and is about 10^{-18}. (See Review of Particle Properties 90).) Nevertheless, the study of the ν_e mass has been pursued in its own right taking advantage of the inner bremsstrahlung process accompanying electron capture,

$$(Z, A) + e^- \rightarrow (Z-1, A) + \nu_e + \gamma \,. \tag{2.10}$$

If the continuous bremsstrahlung spectrum is close in energy to the atomic x-rays, the emission probability is enhanced, and the resulting spectrum has a complicated though, in principle, calculable shape. Much like in

Figure 2.11. Mass difference between tritium and ^{3}He atoms as determined from spectrum endpoint measurements, a mass doublet determination, and an ion cyclotron resonance measurement.

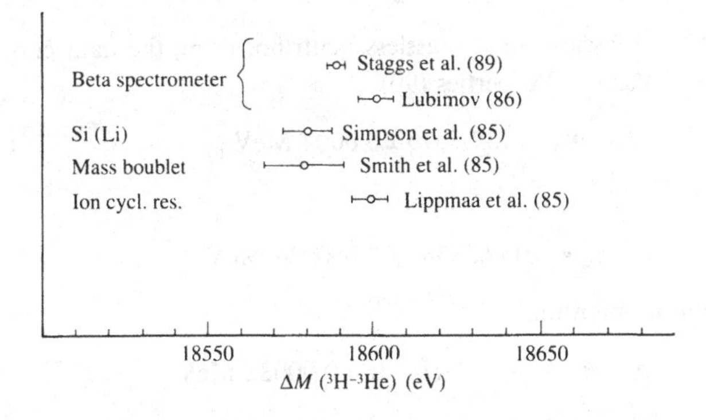

the case of beta decay, a finite neutrino mass will result in changes near the endpoint of the bremsstrahlung spectrum. Studies of the bremsstrahlung with the aim of finding information on the ν_e mass were instigated by De Rujula (81), and experiments were carried out with ^{193}Pt by Jonson et al. (83), and with ^{163}Ho by Yasumi et al. (86), with the (somewhat disappointing) result that the upper limit of the ν_e mass is about 500 eV (90% CL).

Clearly, this technique suffers from many difficulties. The low yield of inner bremsstrahlung and the poorly known atomic branching ratios of the outer shell x-rays are among them, as is the poor energy resolution of presently available efficient photon detectors.

2.2 Muon Neutrino Mass: The Two-Body Decay of the Pion

Muon neutrinos are created in leptonic as well as hadronic processes. The pure leptonic decay of the muon, $\mu \rightarrow e + \nu_e + \nu_\mu$, being a three-body decay, has large phase space for both neutrinos and thus is not well suited for a neutrino mass determination. Instead, the decay of the pion into a muon and a muon neutrino,

$$\pi^+ \rightarrow \mu^+ + \nu_\mu , \tag{2.11}$$

a weak hadronic process governed by simple two-body kinematics, is more appropriate. The process (2.11) has been studied with pions decaying in flight as well as with pions decaying at rest.

First, let us consider the pion decay at rest. A beam of pions is stopped in a target where they can decay. The emerging muons and neutrinos are both monochromatic, and the muon momentum, the only quantity that can be measured in an experiment, is related to the pion mass m_π, muon mass m_μ, and neutrino mass m_ν by the relation

$$p_\mu^2 = \frac{(m_\pi^2 + m_\mu^2 - m_\nu^2)^2}{4m_\pi^2} - m_\mu^2 . \tag{2.12}$$

Evaluating this expression for a massless neutrino, using the data compiled in the Review of Particle Properties (90).

$$m_\pi = 139.5675 \pm 0.0004 \text{ MeV} , \tag{2.13}$$

and

$$m_\mu = 105.658387 \pm 0.000034 \text{ MeV} ,$$

we find a muon momentum of

$$p_\mu (m_\nu = 0) = 29.78986 \pm 0.00032 \text{ MeV} , \tag{2.14}$$

the error being primarily due to the uncertainty in the pion mass. If neutrinos have mass, the muon momentum will be smaller according to (2.12). For example, if $m_\nu = 0.25$ MeV, we find a central value of

$$p_\mu\,(m_\nu = 0.25 \text{ MeV}) = 29.78903 \text{ MeV},$$

a change of only 0.00083 MeV, or a relative change of 2.8×10^{-5}. This fractional change represents the accuracy required in an experimental determination of the muon momentum for pinpointing a neutrino with mass of 0.25 MeV. A somewhat smaller uncertainty is associated with the error in the pion mass determination.

Figure 2.12. Population P_n of final states in the decay of the ^3H atom, $^3H_2$ molecule, methane molecule, and valine molecule.

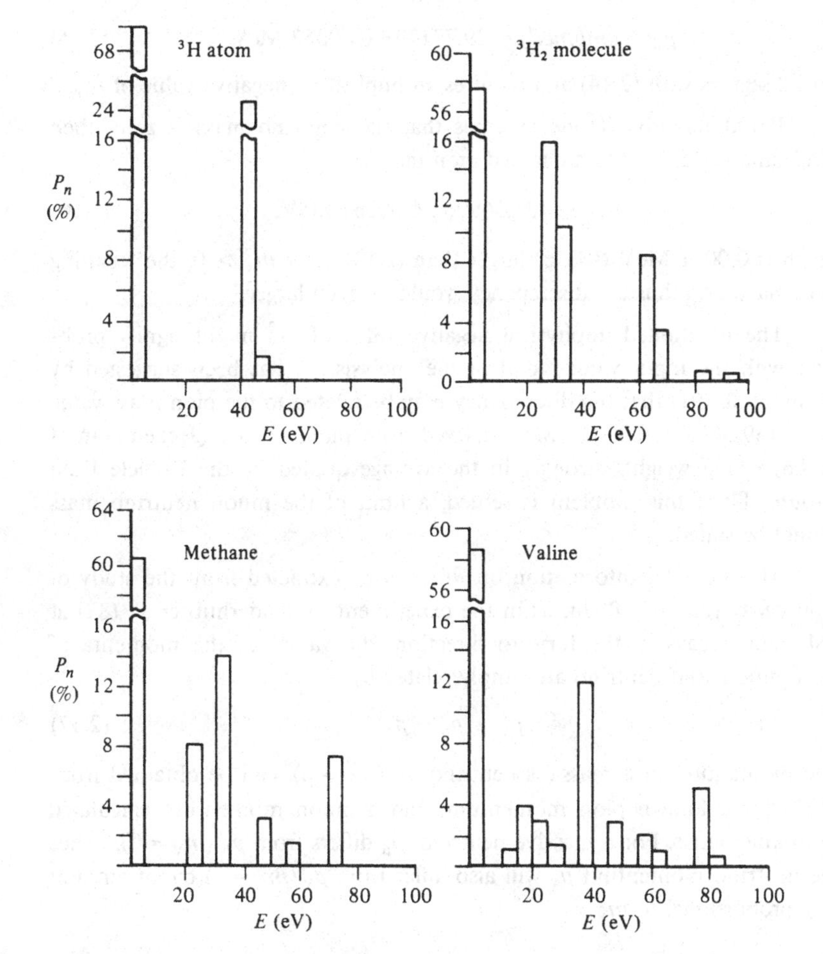

Here, a brief note concerning the charge state of the particles is appropriate. Although the masses quoted in (2.13) have been determined for negative pions and muons, the *CPT* theorem postulates that positive and negative particles possess the same mass. Positively charged particles are more amenable to pion decay experiments, like those described below.

A precise measurement of the decay of a positive pion was carried out at the pion beam at PSI (Paul Scherrer Institute) by Abela et al. (84) and later by Daum et al. (91). The momentum of the μ^+ emerging from the pion stopping target was determined with a magnetic spectrometer, and the following results were obtained by the two groups,

$$p_{\mu^+}\,(Abela\ et\ al.\ 84) = 29.79139 \pm 0.00083\ \text{MeV}\ ,$$

$$p_{\mu^+}\,(Daum\ et\ al.\ 91) = 29.79206 \pm 0.00068\ \text{MeV}\ ,$$

with a combined result of

$$p_{\mu^+}\,(combined) = 29.79179 \pm 0.00053\ \text{MeV}\ . \tag{2.15}$$

This disagrees with (2.14) and requires an unphysical negative value of m_ν^2.

Put differently, if one assumes that the neutrino mass is zero, then (2.12) and (2.15) lead to a charged pion mass of

$$m_\pi^+ = 139.56996 \pm 0.00067\ \text{MeV}, \tag{2.16}$$

which is 0.0021 MeV ($\approx 3\ \sigma$) larger than (2.13). For $m_\nu > 0$, the resulting pion mass, and thus the discrepancy would be even larger.

The mentioned unphysical negative value of m_ν^2 might signify problems with the input values used in the analysis. It has been suggested by Daum et al. (91) that the discrepancy may be related to the pion mass value $m_\pi = 139.56752 \pm 0.00037$ MeV derived from pionic x-rays (Jeckelmann et al. 86) which weights strongly in the average quoted by the Particle Data Group. Until this problem is settled, a limit of the muon neutrino mass cannot be stated.

Alternatively, information on m_{ν_μ} can be extracted from the study of pion decay (2.11) in flight, as in the experiment by Anderhub et al. (82) at PSI. For decays in the forward direction, the values of the momenta of pion, muon, and neutrino are simply related by

$$p_{\nu_\mu} = p_\mu - p_\pi\ . \tag{2.17}$$

The momentum of a massless neutrino, $p_\nu\,(m_\nu = 0)$, can be obtained from (2.17) for a known pion momentum, and a muon momentum calculated from kinematics. For a massive neutrino, p_μ differs from $p_\mu\,(m_\nu = 0)$, hence the neutrino momentum p_ν will also differ from $p_\nu\,(m_\nu = 0)$ by an amount δp_ν, proportional to m_ν^2.

In the PSI experiment, pions with momentum p_π = 350 MeV enter a 180° magnetic spectrometer, instrumented with detectors, and their momenta are measured accurately. Upon leaving the spectrometer, the pions pass through a straight decay section where they decay into muons. These muons return to the very same spectrometer through a beam line with bending magnets. Using one and the same analyzing magnet for pions and muons has the advantage that the systematic errors in the difference $p_\mu - p_\pi$ are much reduced. Each pion and its decay muon are recorded separately and their correlation is identified with the help of a time-of-flight signal. As the decaying pions have relatively large momenta, the errors from the rest mass of the particles (Eq. (2.13)) play only a minor role in determining the neutrino momentum, and the main uncertainties in the final answer stem from errors in determining the particle trajectories, as they enter the spectrometer, as well as from counting statistics. With a final result, $m_{\nu_\mu}{}^2 = -0.14 \pm 0.20$ MeV^2, this experiment provides a mass limit of $m_{\nu_\mu} < 0.50$ MeV (90% CL).

2.3 Tau Neutrino Mass: The Decay of the Tau Lepton

This third neutrino has never been observed directly, in the sense of being responsible for inducing a reaction. At the present time there is no indication that it has finite mass, with a much less restrictive upper limit of its mass, if compared to the muon and, particularly, the electron neutrino. Hence these results are less incisive, and we shall confine our discussion to a brief outline of some pertinent results.

Upper limits of the mass of the tau neutrino, ν_τ, have been derived from studies of the decay of the tau lepton. As in the case of the muon, the tau decays leptonically into a muon or an electron, $\tau \to \mu + \nu_\mu + \nu_\tau$, or $\tau \to e + \nu_e + \nu_\tau$. It also undergoes weak hadronic decay into one or several pions. (For a review of the properties of the tau, see Perl 80.)

The purely leptonic decay involving an electron

$$\tau \to e + \nu_e + \nu_\tau \tag{2.18}$$

has been examined by Bacino et al. (79) at the SPEAR electron-positron storage ring at SLAC (Stanford Linear Accelerator). The electron spectrum was measured with the help of the DELCO detector, a device consisting of spark chambers and Čerenkov counters arranged around the collision region of the electron and positron beams. Near the endpoint, the spectrum, which for $V{-}A$ decay is given by the Michel parameter $\rho = 0.75$, will be lowered if the neutrino mass is finite. No such reduction was seen, and an upper limit of $m_{\nu_\tau} < 250$ MeV could be established.

A similar limit was obtained from a study of the weak hadronic two-body decay,

$$\tau \rightarrow \pi + \nu_\tau , \qquad (2.19)$$

carried out by Blocker et al. (82) at SLAC with the Mark II detector at the electron-positron storage ring SPEAR. The endpoint of the pion energy spectrum of the tau decay (2.19) in flight depends on the masses m_τ, m_π, and m_{ν_τ}. Knowing the former two, the authors derived, from an endpoint measurement, $m_{\nu_\tau} < 250$ MeV.

Somewhat improved limits have been found in a study of the rare tau decay branches, $\tau \rightarrow \pi^+\pi^-\pi^\pm\pi^0\nu_\tau$, and $\tau \rightarrow K^+K^-\pi^+\nu_\tau$ (Matteuzzi et al. 85, Mills et al. 85). However, the best current mass limits come from experimental studies of the tau decay into three and five charged pions and a tau neutrino. The three pion decay mode was investigated by Albrecht et al. (85, 88) with the ARGUS detector at the electron positron storage ring at DESY (Deutsches Electron Synchrotron) at a center-of-mass energy of 10 GeV. A neutrino energy spectrum was obtained from the kinematics of the observed pions. As this spectrum is sensitive to neutrino mass in the region of low energy it was possible to derive a neutrino mass limit of

$$m_{\nu_\tau} < 35 \text{ MeV} \quad (95\% \text{ CL}). \qquad (2.20)$$

A similar limit ($m_{\nu_\tau} < 84$ MeV, 95% CL) was obtained from a study of the decay into five charged pions by Abachi et al. (86) with the High Resolution Spectrometer at the PEP storage ring at SLAC.

3

Neutrino Induced Reactions

As neutrinos are usually detected by means of reactions on nucleons or nuclei, we present in this brief Chapter an evaluation of reaction cross sections. This topic is in some sense auxiliary to our main theme of neutrino mass and mixing, and thus we shall avoid nonessential details. For a more detailed discussion we refer to the books by Commins & Bucksbaum (83) and by Okun (82). The neutral current reactions on nucleons and nuclei are discussed in the reviews by Bilenky & Hosek (82) and by Donnelly & Peccei (79). Some of the related problems are treated in the review articles by Diemoz et al. (86) and by Grenacs (85).

3.1 Charged Current Elastic Reactions

Neutrinos of the lepton flavor l are most simply detected by reactions based on the charged current weak interaction which employ free nucleons at rest as targets. Thus there are two possible "elastic" reactions

$$\nu_l + n \rightarrow p + l^- , \tag{3.1a}$$

$$\bar{\nu}_l + p \rightarrow n + l^+ . \tag{3.1b}$$

(There are no suitable free neutron targets, consequently, (3.1a) can proceed only with bound neutrons, for example, those bound in deuterium.) Analogues of (3.1a) and (3.1b) with nuclear targets are

$$\nu_l + (Z,A) \rightarrow (Z+1,A) + l^- , \tag{3.2a}$$

$$\bar{\nu}_l + (Z,A) \rightarrow (Z-1,A) + l^+ . \tag{3.2b}$$

All these reactions have a threshold equal to the mass m_l of the charged lepton l, plus the difference of the masses of the final and initial nuclear targets. For example, the reaction (3.1b) for electron neutrinos has the threshold E_T = (0.511 + 1.294) MeV, while for muon neutrinos, the threshold is 106.95 MeV. The reaction (3.1a) with electron neutrinos and deuterons as targets has the threshold $E_T = (0.511 + 2.224 - 1.294)$ MeV = 1.441 MeV.

At sufficiently high energies, the excitation of nucleons or nuclei, and the possibility of producing additional particles (e.g., pions, etc.) must be included in the description of the reactions (3.1a) and (3.1b). At very high energies, the regime of deep inelastic collisions,

$$\nu \text{ (or } \bar{\nu}) + N \rightarrow l + X \tag{3.3}$$

will be reached, where X stands for highly excited hadronic matter and N for a nucleon. In deep inelastic collisions the momentum transfer Q^2 is much larger than the nucleon mass and, at the same time, the energy transferred to the hadrons is also larger than the nucleon mass. Deep inelastic collisions of neutrinos with nucleons have been very fruitful in studies of the quark-parton model, scaling phenomena, and in experimental determination of the nucleon structure functions. We, however, shall not describe these problems any further, as they are not directly related to our central theme of neutrino mass and mixing.

The description of the nuclear reactions (3.2a) and (3.2b) is further complicated by the binding of the target nucleons (bound nucleons are not exactly on the mass shell), by the momentum distribution of nucleons in nuclei (Fermi motion), and by the possibility of reactions leading to one of the excited states of the final nucleus.

Interaction with nucleons. We begin our discussion with the most general matrix element of the hadronic weak current, changing a proton of momentum p into a neutron of momentum p', restricted only by the requirement of Lorentz invariance and isotopic spin invariance of the strong interactions. This matrix element is of the form

$$< p' | J_\mu^- | p > = \bar{u}(p') \, [F_1 \gamma_\mu + F_2 \sigma_{\mu\nu} q_\nu + i F_S q_\mu] \, \tau_- \, u(p) \tag{3.4}$$

for the vector part of the current, and

$$< p' | J_{\mu 5}^- | p > = \bar{u}(p') \, [F_A \gamma_5 \gamma_\mu - i F_P \gamma_5 q_\mu - F_T \gamma_5 \sigma_{\mu\nu} q_\nu] \, \tau_- \, u(p) \tag{3.5}$$

for the axial-vector part of the current. The quantities F_i are nucleon form factors, which are all functions of the momentum transfer $q^2 \equiv (p-p')^2$. The presence of form factors is a consequence of the composite nature of nucleons and of the strong interactions between their constituents. The six form factors in (3.4) and (3.5) contain all the information on the semileptonic processes involving nucleons (including beta decay, muon capture, etc.). However, the form factors are difficult to calculate theoretically, and one has to rely to a large extent on experimental data in order to determine their value and their momentum transfer dependence.

The conserved vector current (CVC) hypothesis simplifies the situation considerably for the vector part of the current. According to the CVC theory, the vector part of the weak current and the isovector part of the electromagnetic current form a single isospin triplet of conserved currents. This leads immediately to the results that

$$F_1(0) = 1 \ , \ 2M_p F_2(0) = \mu_p - 1.0 - \mu_n = 3.706 \ . \tag{3.6}$$

Further, because the first two terms in (3.4) are already divergenceless, CVC implies that $F_S = 0$. Moreover, one can use the results of electron scattering experiments to find the Q^2 dependence of the form factors F_1 and F_2. It turns out that the "dipole" formula

$$F_1(Q^2) = (1 + Q^2/0.71 \ \mathrm{GeV}^2)^{-2} \ , \ 2M_p F_2(Q^2) = 3.71 F_1(Q^2) \tag{3.7}$$

is adequate for our purpose (Gaisser & O'Connell 86).

There is no exact analogue of CVC for the axial part (3.5) of the current. However, a simplification is achieved by dividing the individual terms in the current into first and second class currents, according to their transformation properties under G parity, the combination of the charge conjugation transformation C and an isospin rotation $e^{i\pi I_2}$. According to this classification, the already excluded F_S part of the vector current and the F_T part of the axial-vector current are "second class" currents, which are incompatible with the standard model. In fact, experimental data support the assumption that $F_T = 0$ (Grenacs 85). Moreover, the Goldberger-Treiman relation (see, e.g., Commins & Bucksbaum 83) leads to the estimate of the induced pseudoscalar form factor F_P*

$$F_P(Q^2) = 2M_p F_A(Q^2)/(m_\pi^2 - Q^2) \ . \tag{3.8}$$

This relation is also verified, to an accuracy of $\approx$15%, by experiment (Grenacs 85). The contribution of this form factor to the cross section is very small, and vanishes in the extreme relativistic limit. Finally, the neutron beta decay tells us that

$$F_A(0) = -1.261 \pm 0.004 \ . \tag{3.9}$$

The only missing piece of information is the Q^2 dependence of the axial form factor F_A. Here, we assume the same "dipole" form as in (3.7); this is essentially compatible with available data. For F_A the characteristic param-

* In muon capture one often uses a dimensionless quantity $g_P = m_\mu F_P$. At the relevant momentum transfer $Q^2 = -0.9 m_\mu^2$ (for protons) Eq. (3.8) gives $g_P \approx 8.4$.

eter 0.71 GeV^2 in (3.7) should be replaced by $(1.07 \pm 0.06)^2$ GeV^2 (Baker et al. 81); the present world-average value is $(1.032 \pm 0.036)^2$ GeV^2.

Having determined the hadronic current, we can now quote the final formula for the differential cross section of the reactions (3.1a) and (3.1b) (see Walecka 75; Gaisser & O'Connell 86)

$$\frac{d\sigma(\bar{\nu},\nu)}{d\Omega} = \frac{G^2 p_l E_l}{2\pi^2} \frac{1}{1 + 2E_\nu/M_p \sin^2(\theta/2)} \times$$

$$\left[F_1^2 + F_A^2 + Q^2 F_2^2\right]\cos^2(\theta/2) + 2\left[F_A^2\left(1 + \frac{Q^2}{4M_p^2}\right) + \frac{Q^2}{4M_p^2}\mu^2 F_1^2\right]\sin^2(\theta/2)$$

$$\pm \frac{2F_A}{M_p} F_1 \mu \left[Q^2 + \omega^2 \sin^2(\theta/2)\right]^{\frac{1}{2}} \sin(\theta/2) . \qquad (3.10)$$

Here $Q^2 = \vec{q}^2 - \omega^2$, $\omega = E_\nu - E_l$, $\vec{q} = \vec{p}_\nu - \vec{p}_l$, and $\mu = \mu_p - \mu_n = 4.71$. The coupling constant is $G = G_F \cos\theta_C \approx 10^{-5} M_p^{-2}$. (Similar formulae, which include the second class current effects and terms proportional to the charged lepton mass can be found in Llewellyn Smith (72) who, however, uses a slightly different notation for the form factors.)

The final lepton energy E_l, the initial neutrino energy E_ν, and the scattering angle θ (all in the laboratory frame) are related through

$$E_\nu p_l \cos\theta = 0.5 \times (M_f^2 - M_i^2 - m_l^2) + E_l M_i + E_\nu (E_l - M_i) , \qquad (3.11)$$

where M_i (M_f) is the mass of the initial (final) nucleon. If one can neglect the lepton mass as well as the neutron-proton mass difference, formula (3.11) is simplified, and

$$E_l = \frac{E_\nu}{1 + \dfrac{2E_\nu}{M} \sin^2(\theta/2)} .$$

In Figures 3.1 and 3.2 we show the calculated total cross sections (obtained from (3.10) by integrating over all angles) for the electron and muon neutrinos and antineutrinos. The $\pm$ sign associated with the vector-axial-vector interference lowers the cross section for the antineutrino induced reaction, compared with the neutrino induced reaction. The gradual change from the steep quadratic increase with energy near threshold to the linear increase at higher energies, and finally to a constant value is clearly visible.

At the upper end of the energy scale, the elastic reactions (3.1a) and (3.1b) constitute only a part of the inclusive reactions (3.3); the inclusive cross section at high energies is approximately linear with energy E_ν, and between 20 and 250 GeV one obtains $\sigma_{\nu N}^{CC}/E_\nu \approx 0.67 \times 10^{-38}$ GeV^{-1} and

$\sigma_{\bar{\nu}N}^{CC}/E_\nu \approx 0.34\times10^{-38}$ GeV^{-1}. In the simple parton model the ratio of the neutrino to antineutrino charged current inclusive cross sections on isoscalar targets depends on the ratio ε of the antiquark and quark momentum in the nucleon,

Figure 3.1 (a) Comparison of the total cross section for electron neutrinos on neutrons and antineutrinos on protons. The dashed curves are for free nucleons (Eq. (3.10)) and the solid curves are for bound nucleons (Eq. (3.12)). (b) Same for muon neutrinos and antineutrinos. (From Gaisser & O'Connell 86.)

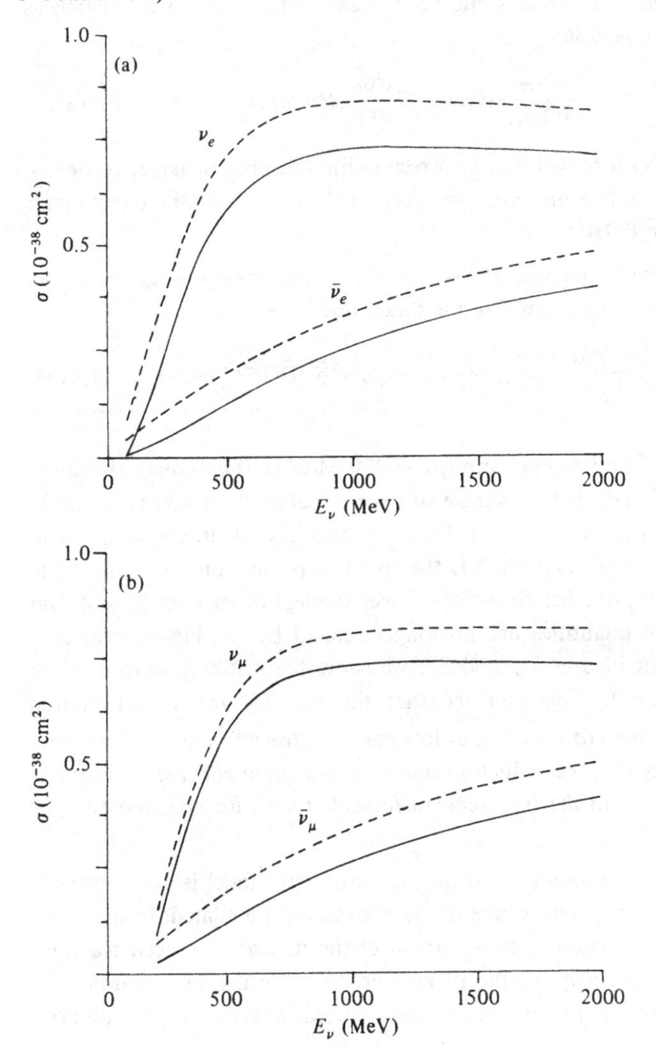

$$\frac{\sigma_{\nu N}^{CC}}{\sigma_{\bar{\nu} N}^{CC}} = \frac{1 + \frac{1}{3}\varepsilon}{\frac{1}{3} + \varepsilon} .$$

Interaction with nuclei. In order to extend the cross section formula (3.10) to the case of nucleons in nuclei, we use the noninteracting Fermi gas model. This model describes quite well electron scattering on nuclei in the quasi-free region and should describe the neutrino induced reactions for energies reasonably far from the threshold and up to several hundred MeV. The doubly differential cross section for (3.2a) and (3.2b) can be written as (Gaisser & O'Connell 86)

$$\frac{d^2\sigma}{d\Omega_l dE_l}(\bar{\nu},\nu) = C\frac{d\sigma_0}{d\Omega_l}R(q,\omega) , \tag{3.12}$$

where $C = Z$ (N) for $\bar{\nu}$ (ν) is the corresponding number of target nucleons, and $d\sigma_0/d\Omega$ is the free nucleon cross section (3.10) without the recoil factor $[1 + 2E_\nu/M_p \sin^2(\theta/2)]^{-1}$.

The response function R in the Fermi gas model is given by an integral over the momentum $\vec{p}$ of the target nucleons

$$R(q,\omega) = \frac{1}{\frac{4}{3}\pi p_F^3}\int \frac{d^3p M_p^2}{E_N E_{N'}}\delta(E_N+\omega'-E_{N'})\Theta(p_F-|\vec{p}|)\Theta(|\vec{p}+\vec{q}|-p_F) , \tag{3.13}$$

where p_F is the Fermi momentum ($p_F = 220$ MeV in the numerical evaluation), $\omega' = \omega - E_B$ (E_B is the average nucleon separation energy, taken as 25 MeV here), E_N, ($E_{N'}$) are the relativistic energies of the nucleon with momentum $\vec{p}$ $(\vec{p} + \vec{q})$, and $\Theta(x)$ is the usual step function. To obtain the total cross section, one has to integrate over the lepton energies E_l and over all angles. These quantities are no longer related by the kinematical constraints (3.11). In Figures 3.1 and 3.2 we show the resulting cross sections per nucleon (C=1). One can see that the binding and Fermi motion drastically affect the cross section at low energies; the effect gradually diminishes with energy and the reduction due to the binding effect is comparable to the uncertainties in the free nucleon formula (3.10) for neutrino energies above 1 GeV.

The above description based on the Fermi gas model is not applicable for energies near threshold, where discrete levels are populated. In that case, however, one can relate the cross section of the neutrino induced reactions to the rate of beta decay for the inverse process. We use this treatment in Chapter 5, where we also present a more careful analysis of the reactions involving deuterons as targets, including the strong interaction of the free final nucleons.

3.2 Neutral Current Reactions

Neutrinos interact with nucleons not only via the charged current weak interaction, but also via the neutral current weak interaction. However, reactions based on the neutral current do not result in the production of charged leptons as the reactions (3.1) and (3.2) do. Instead, they can be recognized by the recoil of the struck nucleon or nucleus, by the excitation of the struck nucleus, or by the production of pions or other hadrons. Assuming that universality holds, the neutral current interaction is expected to be insensitive to the neutrino flavor, and typically cannot give information about the phenomenon of neutrino oscillations. On the other hand, neutral current reactions can give valuable information about the structure of the weak current and, therefore, are important for testing the standard model.

Here, we shall discuss the neutral current neutrino scattering by nucleons and nuclei. The description of the neutral current reaction can be developed along lines essentially similar to the description of the charged

Figure 3.2 Same as Figure 3.1 for low energy electron neutrinos and antineutrinos.

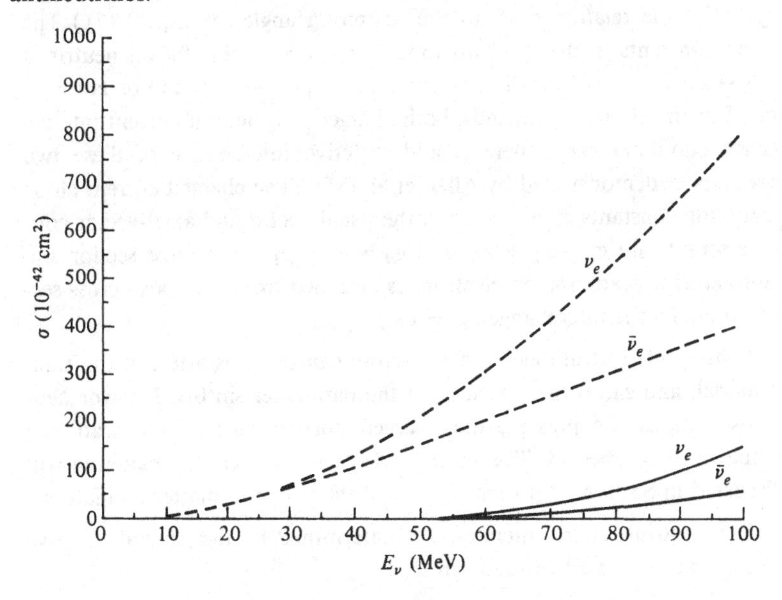

current reactions in Section 3.1. There are, however, two important differences. The first concerns the structure of the nucleon current; while the charged current is by definition an isovector, the neutral current can contain both isovector and isoscalar parts. The second difference is relevant only at relatively low energies or momentum transfers when $qR \leqslant 1$ (R is the nuclear radius and q is the momentum transfer). In this situation coherent elastic scattering occurs and the cross section, being proportional to the square of the number of scatterers, is enhanced.

Neutrinos striking a target will interact not only with the target nucleons (or nuclei) but also with the atomic electrons. Muon and tau neutrinos interact with electrons essentially only via the neutral current weak interaction which is independent of neutrino flavor and, therefore, insensitive to neutrino oscillations. (The charged current reactions $(\nu_\mu, e^-) \rightarrow (\nu_e, \mu^-)$ and $(\nu_\tau, e^-) \rightarrow (\nu_e, \tau^-)$, so called inverse muon and tau decays, are also possible but have a rather high threshold $m_l^2/(2m_e)$, i.e., 11 GeV and 3115 GeV, respectively. Even far above threshold, the cross section for these reactions is much lower than the cross section for the deep inelastic scattering on nucleons.)

The elastic reactions $(\nu_l, e^-) \rightarrow (\nu_l, e^-)$ all have cross sections of the general form

$$\frac{d\sigma}{dT} = \frac{2G_F^2 m_e}{\pi}\left[g_L^2 + g_R^2\left(1 - \frac{T}{E_\nu}\right)^2 - g_L g_R \frac{m_e T}{E_\nu^2}\right],$$

where E_ν is the initial neutrino energy and T is the electron recoil kinetic energy. (For the relation of T to the scattering angle see Eq. (3.17).) The coupling constants in the standard model are $g_R = \sin^2\theta_W$ for all neutrinos and $g_L = \sin^2\theta_W \pm 1/2$ for the electron and, respectively, muon or tau neutrinos. For the electron neutrinos, both charged and neutral current interactions act simultaneously; there is a destructive interference of these two interactions as demonstrated by Allen et al. (85). (The charged current alone has coupling constants $g_L = 1, g_R = 0$; the usual vector and axial-vector coupling constants are $g_V = g_L + g_R$, and $g_A = g_R - g_L$. The cross section for the antineutrino scattering on electrons is obtained from the above cross section formula by the interchange $g_L \leftrightarrow g_R$.)

A study of neutrino electron scattering can serve as a test of the standard model, and can furnish a value of the parameter $\sin^2\theta_W$. If flavor neutrino oscillations are present, the charged current part of the scattering amplitude will be affected. The special role of the $\nu_e e$ elastic scattering will be discussed in Section 5.8 in the context of the neutrino matter oscillations.

The neutrino quark interaction is determined by the general effective quark neutral current J^N of the form

$$(J_\mu^N)_{eff} = \alpha_V^{(0)} V_\mu^s + \alpha_V^{(1)} V_\mu^3 + \alpha_A^{(0)} A_\mu^s + \alpha_A^{(1)} A_\mu^3 - \alpha_{em} J_\mu^{em} \,. \tag{3.14}$$

The currents V^3 and A^3 are the third components of the *strong* isospin current, while V^s and A^s are their isoscalar counterparts. (They are normalized to $(1/2)\tau_3$ and $1/2$, respectively.) The electromagnetic current is $J^{em} = V^3 + V^s$. The coupling constants α depend on the model; in the standard model the current is $J_\mu = J_\mu^3 - 2\sin^2\theta_W J_\mu^{em}$ and, therefore, $\alpha_V^{(0)} = \alpha_A^{(0)} = 0$, $\alpha_V^{(1)} = \alpha_A^{(1)} = 1$, and $\alpha_{em} = 2\sin^2\theta_W$.

The predictions of the standard model can be tested, and the free parameter $\sin^2\theta_W$ determined in the study of deep inelastic charged and neutral current neutrino scattering. If the ratios R are defined as

$$R_{\nu N} = \frac{\sigma_{\nu N}^{NC}}{\sigma_{\nu N}^{CC}} \,, \quad R_{\bar{\nu} N} = \frac{\sigma_{\bar{\nu} N}^{NC}}{\sigma_{\bar{\nu} N}^{CC}} \,,$$

one obtains for isoscalar targets in the lowest approximation

$$R_{\nu N} = \rho \frac{\frac{1}{2} - \sin^2\theta_W + \frac{20}{27}\sin^4\theta_W + \varepsilon\left[\frac{1}{6} - \frac{1}{3}\sin^2\theta_W + \frac{20}{27}\sin^4\theta_W\right]}{1 + \frac{\varepsilon}{3}} \,, \tag{3.15}$$

and

$$R_{\bar{\nu} N} = \rho \frac{\frac{1}{6} - \frac{1}{3}\sin^2\theta_W + \frac{20}{27}\sin^4\theta_W + \varepsilon\left[\frac{1}{2} - \sin^2\theta_W + \frac{20}{27}\sin^4\theta_W\right]}{\frac{1}{3} + \varepsilon} \,. \tag{3.16}$$

Here, ε is the ratio of antiquark momentum to quark momentum in the nucleon, defined above, and $\rho = M_W^2/M_Z^2\cos^2\theta_W$; in the standard model, $\rho = 1$. The study of neutrino and antineutrino deep inelastic scattering is one of the main sources for precise values of $\sin^2\theta_W$.

Neutral current interaction with nucleons. Eq. (3.14) defines the effective quark current. In practice we have to deal with nucleons and, therefore, we must consider an expression analogous to (3.4) and (3.5) for the general matrix element of the nucleon weak neutral current, restricted only by the requirements of Lorentz invariance, time-reversal invariance, and isospin invariance. Leaving out, as before, the second class currents, the proper generalizations of (3.4) and (3.5) are

$$\langle p'\lambda' m_t' | (J_\mu)_T^{M_T} | p\lambda m_t \rangle = \bar{u}(p'\lambda') \left[F_1^{(T)}\gamma_\mu + F_2^{(T)}\sigma_{\mu\nu}q_\nu\right] \langle m_t' | t_{M_T} | m_t \rangle \,, \tag{3.4a}$$

$$<p'\lambda' m_t'|(J_{\mu 5})_T^{M_T}|p\lambda m_t> = \bar{u}(p'\lambda')\,[F_A^{(T)}\gamma_5\gamma_\mu - iF_P^{(T)}\gamma_5 q_\mu]\, <m_t'|t_{M_T}|m_t>, \tag{3.5a}$$

where λ and λ' are the corresponding helicities, and T, M_T are the strong isospin and its projection, respectively. The vector form factors can be obtained from the CVC hypothesis. Thus, $F_1^{(T)} = 1$ for both $T = 0$ and 1, and $F_1^{(0)} + 2MF_2^{(0)} = 0.8795$, the single nucleon isoscalar magnetic moment. The isovector induced pseudoscalar can again be found from the Goldberger-Treiman relation. We are then left with two unknown single nucleon form factors $F_A^{(0)}$ and $F_P^{(0)}$, of which the induced pseudoscalar is irrelevant for ultrarelativistic neutrinos.

In the standard model, and for nucleons composed only of u and d quarks, the axial part of the neutral current is a pure isovector, while the vector part of the neutral current contains an isovector component plus an isoscalar part proportional to $\sin^2\theta_W$. The cross section formula for neutrino and antineutrino elastic scattering on a nucleon can be found in Bilenky & Hosek (82). The fact that these two cross sections are not equal is caused, as in the case of the charged current reactions, by the vector-axial-vector interference and is an indication of parity nonconservation of the neutral current interaction.

In Figure 3.3 we show the total cross sections for neutrino and antineutrino scattering on a nucleon at low and intermediate energies. These cross sections were evaluated with the nucleon form factors given above; in addition, the isoscalar axial form factor $F_A^{(0)}$ was assumed to obey the dipole formula (3.7) (with the substitution $0.89 \leftrightarrow 0.71$), and the static quark model result $F_A^{(0)} = (3/5)F_A^{(1)}(0)$ (see Adler et al. 75) has been used.

The just stated assumption concerning $F_A^{(0)}$ is in disagreement with the standard model where $F_A^{(0)}$ vanishes since the axial current is a pure isovector. On the other hand, recent evidence suggests that the nucleon has some strange quark content. Thus the isospin symmetry is broken and the nucleon acquires an effective coupling $F_A^{(0)}$ that depends on the amount of strangeness in the nucleon.

For completeness, let us give the formulae governing the kinematics of the elastic scattering of massless particles of momentum p on targets of mass M, at rest in the laboratory frame. The recoil kinetic energy is given in terms of the neutrino center-of-mass scattering angle χ as

$$T_{recoil} = \frac{p^2}{2p + M}(1 - \cos\chi)\,. \tag{3.17}$$

In particular, the maximum recoil kinetic energy is

$$T_{recoil}^{Max} = \frac{2p^2}{M + 2p} .$$

In terms of the laboratory scattering angle θ (0° ⩽ θ ⩽ 90°) of the recoiling particle, the recoil kinetic energy is given by the relation

$$T_{recoil} = \frac{2p^2 M \cos^2\theta}{M^2 + 2pM + p^2 \sin^2\theta} ,$$

$$\cos\theta = \frac{p + M}{p}\left(\frac{T_{recoil}}{T_{recoil} + 2M}\right)^{1/2} . \qquad (3.18)$$

In Figure 3.4 we show, as an example, the recoil kinetic energy for a number of nuclear targets. (Note that at low energies, $p << M$, the recoil scales as p^2.)

Figure 3.3 Total neutrino and antineutrino scattering cross section vs. incident energy E_ν for the free nucleon targets. The standard model parameters were used in the calculation.

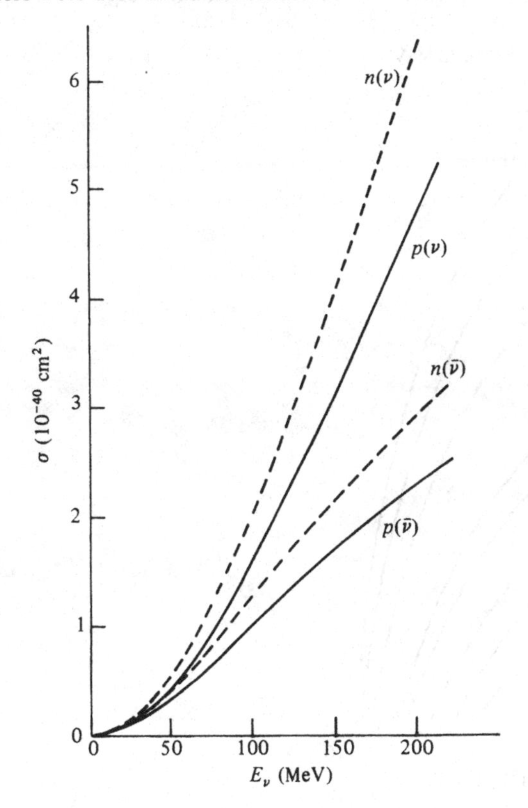

Neutral current interaction with nuclei. In analogy to Eq. (3.12), neutral current neutrino scattering on a nucleus can be written as

$$\frac{d\sigma}{d\Omega} = \sigma_0 F^2(|\vec{q}|, \omega, \theta) , \tag{3.19}$$

where F is the nuclear form factor which depends on the three-vector momentum transfer $\vec{q}$, excitation energy ω, and scattering angle θ. The form factor, in turn, can be expressed in terms of the nuclear matrix elements of operators which are angular momentum and isospin projections of the corresponding hadron currents (see Donnelly & Peccei 79). At low q values, for $q \leqslant p_F$ (Fermi momentum p_F = 200—250 MeV), only two operators are important for neutrino scattering; the inelastic scattering is dominated by the axial-vector dipole operator (analogue of the Gamow-Teller operator in beta decay), and the elastic scattering is dominated by the vector monopole operator (analogue of the Fermi operator in beta decay). It

Figure 3.4 Recoil kinetic energy vs. angle θ between the incident neutrino and the recoiling nucleus, for neutrinos of 100 MeV incident energy.

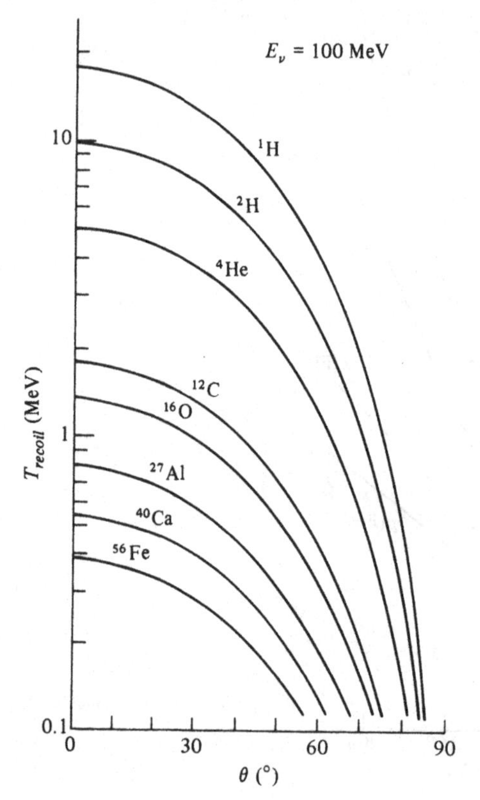

is often possible to deduce the values of the relevant matrix elements from the experimental data on gamma decay and on the charge changing processes such as beta decay or muon capture.

In Figure 3.5 we show the total cross section for the neutrino elastic scattering on a number of nuclei. The calculation, performed by Donnelly (81), is based on the simplest shell model of the nucleus (uncorrelated particles moving in the harmonic oscillator potential); it is assumed that the nuclei remain in their ground states. The Figure shows that nuclear binding has an appreciable effect on the cross section, as one can see by comparing the proton cross section with the cross section for deuterons or ^{4}He. Further inspection of Figure 3.5 shows that, up to an incident neutrino energy of approximately 50 MeV, the scattering is coherent and the resulting cross

Figure 3.5 Total neutrino elastic scattering cross section for the indicated nuclear targets vs. the incident neutrino energy. The standard model parameters were used in the calculation.

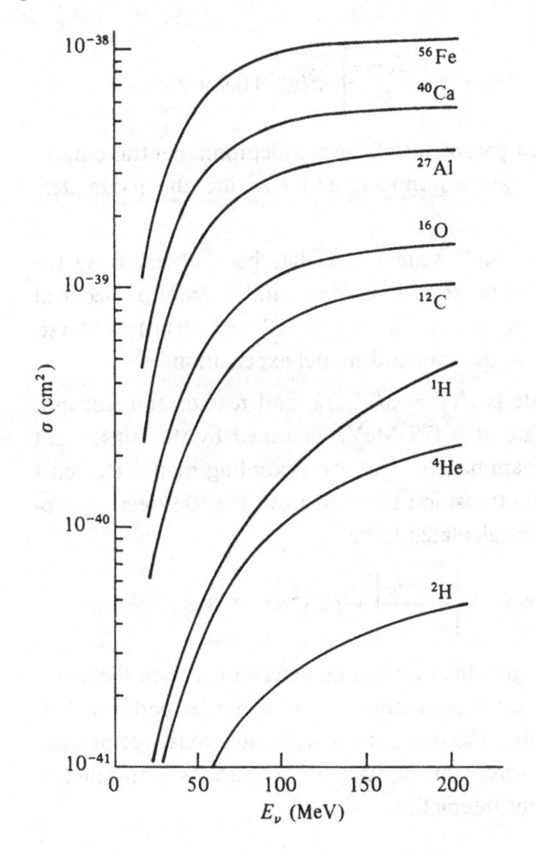

section is proportional to A^2. Above that energy the product qR becomes larger than unity and the coherence is lost over an ever growing interval of scattering angles. Ultimately, at energies somewhat larger than the maximum energy considered (200 MeV), the scattering becomes essentially noncoherent and the cross section becomes proportional to A.

Besides the elastic scattering, the neutrino neutral current interaction can also excite discrete nuclear states. By choosing the appropriate target and nuclear final state, one can determine the corresponding coupling constants $\alpha^{(T)}$ of the hadron current (3.14). Let us illustrate such a program using the excitation of ^{12}C at a meson factory, and of ^{7}Li at a reactor, as examples. For ^{12}C we consider the $I^{\pi},T = 1^{+},1$ state at 15.11 MeV and the predominantly $1^{+},0$ state at 12.71 MeV. The corresponding nuclear matrix elements can be deduced from the beta decay of the analogue states in ^{12}B and ^{12}N. Donnelly & Peccei (79) then calculate the following cross sections:

$$\sigma(\omega = 15.11 \text{ MeV}) = 1.1\times10^{-38} \left(\frac{E_{\nu}-\omega}{M} \right)^2 (\kappa\alpha_A^{(1)})^2 \text{ cm}^2 ,$$

and

$$\sigma(\omega = 12.71 \text{ MeV}) = 0.9\times10^{-38} \left(\frac{E_{\nu}-\omega}{M} \right)^2 \kappa^2(\alpha_A^{(0)}+0.06\alpha_A^{(1)})^2 \text{ cm}^2 .$$

Here κ is the overall strength parameter of the semileptonic neutral current interaction ($\kappa = 1$ in the standard model) and $\alpha_A^{(T)}$ are the parameters defined in (3.14).

Excitation of the 15.11 MeV state in ^{12}C has been observed by the KARMEN Collaboration (Drexlin et al. 91). For ν_e and $\bar{\nu}_{\mu}$ from μ^+ decay at rest the flux averaged cross section is $\sigma_{NC} = [10.8\pm5.1(\text{stat})\pm1.1(\text{syst})]\times10^{-42}\text{cm}^2$, in agreement with the standard model expectation.

For ^{7}Li the ground state is $I^{\pi},T = 3/2^{-},1/2$, and reactor antineutrinos could excite the $1/2^{-},1/2$ state at 0.478 MeV, identified by the subsequent emission of the 0.478 MeV gamma ray. The corresponding matrix elements are extracted from the gamma transition rate and from the ^{7}Be electron capture rate. The cross section is calculated to be

$$\sigma(\omega = 0.478 \text{ MeV}) = 6.9\times10^{-38} \left(\frac{E_{\nu}-\omega}{M} \right)^2 \kappa^2(0.32\alpha_A^{(0)}-0.33\varphi\alpha_A^{(1)})^2 \text{ cm}^2 ,$$

where $\varphi=\pm1$ is the relative sign which cannot be determined from the available input. If these and similar experiments could indeed be performed, it would be possible to determine the otherwise inaccessible axial-vector coupling constants and put severe constraints on the various nonstandard models of weak neutral current interaction.

Coherent scattering of relic neutrinos. In Chapter 7 we describe the relic neutrino sea which permeates the whole universe. A demonstration of the existence of this sea, and a measurement of its density, are problems of extraordinary importance. Background neutrinos have very low energy; their effective temperature is only about 1.9 K. Thus, these neutrinos could interact coherently over large distances, and several proposals have been made to utilize this coherence.

The interaction of low energy neutrinos with matter is usually described in terms of the index of refraction n for the propagation of the neutrino wave. This description is justified, because the neutrino wavelength is large compared with the interatomic spacing. The index of refraction is given in terms of the forward scattering amplitude as

$$n_{\nu,\bar{\nu}} = 1 + \frac{2\pi}{p^2}\sum_a N_a f^a_{\nu,\bar{\nu}} \,, \tag{3.20}$$

where N_a is the number density of scatterers of the type a. The forward scattering amplitude, in turn, is given by (Langacker et al. 83)

$$f^a_{\bar{\nu},\nu}(0) = \pm\frac{G_F E}{\pi\sqrt{2}} K(p,m_\nu)\, g^a_V \,, \tag{3.21}$$

where we have left out the term corresponding to the polarization of the scatterers. In (3.21) the upper (lower) sign refers to antineutrinos (neutrinos) and the function $K(p,m_\nu) = 1$ for $m_\nu = 0$ and $K(p,m_\nu) = 1/2$ for nonrelativistic neutrinos. The coupling constant g_V is the vector coupling constant in the effective Lagrangian. For protons and any neutrinos, $g^p_V = 1/2 - 2\sin^2\theta_W$, while for neutrons, $g^n_V = -1/2$. For electrons, we have to distinguish between electron neutrinos, where an interference with the charged current interaction occurs, and $g^e_V = 1/2 + 2\sin^2\theta_W$, and all other neutrinos where $g^e_V = 2\sin^2\theta_W - 1/2$. Using the nominal value $\sin^2\theta_W = 1/4$, we arrive at the formula for the index of refraction

$$n_{\bar{\nu}_e,\nu_e} - 1 = \pm\frac{G_F}{\sqrt{2}}\frac{EKN}{p^2} \times (3Z - A) \,, \tag{3.22}$$

where N is the atomic number density of the target, and

$$n_{\bar{\nu}_l,\nu_l} - 1 = \pm\frac{G_F}{\sqrt{2}}\frac{EKN}{p^2} \times (Z - A) \,, \tag{3.23}$$

for ν_μ and ν_τ.

The proposed experiments (see Langacker et al. (83) for a list of references) suggested measurement of the energy, momentum, or angular momentum transferred from the neutrino sea to a macroscopic target as it moves through the sea together with the motion of the Earth through the galaxy. However, Cabibbo & Maiani (82) and Langacker et al. (83) have shown that, in general, the transfer of energy, momentum, and (for isotropic targets) angular momentum vanishes to first order in weak interaction for a homogeneous (when time-averaged) incident neutrino beam. There is one exception to this conclusion, involving the use of a polarized target (Stodolsky 75). However, the expected effect appears to be so small that it is doubtful that it can ever be observed.

4

Heavy Neutrinos and Neutrino Decay

Several kinematic tests leading to limits of neutrino mass were discussed in Chapter 2. There, the conclusion was reached that the masses of the muon and tau neutrinos are below their present detection limits of 0.25 MeV and 35 MeV, respectively, and that the electron neutrino mass is at most 9 eV.

In this Chapter we shall explore the possibility that weak decays are accompanied not only by the dominant light (or massless) neutrinos described in Chapter 2, but also by one or more subdominant heavy neutrinos. These heavy neutrinos may well be heavier than the present detection limit. Thus a weak decay may proceed by a dominant branch, as well as by a subdominant heavy neutrino branch.

The basis for these ideas is neutrino mixing. A weak eigenstate ν_l ($l = e, \mu, \tau, \ldots$) may be a superposition of mass eigenstates ν_i ($i = 1, 2, 3, \ldots, N$), each with different mass m_{ν_i},

$$\nu_l = \sum_{i=1}^{N} U_{li} \nu_i \ , \qquad (4.1)$$

where U_{li} is a unitary mixing matrix. The dominant state, referred to above, may be (for example) $\nu_i \approx \nu_e$, and the subdominant heavy neutrino state $\nu_i = \nu_H$ is coupled to ν_e through (4.1) with a coupling strength $|U_{eH}|^2 << 1$. (There are restrictions on the number N in Eq. (4.1). The measured width of the intermediate neutral vector boson Z^0 implies that $N_\nu = 2.9 \pm 0.1$ (Dydak 90) for the number of weak doublet neutrinos with masses $m_\nu \leqslant M_Z/2$. Consideration of the primordial nucleosynthesis restricts $N_\nu \leqslant 3.4$ for the number of neutrinos at equilibrium at the neutrino decoupling temperature of the early universe (see Chapter 7).)

As a result of this admixture, several phenomena are expected to occur. Evidence for a heavy neutrino could be found in the lepton spectra accompanying two-body weak decays (as, for example, the kaon decay $K \rightarrow e\nu$), as well as three-body weak decays (such as nuclear beta decay

$Z \rightarrow (Z+1)+e^-+\bar{\nu}_e$). Also, a heavy neutrino may undergo spontaneous decay through various modes, including the decays $\nu_H \rightarrow \nu_1\nu_2\nu_3$, $\nu_H \rightarrow \nu_i\gamma$, $\nu_H \rightarrow \nu_i e^+e^-$, etc., some of which may be detectable by experiment. We note that neutrino decay in this context is closely related to flavor changing transitions between leptons, an example of which is the transition $\mu \rightarrow e\gamma$.

At the present time, there is no confirmed evidence for heavy neutrinos. A large number of experiments, some of which are described in the following Sections, provide stringent limits for the mixing parameters $|U_{eH}|^2$ and $|U_{\mu H}|^2$.

A discussion of subdominant heavy neutrinos and their effects on two- and three-body decays is presented in Section 4.1. Section 4.2 is concerned with the decay of neutrinos. Experimental evidence is given in each Section and summarized in two overview figures (Figures 4.5 and 4.6). As for reviews on the theoretical aspects, we mention the papers by Shrock (80, 81), while much of the phenomenology up to 1983 is documented in Boehm & Vogel (84).

4.1 The Admixture of Heavy Neutrinos

4.1.1 The two-body decay

We begin this Section with the two-body decay of a charged pseudoscalar meson M, such as a charged π, K, F, D or B meson, into a charged lepton and a neutrino. Take the decay of the kaon, $K \rightarrow e\nu_e$, where ν_e is the usual light (or massless) electron neutrino. By virtue of energy and momentum conservation, the electrons emitted in the decay of a K at rest are monochromatic and the observed electron spectrum exhibits but a single peak. A heavy neutrino branch will give rise to an extra electron peak at an energy below this main peak, the energy difference being related to the heavy neutrino mass. The situation is illustrated in the schematic Figure 4.1. By measuring the charged lepton momentum (or energy), one can determine in a straightforward way the mass of the admixed heavy neutrino. The branching ratio (height of the small peak in Figure 4.1) allows us to determine the mixing parameters $|U_{eH}|^2$.

In a two-body decay of a pseudoscalar meson M into a charged lepton and neutrino,

$$M^- \rightarrow l^- + \bar{\nu}_l, \qquad M^+ \rightarrow l^+ + \nu_l\,, \tag{4.2}$$

the momenta and energies of the leptons as well as the decay rate of M depend on the lepton masses m_l, and m_ν through the ratios

$$\delta_l = \left(\frac{m_l}{m_M}\right)^2, \quad \delta_\nu = \left(\frac{m_\nu}{m_M}\right)^2 . \tag{4.3}$$

We assume that δ_l is known; our goal is to determine δ_ν.

In the coordinate frame where M is at rest, the charged lepton l and the neutrino ν_l are emitted back to back, each with the same momentum p. We have (in units $\hbar = c = 1$)

$$m_M = E_l + E_\nu \equiv (m_l^2 + p^2)^{\frac{1}{2}} + (m_\nu^2 + p^2)^{\frac{1}{2}},$$

and, consequently

$$\begin{aligned} E_l &= (m_M^2 + m_l^2 - m_\nu^2)/2m_M \\ E_\nu &= (m_M^2 + m_\nu^2 - m_l^2)/2m_M \end{aligned} , \qquad (4.4)$$

as well as

$$p = \frac{1}{2m_M}[\, m_M^4 + m_l^4 + m_\nu^4 - 2m_M^2 m_l^2 - 2m_M^2 m_\nu^2 - 2m_l^2 m_\nu^2\,]^{\frac{1}{2}}. \qquad (4.5)$$

The expressions (4.4) and (4.5) can also be written in a more compact form using (4.3) and the notation

$$\lambda(x,y,z) \equiv x^2 + y^2 + z^2 - 2yz - 2zx - 2xy\, . \qquad (4.6)$$

This gives

$$E_l = \frac{m_M}{2}(1 + \delta_l - \delta_\nu),\ E_\nu = \frac{m_M}{2}(1 + \delta_\nu - \delta_l)\ , \qquad (4.4')$$

$$p = \frac{m_M}{2}[\lambda(1,\delta_l,\delta_\nu)]^{\frac{1}{2}}\, . \qquad (4.5')$$

From a measurement of p we can determine m_ν from the relation

$$m_\nu^2 = m_M^2 + m_l^2 - 2m_M(p^2 + m_l^2)^{\frac{1}{2}}\, . \qquad (4.7)$$

Figure 4.1. Illustration of the effect of a heavy neutrino ν_H on the momentum spectrum of positrons in K^+ decay.

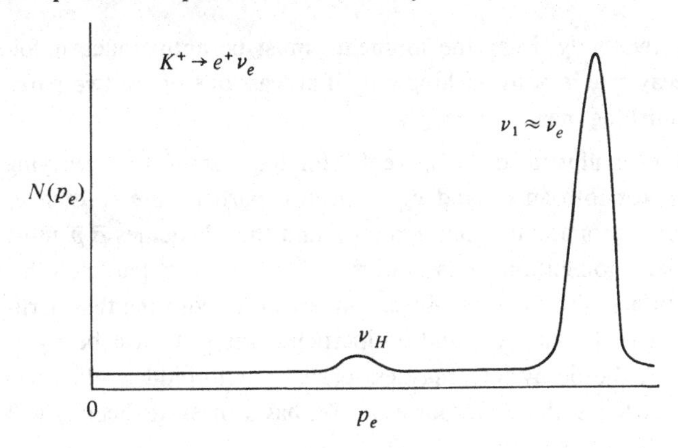

If the neutrino ν_l is in a mixed state described by (4.1), then the N components contribute to the decay incoherently. (This situation is in contrast to the case of neutrino oscillations as discussed in Section 5.2). The lepton spectrum will exhibit one or more extra peaks, as illustrated in Figure 4.1 and the momenta p_H and masses m_{ν_H} for each heavy neutrino are related by (4.7). For the energy difference between the main peak ($m_\nu = 0$) and the extra peak due to m_{ν_H} one finds $\Delta E = m_{\nu_H}^2/2m_M$.

The intensity of each component is proportional to the mixing parameter $|U_{lH}|^2$ and depends on the mass m_{ν_H}. Shrock (80, 81) has pointed out that a significant enhancement of the heavy neutrino branch may occur in some cases.

The rate of decay of $M \to l + \nu$ involving a massive but unmixed neutrino has been evaluated by Konopinski (66) and is given by

$$\Gamma(M \to l + \nu) = G_F^2 \frac{f_M^2}{2\pi} pE_l E_\nu (1 + \vec{v}_l \cdot \vec{v}_\nu) , \tag{4.8}$$

where G_F is the usual Fermi coupling constant, f_M is the coupling constant for the decay of the meson M (which is independent of the lepton flavor and lepton mass), and $\vec{v}_l$ and $\vec{v}_\nu$ are the charged lepton and neutrino velocities, respectively.

In our compact notation we have

$$\Gamma(M \to l + \nu) = G_F^2 \frac{f_M^2}{2\pi} \frac{m_M^3}{4} \lambda^{\frac{1}{2}}(1,\delta_l,\delta_\nu)[\delta_l + \delta_\nu - (\delta_l - \delta_\nu)^2] . \tag{4.9}$$

The neutrino mass dependence enters through the "phase-space factor" $pE_l E_\nu$, but more significantly, through the term $(1 + \vec{v}_l \cdot \vec{v}_\nu)$. This term reflects the correlation between the charged lepton and neutrino momenta in a form characteristic for a $0 \to 0$ transition. It is often referred to as the "helicity factor".

Since in a two-body decay the momenta must be antiparallel, it follows that the decay rate is nonvanishing only if at least one of the two particles has a nonvanishing mass, implying $v < 1$.

The situation is illustrated in Figure 4.2 for the case of a π^+ decaying into a μ^+ and ν_μ (or into an e^+ and ν_e). The two particles are emitted in opposite directions (momentum conservation), and their helicities $\vec{\sigma} \cdot \vec{p}$ must be equal (angular momentum conservation). For massless particles this situation is in conflict with the laws of weak interaction, requiring that particles (the ν_μ) must be left-handed, and antiparticles (the μ^+) must be right-handed. Therefore, the decay into massless particles is forbidden. If, however, one of the particles, the μ^+ in our example, has a mass, its helicity will deviate from the full value in the measure by which its velocity v differs from the velocity of light. The helicity factor $(1 + \vec{v}_l \cdot \vec{v}_\nu)$ states this conflicting requirement.

We note that, while (4.8) and (4.9) were derived in standard $V-A$ theory, the helicity factor $(1 + \vec{v}_l \cdot \vec{v}_\nu)$ is a general feature of any combination of V and A couplings with an arbitrary degree of parity nonconservation.*

In an experiment, one compares the area of the extra peak (or a limit of this quantity) to the area of the observed main peak corresponding to $m_\nu = 0$. From (4.9) it follows that the ratio of these areas, i.e., the rate of $M \to l + \nu_H$ relative to the conventional decay with massless ν is given by

$$R_{lH} = \bar{\rho}_{eH} |U_{lH}|^2 \, ,$$

where

$$\bar{\rho}_{lH} = \frac{\lambda^{\frac{1}{2}}(1,\delta_l,\delta_{\nu_H})(\delta_l + \delta_{\nu_H} - (\delta_l - \delta_{\nu_H})^2)}{\delta_l(1 - \delta_l)^2} \, . \tag{4.11}$$

In Figures 4.3 and 4.4 we show the kinematic factor $\bar{\rho} \equiv \bar{\rho}_{lH}$. The enhanced sensitivity to decays involving positrons (or electrons) plus a heavy neutrino is obvious.

Let us now review the experimental situation. Several searches for

Figure 4.2. Illustration of helicity suppression in π^+ decay. The quantities p_ν and p_l are the neutrino and lepton momenta respectively, and σ is their spin. The Figure shows how momentum and angular momentum conservation are in conflict with the requirement of a left-handed particle and right-handed antiparticle imposed by weak interaction.

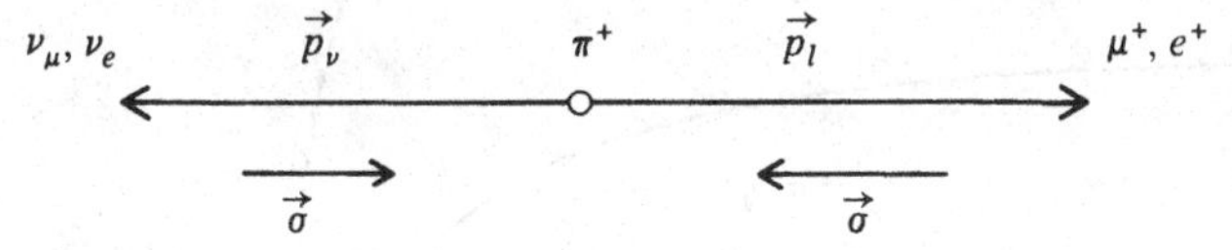

ν_μ, ν_e $\vec{p}_\nu$ π^+ $\vec{p}_l$ μ^+, e^+

$\vec{\sigma}$ $\vec{\sigma}$

---

*In order to verify whether or not the weakly admixed heavy neutrino obeys the standard $V-A$ theory, it would be necessary also to measure the longitudinal polarization of the lepton. In $V-A$ theory this polarization P is given by

$$P = \frac{\pm p(E_l - E_\nu)}{E_\nu E_l - p^2} = \frac{\pm\lambda^{\frac{1}{2}}(1,\delta_l,\delta_\nu)(\delta_l - \delta_\nu)}{\delta_l + \delta_\nu - (\delta_l - \delta_\nu)^2} \, . \tag{4.10}$$

The $\pm$ signs refer to lepton or antilepton, respectively. For $m_\nu = 0$ we have $P = \pm 1$, independent of m_l. For non $V-A$ theories (non left-handed couplings) the polarization will differ from (4.10).

extra peaks in the energy spectra of muons and electrons from pion and kaon decay have been conducted. Up to now, no such peaks have been found and, consequently, there is no evidence for heavy neutrinos. In Figures 4.5 and 4.6 we show the upper limits for the mixing coeeficients $|U_{eH}|^2$ and $|U_{\mu H}|^2$. Below, we shall discuss briefly how this information was obtained.

We begin with the pion decays $\pi \to \mu\nu$ and $\pi \to e\nu$. Abela et al. (81) have studied the decay $\pi^+ \to \mu^+\nu_\mu$ with the pion beam of the PSI ring-cyclotron. The μ^+ spectrum was measured with a plastic scintillation counter. A peak search below the main peak at $T_\mu = (m_\pi - m_\mu)^2/2m_\pi = 4.1$ MeV associated with known light neutrino ν_μ has not revealed a heavy neutrino in the mass range of $5 < m_{\nu_H} < 30$ MeV, with an upper limit for the coupling strength $|U_{\mu H}|^2$ shown in Figure 4.6. The heavy neutrino mass region was later extended down to a lower limit of

Figure 4.3. Kinematical factor $\bar{\rho}$ (4.11) for kaon and pion decays involving muons in the final state. The maximum neutrino mass in each channel is $m_\nu^{Max} = m_M - m_l$.

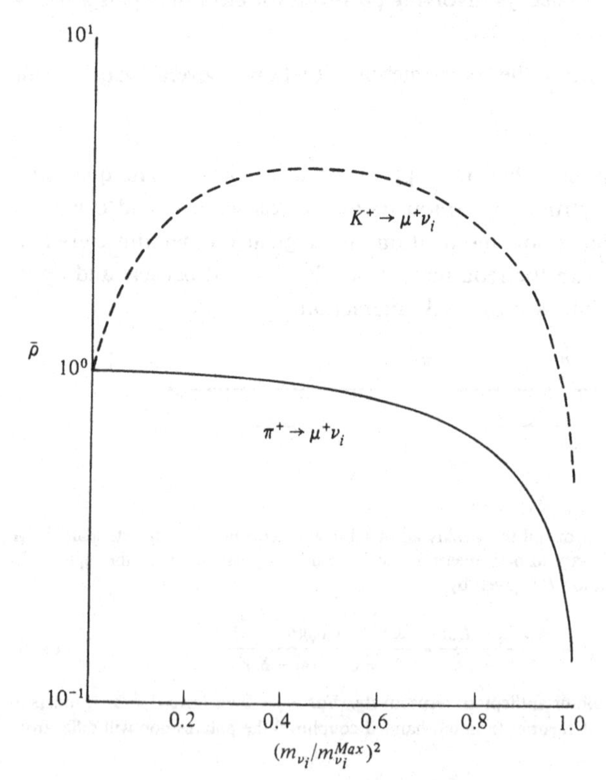

$m_H > 2.5$ MeV by Minehart et al. (84) using a Ge detector, again without finding a peak. Figure 4.6 summarizes the limits of $|U_{\mu H}|^2$ for this experiment as well.

A similar peak search has been carried out for the decay $\pi^+ \to e^+\nu_e$ with the pion beams from the TRIUMF cyclotron at Vancouver (Canada) and the PSI ring-cyclotron. In an experiment by Bryman et al. (83), the positive pions were stopped and allowed to decay in a scintillation counter telescope. The positron spectrum was measured with a large NaI scintillation counter. A complication arose because, on account of the mentioned helicity factor, the branch $\pi^+ \to e^+\nu_e$ is strongly suppressed. As a result, the $\pi^+ \to e^+\nu_e$ signal is superimposed on a large positron background from the $\mu^+ \to e^+\bar{\nu}_\mu\nu_e$ decay. The relatively long lifetime of the μ^+ (2.2 μs), however, makes it possible to reduce the μ^+ decay background by timing. The posi-

Figure 4.4. Same as Figure 4.3 but for positrons in the final state.

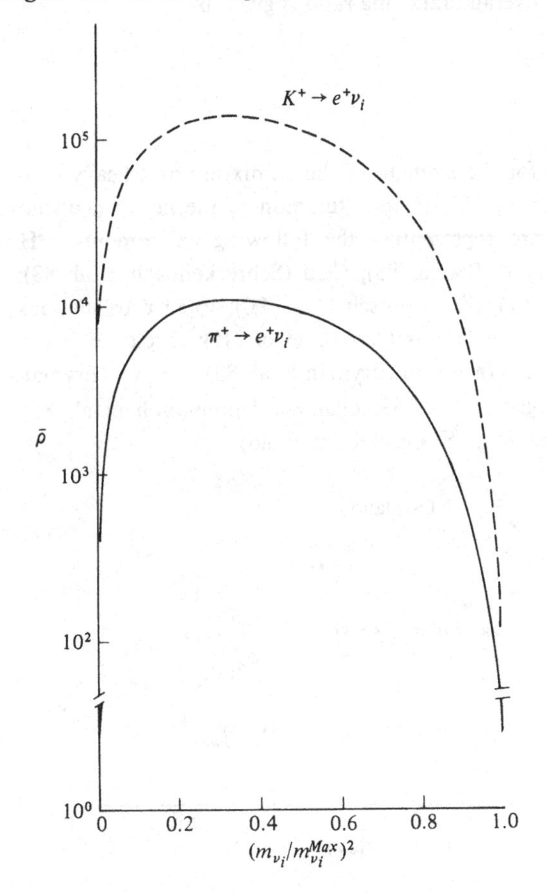

tron spectrum in the NaI thus is recorded as a function of time with respect to the pion stop signal in the telescope. Figure 4.7(a) shows the "prompt" spectrum (2—22 ns after pion stop), with the electron peak at 66 MeV superimposed on the muon decay spectrum. The delayed spectrum (135—160 ns), representing muon decay only, was subtracted from the prompt spectrum to obtain the corrected prompt spectrum shown in Figure 4.7(b). Following this procedure, no peaks corresponding to heavy neutrinos were seen in the mass range between 4 MeV and 135 MeV, with limits of $|U_{eH}|^2$ shown in Figure 4.5. Azuelos et al. (86) and De Leener-Rosier et al. (86) have carried out a similar experiment.

Of interest also is the study of the branching ratio in two competing two-body decays. The relative rate of two decays, such as $\pi \rightarrow e\nu$ and $\pi \rightarrow \mu\nu$, is a sensitive gauge for heavy neutrino admixture, as will be briefly outlined below.

If, as in the case of the pion decay, the meson M can decay into several lepton flavors, the overall branching ratio is given by

Figure 4.5. Upper limits for the strength of the admixture of a heavy neutrino to an electron neutrino, $|U_{eH}|^2$, as a function of the heavy neutrino mass m_{ν_H}. The curves are representing the following experiments: ^{3}H (Simpson 81); ^{35}S (Markey & Boehm 85); ^{64}Cu (Schreckenbach et al. 83); ^{63}Ni (Hetherington et al. 87); ^{20}F (Deutsch et al. 90); $\nu_e \rightarrow x$ oscillations (Zacek et al. 86); $\nu_\mu \rightarrow \nu_e$ oscillations (Ahrens et al. 85); reactor $\nu \rightarrow ee\nu$ (Oberauer et al. 87); $(\pi \rightarrow e\nu)/(\pi \rightarrow \mu\nu)$ (Bryman et al. 83); $\pi \rightarrow e\nu$ (Bryman et al. 83); CHARM (Bergsma et al. 83; Gall 84; Dorenbosch et al. 86); BEBC (Cooper-Sarkar et al. 85); PS (Bernardi et al. 86).

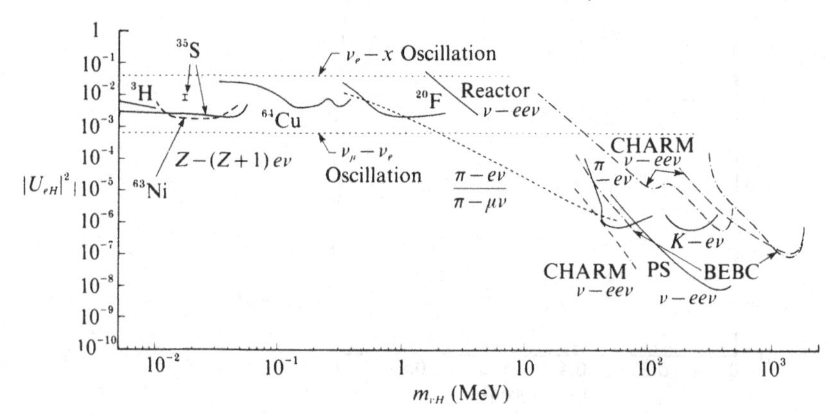

$$R(M) = \frac{\sum_{i=1}^{N} |U_{l_1 i}|^2 \lambda^{\frac{1}{2}}(1,\delta_{l_1},\delta_{\nu_i})[\delta_{l_1}+\delta_{\nu_i}-(\delta_{l_1}-\delta_{\nu_i})^2]\times\Theta(m_M-m_{l_1}-m_{\nu_i})}{\sum_{i=1}^{N} |U_{l_2 i}|^2 \lambda^{\frac{1}{2}}(1,\delta_{l_2},\delta_{\nu_i})[\delta_{l_2}+\delta_{\nu_i}-(\delta_{l_2}-\delta_{\nu_i})^2]\times\Theta(m_M-m_{l_2}-m_{\nu_i})} . \quad (4.12)$$

In (4.12) the summation is over all massive neutrinos which kinematically could participate in the decay, that is, $m_{\nu_i} \leqslant m_M - m_l$. We should stress that (4.12) is valid only if universality holds, i.e., if l_1 and l_2 and their associated neutrinos have the same weak interaction couplings.

In Bryman's experiment, this branching ratio can easily be obtained from the ratio of the areas of the prompt and delayed spectra and was found to be $R = (1.218 \pm 0.014) \times 10^{-4}$. This result is in agreement with R_{th} for the case where there are no heavy neutrinos. The calculated branching ratio R_{th}, including radiative corrections, is given by (Kinoshita 59)

$$R_{th} = R_o(1 - 16.9\ \alpha/\pi) = 1.233 \times 10^{-4} \quad (4.13)$$

Figure 4.6. Upper limits for the strength of the admixture of a heavy neutrino to a muon neutrino, $|U_{\mu H}|^2$, as a function of the heavy neutrino mass $m_{\nu H}$, $\pi \rightarrow \mu\nu$ (Abela et al. 81, full curve; Minehart et al. 84, dashed curve); $\mu^-\ {}^3He$ (Deutsch et al. 83); $K \rightarrow \mu\nu$ (Yamazaki 84); BEBC (Cooper-Sarkar et al. 85); PS (Bernardi et al. 86).

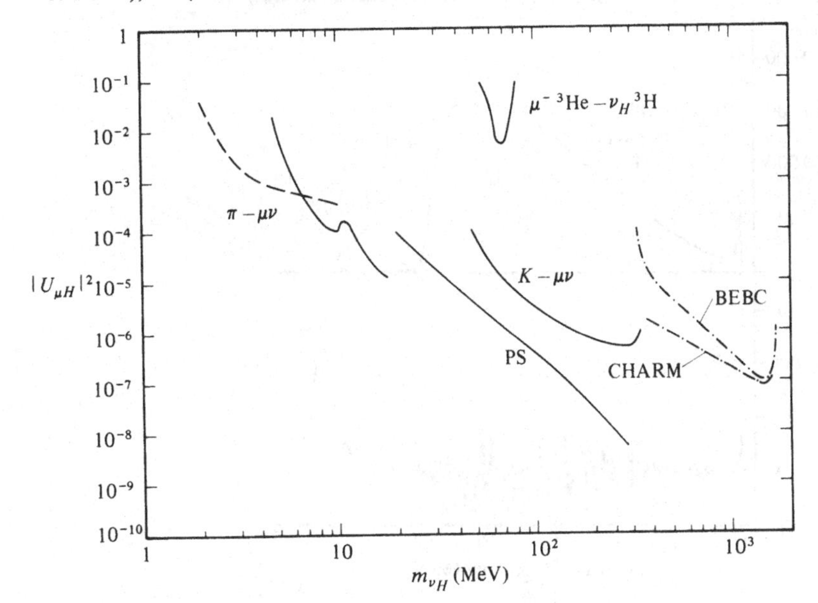

with

$$R_o = \frac{\Gamma(\pi - e\nu_e)}{\Gamma(\pi - \mu\nu_\mu)} = \frac{m_e^2}{m_\mu^2}\left(\frac{m_\pi^2 - m_e^2}{m_\pi^2 - m_\mu^2}\right)^2 = 1.284 \times 10^{-4} . \qquad (4.14)$$

Limits for the admixture $|U_{eH}|^2$ can be obtained from (4.12), taking $|U_{\mu H}|^2$ as derived from the peak search. These upper limits in the range $2.5 < m_{\nu_H} < 30$ MeV, again, are shown in Figure 4.5. For $m_{\nu_H} > 35$ MeV, the $\pi \to \mu\nu$ channel is kinematically forbidden, while $\pi \to e\nu$ is still allowed, and the test remains applicable up to $m_{\nu_H} \leqslant 55$ MeV.

The two-body decays of the charged kaon, $K^+ \to \mu^+\nu_\mu$ and $K^+ \to e^+\nu_e$ were studied extensively by a group at the KEK Laboratory in Japan (Yamazaki et al. 84). The laboratory's proton synchrotron of 12 GeV beam energy produced a kaon beam of 550 MeV. This beam passed through a degrader and then was stopped in a stack of plastic scintillation counters. The decay muons and electrons were momentum analyzed ($\Delta p/p \approx 10^{-2}$) with the help of a magnetic field and several planes of wire counters. Upon exiting the magnetic spectrometer, the particles enter a range counter telescope which, together with a time-of-flight signal, serves to

Figure 4.7. Positron energy spectrum from $\pi \to \mu\nu$, $\mu \to e\bar{\nu}\nu$ and $\pi \to e\nu$ decays. The upper figure (a) shows a prompt spectrum. Upon subtraction of the delayed $\mu \to e\bar{\nu}\nu$ spectrum the pure $\pi \to e\nu$ spectrum is obtained and depicted in the lower figure (b). (From Bryman et al. 83.)

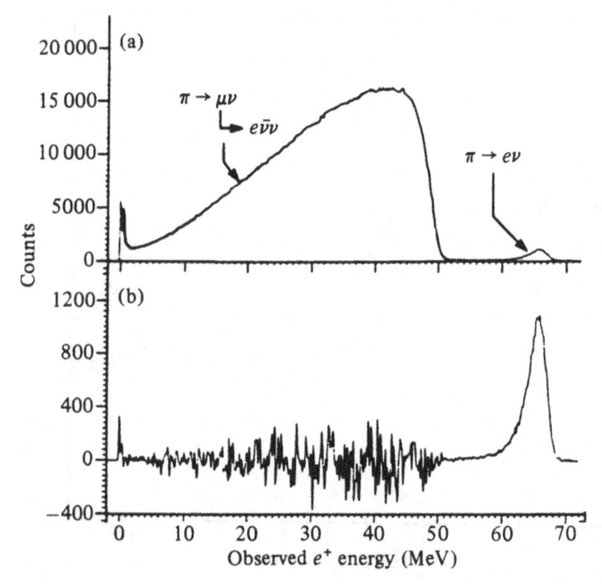

identify the muons or electrons. The experimental arrangement is shown in Figure 4.8. NaI gamma counters serve to reduce the backgrounds from $K^+ \rightarrow \mu^+\nu\pi^o$ and $K^+ \rightarrow \mu^+\nu\gamma$ decays. The momentum spectrum for the muons is shown in Figure 4.9. There is no evidence for a peak in the momentum interval between 110 and 230 MeV and the corresponding upper limit for $|U_{\mu H}|^2$ obtained from a peak fitting analysis and from the kinematic factor $\bar{\rho}$ (Figure 4.3) is exhibited in Figure 4.6. In a similar way an upper limit for $|U_{eH}|^2$ was obtained and is displayed in Figure 4.5.

Muon capture, $\mu^- + ({}_ZA)_i \rightarrow \nu_\mu + ({}_{Z-1}A)_f$, is another example of a weak process with a two-body final state. If the final neutrino has mass m_ν, then its momentum, and the momentum of the final recoiling nucleus, is given by

$$p(m_\nu) = (\Delta^2 - m_\nu^2)^{\frac{1}{2}}[(1 - \Delta/2M)^2 - (m_\nu/2M)^2]^{\frac{1}{2}} , \qquad (4.15)$$

where Δ is the available energy and M is the initial mass,

$$\Delta = M(A_i) + m_\mu - E_{binding} - M(A_f) \equiv M - M(A_f) . \qquad (4.16)$$

Figure 4.8. Experimental arrangement for the study of secondary positron peaks in $K^+ \rightarrow e^+\nu$ decay. (From Hayano et al. 82.)

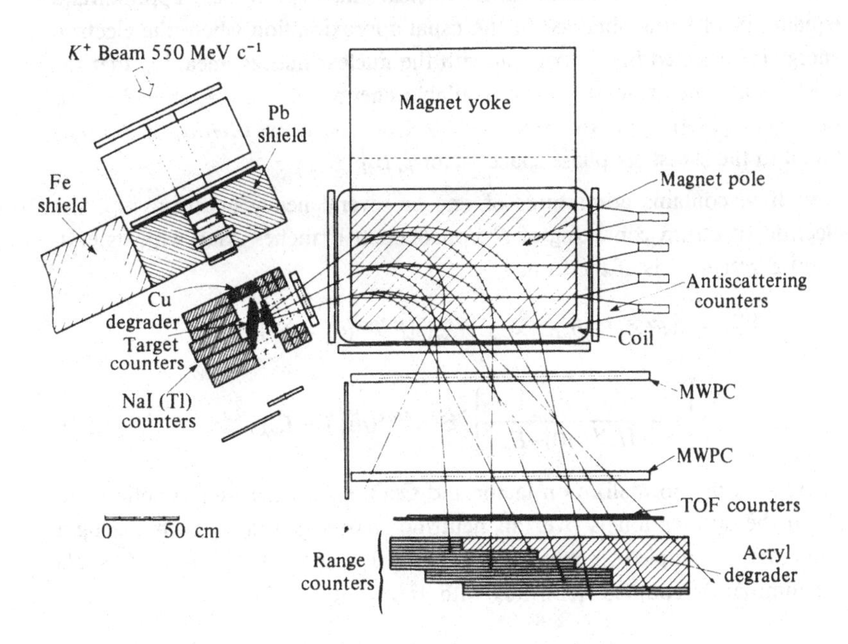

The intensity of the branch with neutrino of mass m_{ν_H}, relative to the muon capture with $m_\nu = 0$, is

$$R = |U_{\mu H}|^2 \frac{p^2[m_{\nu_H})(1 - p(m_{\nu_H})/M]}{p^2(0)[1 - p(0)/M]} . \tag{4.17}$$

Deutsch et al. (83) have analyzed muon capture in ^3He from this point of view and have extracted limits for the mixing parameter $|U_{\mu H}|^2$ (see Figure 4.6).

4.1.2 *Nuclear beta decay as a three-body decay*

Next, we consider the effect of heavy neutrinos on the spectrum shape of the allowed nuclear beta decays

$$({}_ZA)_i \rightarrow ({}_{Z+1}A)_f + e^- + \bar{\nu}_e , \tag{4.18}$$

or its analogues with positron emission. For ground-state to ground-state transitions and neutrino mass m_ν, the maximum electron energy (endpoint energy) is obtained when the recoiling final nucleus and neutrino move as one body of mass $M_f + m_\nu$. In that case

$$E_e^{Max}(m_\nu) = \frac{M_i^2 + m_e^2 - (M_f + m_\nu)^2}{2M_i} \approx (M_i - M_f - m_\nu) . \tag{4.19}$$

Here M_i (M_f) are the initial (final) nuclear masses. The last approximate equality in (4.19) is obtained in the usual approximation where the electron energy is neglected in comparison with the nuclear masses, meaning that the electron and neutrino share the available energy, $E_\nu + E_e = M_i - M_f$. In this (very good) approximation the electron spectrum dN/dE_e is proportional to the statistical phase space factor $p_e E_e p_\nu E_\nu$.

If ν_l contains admixtures of one or several heavy neutrinos ν_H, the electron spectrum consisting of N noncoherent branches, each with its endpoint $E^{Max}(m_{\nu_H})$, is of the form

$$\frac{dN}{dE_e} = AF(Z_1E_e)p_eE_e \times \sum_{i=1}^{N} |U_{ei}|^2[E^{Max}(0) - E_e]^2$$

$$\times \left\{1 - \frac{m_{\nu_i}^2}{[E^{Max}(0) - E_e]^2}\right\}^{\frac{1}{2}} \Theta(E^{Max}(m_{\nu_i}) - E_e) , \tag{4.20}$$

where A is the normalization factor and $\Theta(x)$ is the usual step function. In (4.20) the summation is over all neutrino mass eigenstates, both the light dominantly coupled neutrinos $|U_{ei}|^2 \approx 1$, and the heavy ($i = H$) subdominantly coupled neutrinos with $|U_{eH}|^2 << 1$. The spectrum, therefore, has discontinuities in slope ("kinks") at each endpoint energy $E^{Max}(m_{\nu_H})$. All beta decays are expected to display such kinks independent of the initial and final nucleus. The position of the kink (its distance from the endpoint $E^{Max}(0)$) determines m_{ν_H}; the step size in the slope of the spectrum determines $|U_{eH}|^2$.

In Figure 4.10 we show a schematic example of such a spectrum in the form of the normalized difference between the electron spectrum with and without the admixture of one heavy neutrino. In constructing the Figure, we assumed that the largest endpoint $E^{Max}(0)$ is known, being the same for

Figure 4.9. Momentum spectrum of all charged particles (upper curve) and of the μ^+ only (lower curve). (From Yamazaki et al. 84)

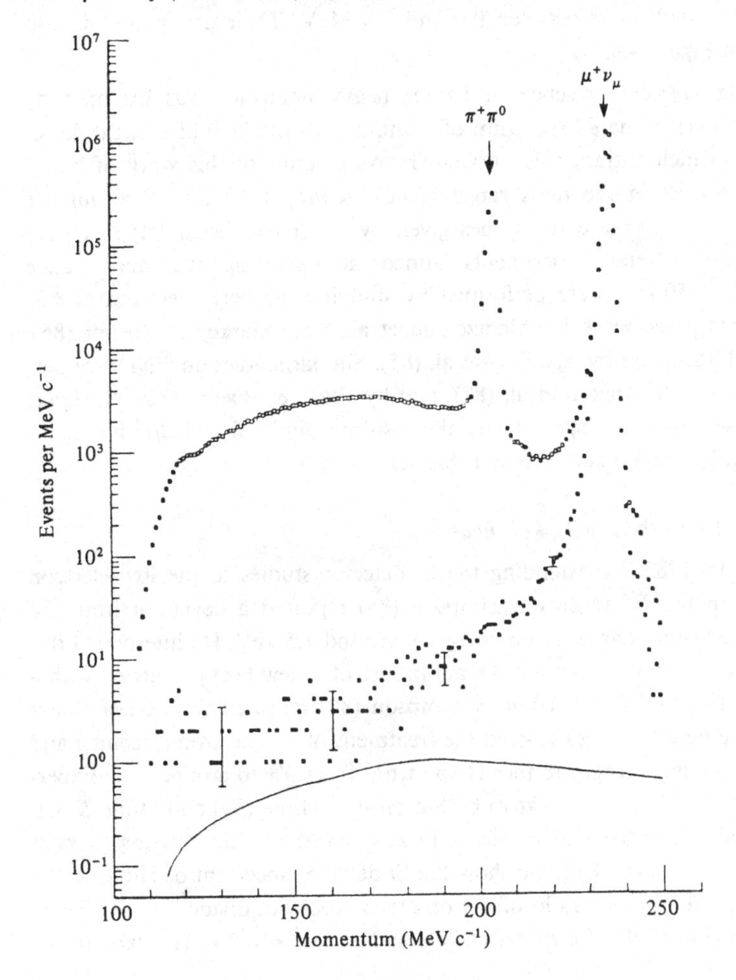

both spectra. In the analysis of the experimental data, this may not be the case and the corresponding plot can look different in the vicinity of the end-point.

An experimental study of the spectral shape in e^+ and e^- decays of ^{64}Cu carried out by Schreckenbach et al. (83) has not revealed such a kink, and the resulting upper limit for $|U_{eH}|^2$ is shown in Figure 4.5. This experiment was performed at the ILL (Laue-Langevin Institute) at Grenoble using a magnetic double focusing spectrometer with the source at the reactor core (see Section 5.4). Both the e^+ and e^- spectra were found to be represented by horizontal lines in the energy range of 75—600 keV, if plotted as the relative deviation from the "Kurie plot". The resulting heavy neutrino mass probed in this experiment was $30 < m_{\nu_H} < 460$ keV. Recently, Deutsch et al. (90) have searched for a kink in the beta spectrum of ^{20}F associated with a neutrino with mass between 0.4 and 2.8 MeV. Their upper limit is also shown in Figure 4.5.

The possible presence of lighter heavy neutrinos was explored by Simpson (81) in the e^- spectrum of tritium with the help of a Si(Li) detector, into which tritium was implanted. As a result of this work, if heavy neutrinos exist in the mass range 100 eV $< m_{\nu_H} <$ 10 keV, their mixing strength cannot exceed the values given by the line (labeled ^{3}H) in Figure 4.5. Also, several experiments aimed at exploring the mass range $5 < m_{\nu_H} < 50$ keV were performed by studying the beta spectrum of ^{35}S. We mention the work by Altzitzoglou et al. (85), Markey & Boehm (85), Ohi et al. (85), and by Apalikov et al. (85). Similar studies on ^{63}Ni were performed by Hetherington et al. (87), and by Wark & Boehm (86). No kinks were observed in the spectra and the resulting limits for $|U_{eH}|^2$ are again shown in Figure 4.5 and also in Table 4.1.

4.1.3 *A hypothetical 17 keV neutrino*

In 1985, by extending the Si detector studies to the low electron energies in the ^{3}H spectrum, Simpson (85) reported a deviation from the Fermi spectrum at an electron energy of around 1.5 keV. He interpreted this "kink" as originating from a $\approx$3% admixture of a new heavy neutrino with a mass of 17 keV. Later, Hime & Simpson (89) repeated this measurement with a Ge detector. They revised the treatment of the electron screening and thus the mixing strength in the ^{3}H spectrum from 3% to around 1%. Experiments with ^{35}S (E_0 = 167 keV) by Simpson & Hime (89) and Hime & Jelley (91) also reported evidence for a 17 keV neutrino with a mixing strength of 0.84%. In Figure 4.11, we show the Si detector spectrum of Hime & Jelley (91) presented as a ratio of the observed spectrum divided by the Fermi spectrum (Eq. (4.20)) for m_ν = 17 keV and $|U_{eH}|^2$ = 0.9%. The experimental spectrum is normalized to the Fermi spectrum over the upper part (about 15 keV) of the spectrum which is unaffected by the kink. Evidence for a 17 keV neutrino was also reported by Sur et al. (91) from a measurement of the beta spectrum from ^{14}C implanted into a Ge detector, as well as by Zlimen et al. (91) from a measurement of the bremsstrahlung spectrum of ^{71}Ge.

Figure 4.10. Schematic example of the effect of heavy neutrino mixing on the electron spectrum in beta decay. For illustration an unrealistically large value $|U_{eH}|^2 = 0.10$ was chosen. The upper part shows the expected spectrum $N(E)$, in arbitrary units. The dashed curve indicates extrapolation of the spectrum $N_{ext}(E)$ assuming the allowed shape without massive neutrinos and using the part of the spectrum above the discontinuity at $E_0 - m_\nu$ as the basis of the fit. The lower part of the Figure shows the ratio between the mixed (true spectrum) and the extrapolated spectrum.

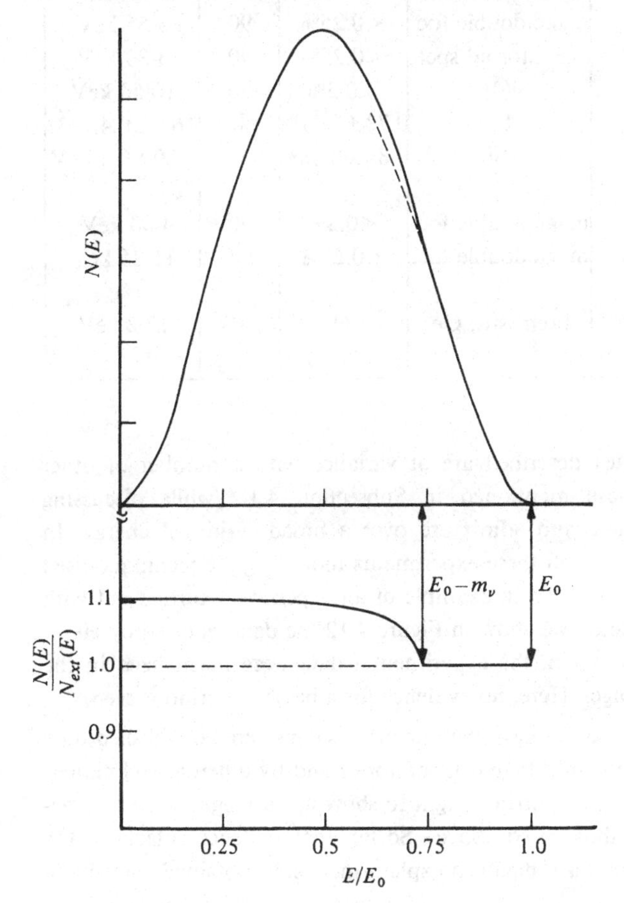

Table 4.1 *Results from searches for heavy neutrinos, including a hypothetical 17 keV neutrino.*

Experiment	Nucleus	Spectrometer	$\vert U_{eH}\vert^2$	CL(%)	m_ν or range explored
Simpson(85)	^3H	Si	≈3%	-	17.1 keV
Hime-Simpson(89)	"	Ge	0.6-1.6%	-	16.9±0.1 keV
Bahran-Kalbfleisch(91)	"	proport.counter	<0.4%	90	≈ 17 keV
Sur et al.(91)	^{14}C	implanted in Ge	1.4±0.5%	-	17±2 keV
Altzitzoglou et al.(85)	^{35}S	magn.lens spec.	<0.4%	99	5-50 keV
Markey-Boehm(85)	"	magn.double foc.	<0.25%	90	5-55 keV
Apalikov et al.(85)	"	magn.toroid spec.	<0.25%	90	5-80 keV
Ohi et al.(85,86)	"	Si(4π)	<0.3%	90	10-50 keV
Simpson-Hime(89)	"	Si	0.73±0.11%	-	16.9±0.4 keV
Hime-Jelley(91)	"	Si	0.84±0.08%	-	17.0±0.4 keV
Hetherington(87)	^{63}Ni	magn.double foc.	<0.3%	90	4-40 keV
Wark-Boehm(86)	"	magn.double foc.	<0.25%	90	11-25 keV
Zlimen et al.(91)	^{71}Ge	bremsstr., Ge	1.6%	-	17.2 keV

The findings just described are at variance with a number of other results including those mentioned in Subsection 4.1.2 while discussing searches for heavy neutrino admixture over a broad region of energy. In Table 4.1 we summarize all these experiments indicating the technique used and the results obtained. As an example of an experiment carried out with a magnetic spectrometer, we show in Figure 4.12 the data reported by Hetherington et al. (87). Again, the experimental data were normalized in the the upper 15 keV range. Here, no evidence for a heavy neutrino is seen.

As for all of physics, a new phenomenon is considered established only if it can be reproduced in different laboratories and by different techniques. The hypothetical 17 keV neutrino ought to show up in magnetic spectrometer experiments like those cited above. So far, that evidence is lacking. On the other hand, it remains difficult to explain the results obtained with the Si

Figure 4.11. Si-detector beta spectrum of ^{35}S from Hime & Jelley (91) plotted as ratio of experimental spectrum to Fermi spectrum for $m_\nu = 17$ keV and $|U_{eH}|^2 = 0.9\%$. The experimental spectrum is normalized to the Fermi spectrum over the region above 150 keV. The endpoint energy is 167 keV. The smooth curve corresponding to the admixture given agrees with the data. The horizontal line indicates zero admixture.

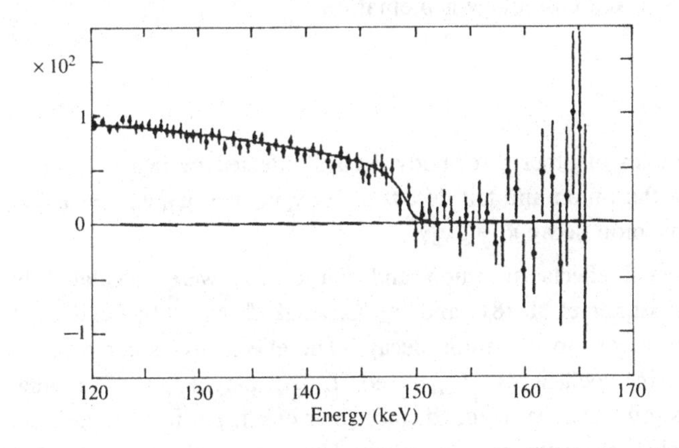

Figure 4.12. Magnetic spectrometer data for the beta spectrum of ^{63}Ni from Hetherington et al. (87) plotted as ratio of experimental spectrum to Fermi spectrum normalized over the upper 15 keV. The horizontal line corresponds to zero admixture, while the dotted line shows the expected shape for a 3% admixture of a 17 keV neutrino. The endpoint energy is 66.9 keV. The data agree with zero admixture.

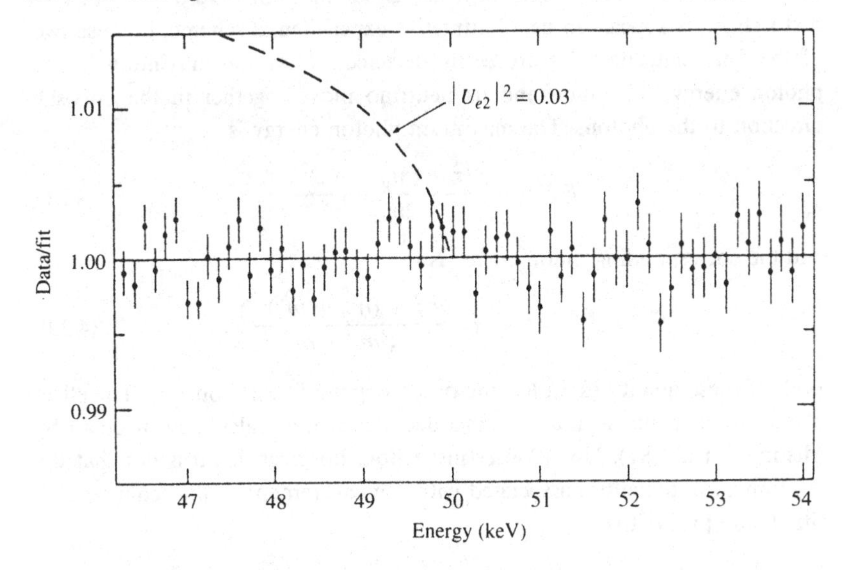

and Ge detectors. To overcome the experimental problems hiding or simulating this small effect is a difficult task. It illustrates a challenge in neutrino physics aimed at finding important evidence for new physics beyond the Standard Model. Further work now in progress in several laboratories should help resolve the present disparity. Until an unambiguous confirmation of the 17 keV neutrino is provided, a discussion of its impacts on particle physics and cosmology is premature.

4.1.4 *Other three-body decays*

Examples of other three-body decays, affected by heavy neutrino admixtures, are the muon and tau decays $\mu \rightarrow e\nu_e\nu_\mu$, $\tau \rightarrow \mu\nu_\mu\nu_\tau$, $\tau \rightarrow e\nu_e\nu_\tau$, and the radiative pion decay $\pi \rightarrow \mu\nu_\mu\gamma$.

The expected effects in muon and tau decays were calculated by Shrock (81), Missimer et al. (81) and by Kalyniak & Ng (82); Dixit et al. (83) applied the theory to the muon decay. The effects are rather complex because two mixing coefficients, $|U_{eH}|^2$ and $|U_{\mu H}|^2$, play a role. The radiative corrections must also be included since their effect, particularly near the end of the Michel spectrum, is substantial. Here, it should be noted that, unlike in the case of nuclear beta decay, it is necessary to distinguish the effects of the heavy neutrino admixtures from the effects associated with possible nonstandard weak interaction couplings, for example, due to the right-handed currents.

The radiative pion decay $\pi \rightarrow \mu\nu_\mu\gamma$ has long been recognized as a possible indicator of muon neutrino mass (Goldhaber 63; Bardin et al. 71). At first sight, this appears to be an attractive experimental scheme, because two of the final state particles are easily detected. Near the maximum of the photon energy, the muon and its neutrino move together in the opposite direction to the photon. The maximum photon energy is

$$E_\gamma^{Max} = \frac{m_\pi^2 - (m_\mu + m_\nu)^2}{2m_\pi}, \tag{4.21}$$

and the corresponding muon energy is

$$E_\mu^{(E_\gamma - Max)} = m_\mu \frac{m_\pi^2 + (m_\mu + m_\nu)^2}{2m_\pi(m_\mu + m_\nu)}. \tag{4.22}$$

Both of these quantities, in leading order, depend linearly on m_ν. The effect of the muon neutrino mass on this decay has been calculated in detail by Missimer et al. (81). Upon closer inspection, however, it turns out that the experimental difficulties associated with a measurement of the required sensitivity are prohibitive.

4.1.5 Very heavy neutrinos

What happens if a heavy neutrino has a mass larger than several GeV, so that it cannot be produced in decays involving known leptons or hadrons? It has been pointed out by Gronau et al. (84) that even in this case we can obtain meaningful limits for the corresponding mixing parameters. The arguments are as follows: due to the mixing with the hypothetical very heavy neutrino the weak interaction coupling strength of all conventional light left-handed neutrinos ν_{lL} will be reduced, and the effect of the coupling can be taken into account by a substitution

$$\nu_{lL} \rightarrow \nu_{lL}(1 - |U_{lH}|^2/2) \tag{4.23}$$

in the formulae for the decay matrix elements. When comparing the muon lifetime and the rate of superallowed Fermi-type nuclear beta decays, the reduction factor (4.23) for $l = e$ drops out and one is left with the relation

$$\frac{|V_{ud}^{KM}|}{1 - |U_{\mu H}|^2/2} = 0.9737 \pm 0.0025 ,$$

where V_{ud}^{KM} is the matrix element of the Kobyashi-Maskava matrix, describing the mixing of quarks in analogy to (4.1). From unitarity of the Kobyashi-Maskava matrix, and from studies of processes involving charmed quarks, one concludes that $|V_{ud}^{KM}| \geqslant 0.9748$ and, therefore, $|U_{\mu H}|^2 \leqslant 8 \times 10^{-3}$. Going back to the branching ratio of pion decay into muons and electrons discussed previously, and using the above $|U_{\mu H}|^2$, we find $|U_{eH}|^2 \leqslant 4.3 \times 10^{-2}$. For tau neutrinos, one can compare the theoretical and experimental values of the lifetime and one obtains $|U_{\tau H}|^2 \leqslant 0.3$. With improved accuracy of the experimental lifetime, this limit can be substantially improved.

4.2 Neutrino Decay

The existence of subdominantly coupled heavy neutrinos, discussed in the previous Section, necessarily leads to neutrino instability. Here, we discuss the information on the mass and mixing of such heavy neutrinos, as it was obtained from the study of neutrino decay.

If subdominantly coupled heavy neutrinos exist, they will be produced by any neutrino source, provided the appropriate kinematic criteria are met. Pseudoscalar mesons, for example, will produce heavy neutrinos ν_H at the relative rate (4.11). (These heavy neutrinos will have momenta (4.5) in the meson rest frame. However, if the decaying meson moves in the laboratory with velocity v_M in the direction of the neutrino beam, the laboratory velocity of the neutrino ν_H is given by

$$v_{Lab} = \frac{p_\nu^{(c)} + v_M E_\nu^{(c)}}{E_\nu^{(c)} + v_M\, p_\nu^{(c)}}\,, \tag{4.24}$$

where $p_\nu^{(c)}$ ($E_\nu^{(c)}$) is the momentum (4.3) (energy (4.4)) of the neutrino in the meson rest frame.)

Similarly, if the neutrino source is nuclear beta decay (the initial nuclei are assumed to be at rest in the laboratory), the branching ratio for emission of ν_H with energy E_ν is

$$R(E_\nu) = |U_{eH}|^2 \left(1 - \frac{m_{\nu_H}^2}{E_\nu^2}\right)^{\frac{1}{2}} \Theta(E_\nu - m_{\nu_H})\,. \tag{4.25}$$

In both cases, the production is suppressed, in comparison to the dominant mode, by the small mixing coefficient $|U_{lH}|^2$.

Let us consider the decay of ν_H. If ν_H is heavier than twice the electron mass ($m_{\nu_H} > 2m_e$), the dominant mode of the decay will be

$$\nu_H \to e^+ + e^- + \nu_i\,, \tag{4.26}$$

described by the Feynman graphs in Figure 4.13. Here, ν_i is a light neutrino which couples strongly to electrons $|U_{ei}|^2 \approx 1$. We assume that its mass m_{ν_i} is negligible compared to the electron mass m_e (see Chapter 2). In the coordinate system, where ν_H is at rest, the Feynman graphs in Figure 4.13 are easily evaluated and give the decay rate (Shrock 81)

$$\Gamma^{CM} = \frac{G_F^2}{192\pi^3}\, m_{\nu_H}^5 |U_{eH}|^2\, h\left(\frac{m_e^2}{m_{\nu_H}^2}\right). \tag{4.27}$$

Figure 4.13. Feynman graphs describing $\nu_H \to e^+ + e^- + \nu_i$ decay: (a) is for Dirac neutrinos; for Majorana neutrinos graphs (a) and (b) must be added.

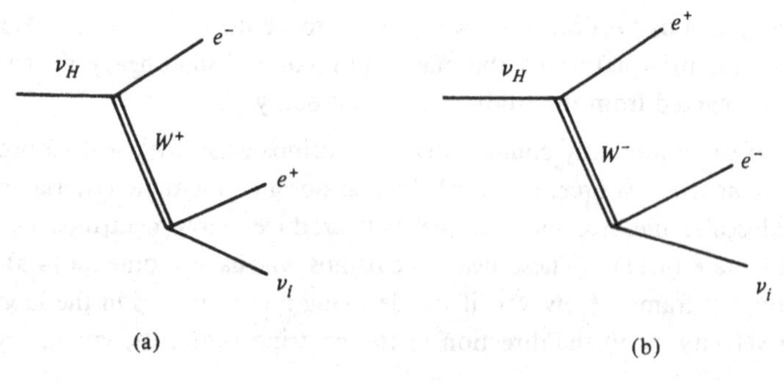

Here, in relevant units,

$$G_F^2/192\pi^3 = 3.5 \times 10^{-5}\ \text{MeV}^{-5}\ \text{s}^{-1}\ .$$

The phase space factor $h(a = m_e^2/m_{\nu_H}^2)$ has been evaluated by Shrock (81) and is equal to

$$h(a) = (1 - 4a)^{\frac{1}{2}}(1 - 14a - 2a^2 - 12a^3)$$
$$+ 24a^2(1 - a^2)\ln\frac{1 + (1 - 4a)^{\frac{1}{2}}}{1 - (1 - 4a)^{\frac{1}{2}}}\ . \tag{4.28}$$

As one could guess , $h(a) \to 0$ for $a \to 4$ (i.e., $m_{\nu_H} \to 2m_e$) and $h(a) \to 1$ for $a \to 0$ (i.e., $m_{\nu_H} >> m_e$). The rate of the decay is the same for Dirac and Majorana neutrinos. For Majorana neutrinos the two graphs in Figure 4.13 do not interfere, because the final neutrino ν_i is left-handed for process (a) and right-handed for process (b). For the same reason, the angular distribution of the electron with respect to the polarization of ν_H can distinguish between the Dirac and Majorana cases (Li & Wilczek 82).

If the e^+e^- pairs are observed in a detector of length L_{det}, the probability of decay (4.26) in the detector is

$$P = \frac{m_{\nu_H}}{E_\nu}\,\Gamma^{CM} \times L_{det}\ , \tag{4.29}$$

where the factor m_ν/E_ν arises from the relativistic time dilatation. Analysis of actual experiments has to take into account the energy and angular distribution of the e^+e^- pairs. The general formalism developed by Shrock (81) has been applied to the reactor neutrinos by Vogel (84a) and Oberauer et al. (87), while Toussaint & Wilczek (81) considered the decay of solar neutrinos. In the former case, a limit for the mixing parameter U_{eH} is deduced from the experimental limit for the number of e^+e^- pairs, created in a neutrino detector near a nuclear reactor. In the second case, a more stringent, but more model dependent limit on U_{eH} is deduced from the experimental limit for the positron flux in interplanetary space. (Bahcall et al. (72) considered the decay of solar neutrinos into a lighter neutrino ν' and a massless scalar or pseudoscalar particle φ, and obtained a decay rate formula depending on the neutrino mass and the corresponding coupling constant.)

In general, formulae (4.28), (4.29) in conjunction with (4.25) or (4.11) allow calculation of the expected number of e^+e^- pairs for each value of the mass m_{ν_H}. That number is proportional to $|U_{eH}|^4$, if the original source involved electrons, and to $|U_{eH}|^2|U_{\mu H}|^2$, if the original source involved muons. More complicated decays, such as $\nu_H \to \mu^+e^-\nu_L$, $\nu_H \to e^-\pi^+$, etc. have been considered by Shrock (81).

If $m_{\nu_H} \leqslant 2m_e$, the decay (4.26) is energetically forbidden. While its analogue with three neutrinos in the final state is possible, its observation is obviously difficult. A decay of this type, however, could be important in astrophysics.

Another possibility is the decay $\nu_H \to \nu_i + J$, where J is the massless majoron (Chikashige et al. 81). The decay rate in that case depends on unknown parameters that determine the majoron-neutrino coupling. Relatively short lifetimes can be obtained in this way (Glashow 91).

Experimental evidence for heavy neutrinos ν_H has been sought in several studies at high energy accelerators. Heavy neutrinos may be produced in the decay of pions or kaons in competition with the well-known two-body and three-body decays into light neutrinos. The rate of production of heavy neutrinos is proportional to the mixing strengths $|U_{eH}|^2$ or $|U_{\mu H}|^2$ (depending on whether ν_H is produced with an electron or a muon). The production rate is also proportional to a kinematic factor which includes the helicity suppression for two-body decays discussed above.

Heavy neutrinos may also be produced in the decay of heavy charmed mesons, such as the D or F mesons. In a beam dump experiment, where the meson production target is thick, so that all mesons are stopped, the π and K mesons are absorbed by target nuclei before they can decay. This greatly suppresses the "conventional" neutrino flux and enhances the fraction of neutrinos produced in the decay of the short lived D and F. Again, the helicity suppression for the decay into light neutrinos, $D \to l + \nu_e$, helps to favor the decay into a heavy neutrino, $D \to l + \nu_H$. In analyzing the results from the F decay, $F \to \tau + \nu_\tau$, the assumption is being made that the ν_τ couples dominantly to the mass eigenstate ν_H. The mixing strength associated with the production, then, is $|U_{\tau H}|^2 \approx 1$.

The heavy neutrino thus produced may decay through one or several of the channels $\nu_H \to ee\nu$, $\mu\mu\nu$, $e\mu\nu$, $e\pi$ and $\mu\pi$ with a decay rate proportional to $|U_{eH}|^2$, etc. The result of an experimental search for charged particle pairs is thus proportional to $|U_{eH}|^4$, $|U_{eH}|^2|U_{\mu H}|^2$, or $|U_{\mu H}|^4$, except for the case of $\nu_\tau = \nu_H$, mentioned above, where the experimental result is proportional to $|U_{eH}|^2$, $|U_{eH}||U_{\mu H}|$, or $|U_{\mu H}|^2$.

As an example of the accelerator based searches for heavy neutrino decay, we show in Figure 4.14 the schematic arrangement of the CERN CHARM experiment (Bergsma et al. 83; Gall 84; Dorenbosch et al. 86). The 400 GeV proton beam from the SPS at CERN strikes a thin target producing a beam of π and K mesons. The ν_H beam, resulting from the process $\pi,K \to \nu_H e(\nu_H\mu)$, enters a detector region, after having been "cleansed" from all other particles by a shield of iron and earth. The reaction

$\nu_H \rightarrow e^+e^-\nu_e$ can now be studied with the help of several detector planes in the decay region and a calorimeter. The reconstruction of e^+e^- trajectories allowed identification of candidate events. No valid e^+e^- events were found, and the result was interpreted as an upper limit for $|U_{eH}|^4$. In terms of $|U_{eH}|^2$ this limit is around 10^{-5}—10^{-6}, as shown in Figure 4.5. The mass range for ν_H is limited by the kaon mass.

The CHARM group also made use of the beam dump concept and obtained upper limits for the mixing strengths. These limits are shown in Figures 4.5 and 4.6. Clearly, the experiment has greater sensitivity if the assumption is made that $\nu_\tau = \nu_H$. In this case, however, the range of validity for m_{ν_H} cannot exceed $m_{\nu_\tau} < 35$ MeV (Albrecht et al. 88).

We note that the distance between ν_H production and the CHARM detector was 900 m. Neutrino decay may appreciably attenuate the ν_H beam when the mixing is large ($|U_{eH}|^2 > 10^{-3}$), because the decay length is then shorter than the distance to the detector.

Production and decay of heavy neutrinos have also been studied with the CERN BEBC bubble chamber (Cooper-Sarkar et al. 85). No events were seen that satisfy the required kinematic constraints and the resulting upper limits for $|U_{eH}|^2$ and $|U_{\mu H}|^2$ are also given in Figure 4.5 and 4.6.

4.2.1 Radiative neutrino decay

For a heavy neutrino with mass below the threshold for e^+e^- decay (4.26), the radiative decay

$$\nu_H \rightarrow \nu_i + \gamma \tag{4.30}$$

Figure 4.14. Schematic illustration of the experimental search for the formation of a heavy neutrino and its decay into a e^+e^- pair, such as the work performed by the CHARM collaboration (Bergsma et al. 83).

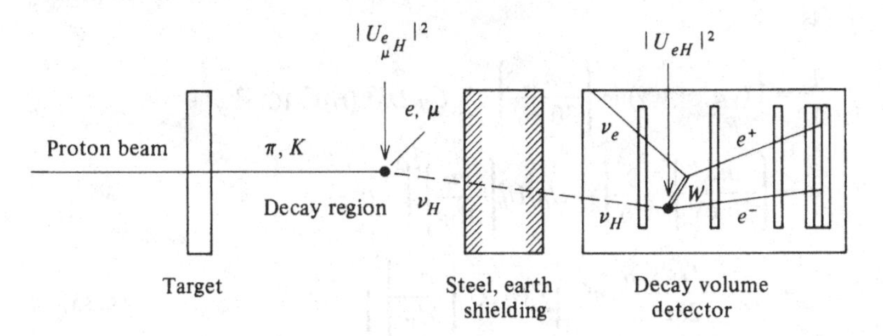

is essentially the only viable candidate for direct observation. The decay rate for the mode (4.30) is given by

$$\Gamma(\nu_H \to \nu_i \gamma) = \frac{1}{8\pi}\left(\frac{m_{\nu_H}^2 - m_{\nu_i}^2}{m_{\nu_H}}\right)^3 \left(|a|^2 + |b|^2\right), \tag{4.31}$$

where a and b are the coefficients appearing in the most general effective matrix element for the decay (4.30) compatible with gauge invariance,

$$M = \varepsilon^{\mu} q^{\nu} \overline{\nu}_i \sigma_{\mu\nu}(a + b\gamma_5)\nu_H . \tag{4.32}$$

Here ε^{μ} and q^{ν} are the polarization vector and momentum of the photon, respectively.

The coefficients a and b can be calculated within the Standard Model by evaluating the Feynman graphs in Figure 4.15. Pal & Wolfenstein (82) obtained

$$a_D = -\frac{eG_F}{8\sqrt{2}\pi^2}(m_{\nu_H} + m_{\nu_i})\sum_e U_{li} U_{lH}^* F(r_l) ,$$

$$b_D = (m_{\nu_H} - m_{\nu_i})/(m_{\nu_H} + m_{\nu_i}) a_D , \tag{4.33}$$

where U is the mixing matrix in (4.1), and the sum is over all charged leptons l. The function $F(r_l)$ is a smooth function of the variable $r_l = (m_l/M_W)^2$ ($M_W \approx 81$ GeV is the mass of the intermediate vector boson); for small r_l we obtain

$$F(r) \cong -\frac{3}{2} + \frac{3}{4} r . \tag{4.34}$$

Equation (4.33) is valid when both ν_H and ν_i are Dirac neutrinos (hence subscript D in (4.33)). For Majorana neutrinos, one has instead $a_M = 0$, $b_M = 2b_D$ or $a_M = 2a_D$, $b_M = 0$, depending on the relative CP phase of the neutrinos ν_H and ν_i (see Chapter 6 for a discussion of the CP phase of neutrinos). Restricting ourselves to the case of Dirac neutrinos and CP conserving theories and, in addition, assuming that $m_{\nu_H} >> m_{\nu_i}$, we obtain in practical units

$$\frac{1}{\tau} = \Gamma(\nu_H \to \nu_i \gamma) = \left(\frac{m_{\nu_H}}{30 \text{ eV}}\right)^5 |\sum_l U_{lH} U_{li}^* F(r_l)|^2 \, 10^{-22} \text{ y}^{-1}$$

$$= \left(\frac{m_{\nu_H}}{30 \text{ eV}}\right)^5 \frac{9}{16} \left|\sum_l U_{lH} U_{li}^* \left(\frac{m_l}{M_W}\right)^2\right|^2 10^{-22} \text{ y}^{-1}$$

$$= \frac{9}{16} \frac{G_F^2}{128\pi^3} \frac{\alpha}{\pi} m_{\nu_H}^5 \left|\sum_l U_{lH} U_{li}^* \left(\frac{m_l}{M_W}\right)^2\right|^2 . \tag{4.35}$$

If we further assume that the tau is the heaviest charged lepton in existence, and that its contribution dominates the sum in (4.35), we obtain

$$\Gamma \approx \left[\frac{m_{\nu_H}}{30 \text{ eV}} \right]^5 |U_{\tau H} U^*_{\tau i}|^2 \, 10^{-29} \text{ y}^{-1} . \tag{4.36}$$

(The neutrino lifetime τ ($\tau = 1/\Gamma$) for maximum mixing $U_{\tau H} = U_{\tau i} = 1$ is shown in Figure 4.16.)

In the Standard Model, radiative decay is characterized by a very long lifetime, much longer than the age of the universe. The additional suppression, apparent when one compares (4.35) and (4.36), is related to the "GIM cancellation" factor $r_l^2 = (m_l/M_W)^4 << 1$. Indeed, owing to the unitarity of the mixing matrix U, the constant term in F (4.34) does not contribute in the Standard Model. The only way to avoid this suppression is the introduction of a new generation of very heavy leptons or the introduction of sterile weak interaction singlet neutrinos.

Another way to speed up the radiative decay is the introduction of models in which right-handed weak interactions occur. In this case (see, e.g., Enquist et al. 83) the GIM cancellation is absent and, moreover, the $m^5_{\nu_H}$ dependence of the rate is replaced by $m^3_{\nu_H} m_l^2$, a much larger quantity. Thus, observation of the radiative neutrino decay would be a powerful argument in favor of the existence of right-handed weak interactions.

In order to detect photons from the radiative neutrino decay, or to establish a limit for the corresponding lifetime, it is necessary to take into account the photon angular distribution, which represents another signature for the identity of the neutrinos (Dirac vs. Majorana).

In the center-of-mass system of ν_H, the only preferred direction is the neutrino polarization; the photon angular distribution with respect to the

Figure 4.15. Feynman graphs describing neutrino radiative decay $\nu_H \rightarrow \nu_i + \gamma$. The charged lepton is denoted by l and the intermediate vector boson by W.

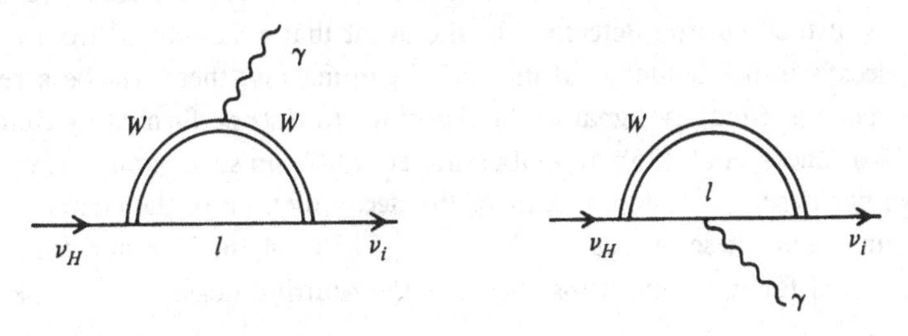

polarization direction is calculated to be (Shrock 82; Li & Wilczek 82)

$$dN = 1/2(1 + \alpha\cos\theta)d\cos\theta, \qquad |\alpha| \leqslant 1\,, \tag{4.37}$$

where $\alpha = 0$ for Majorana neutrinos and $\alpha = 1$ for left-handed Dirac neutrinos ν_H and massless final neutrinos ν_i (in that case the photons are preferably emitted "backward").

Taking into account (4.37) and the relativistic time dilatation, we find the resulting photon spectrum in the laboratory (for $m_{\nu_i} \cong 0$)

$$\frac{dN}{dE_\gamma} = m_{\nu_H}\Gamma^{CM}\int_{E_\gamma}^{\infty}\frac{1 - \alpha + 2\alpha E_\gamma/E_\nu}{E_\nu^2}N(E_\nu)dE_\nu\,, \tag{4.38}$$

where $N(E_\nu)$ is the spectrum of the decaying neutrinos ν_H, and Γ^{CM} is the center-of-mass decay rate (4.31). Experimental values (or limits) of the photon flux, therefore, can be used to establish a limit of the product $m_{\nu_H}\,\Gamma^{CM}$.

The radiative decay rate (4.35) is a fast increasing function of the mass of the initial neutrino m_{ν_H}. However, it turns out (Nieves 83) that a decay with two photons in the final state

$$\nu_H \rightarrow \nu_i + \gamma + \gamma \tag{4.39}$$

increases even faster and is free of the "GIM cancellation" mentioned earlier. Assuming, as before, that the final neutrino ν_i is very light and also that $\nu_i \approx \nu_e$, one obtains the rate

$$\Gamma(\nu_H \rightarrow \nu_i\gamma\gamma) = (2 \times 10^{11}\ \mathrm{s})^{-1}\left(\frac{m_{\nu_H}}{1\ \mathrm{MeV}}\right)^9 |U_{eH}U_{ei}^*|^2\,, \tag{4.40}$$

which is faster than the single photon mode for $m_{\nu_H} \geqslant 0.1$ MeV. This example shows that in the study of the neutrino decay, one has to keep in mind other less obvious decay channels. In cosmological and astrophysical applications neutrino decays involving as yet unobserved particles, such as light scalars or pseudoscalars, majorons (see above), etc. are often considered; these decay modes are not discussed here.

Experimental searches for $\nu \rightarrow \nu\gamma$ and $\nu \rightarrow \nu\gamma\gamma$ decays have been conducted for ν_e and ν_μ neutrinos, using as the gamma ray detector certain elements of the neutrino detector. In the event that a $\bar{\nu}_e$ emitted from a reactor decays into a lighter neutrino and a gamma ray, there will be a reactor associated gamma ray signal in the detector. In tests performed by Reines et al. (74), Zacek et al. (86), and Oberauer et al (87) no such gamma ray events have been seen. The lower limit of the decay lifetime in the center-of-mass system from these works is shown in Figure 4.16. Similar tests were conducted for ν_μ decay, using the LAMPF neutrino detector of Frank et al.

(81); the lifetime limits are also shown in Figure 4.16. The pertinent curves extend to the experimental mass limits of 35 eV and 250 keV, respectively. The Figure also shows a limit for τ_{ν_e} derived from considerations of the solar neutrino flux (Raffelt 85). The line, denoted τ_ν(astr), gives the lower limit of the radiative lifetime as required by the experimental limits on photon fluxes in interstellar space, as well as by other considerations, such as supernova energetics and emission, etc. (Turner 81). The line extends to 200 eV, the largest neutrino mass compatible with stability of the universe (see Chapter 7). Absence of γ-ray emission during the SN1987A neutrino burst was used to obtain stringent limits on the radiative decay of neutrinos

Figure 4.16. Neutrino radiative lifetime τ in the center-of-mass system as a function of neutrino mass. The curve labeled $\tau_{\nu \to \nu' + \gamma}$ (th) is the calculated curve for maximum mixing. The experimental limits are τ_{ν_e} (solar), τ_{ν_e} (exp) from reactor experiments, and τ_{ν_μ} (exp) from muon beam experiments. The line labeled τ_ν (astr) is obtained from analysis of photon fluxes in space. For references see text.

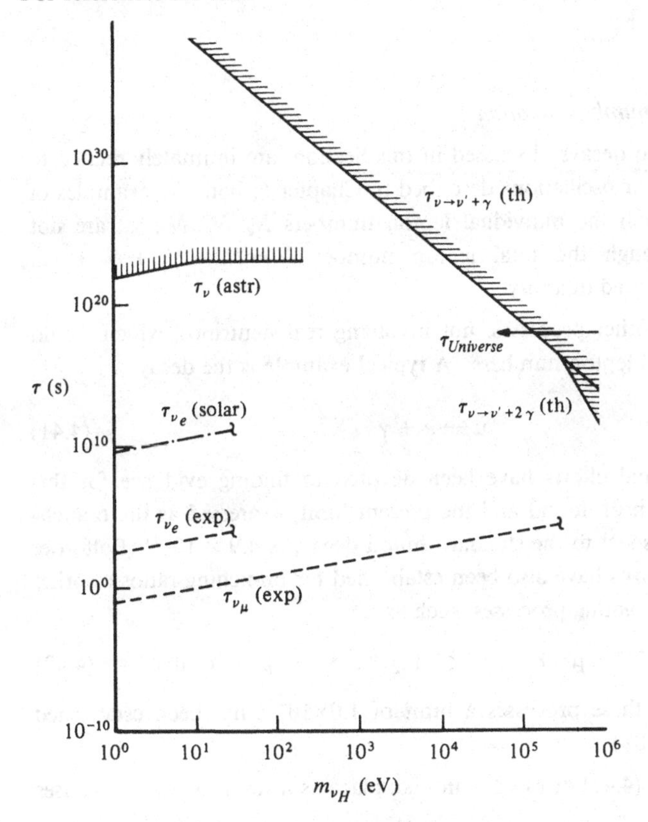

(see Section 7.2). An even more stringent limit on the radiative decay rate is obtained when one makes use of the relation with the photon-neutrino coupling effect on the neutrino electron scattering (Raffelt 89, see also Section 7.2).

The decay (4.30), when the initial and final neutrinos ν_H and ν_i are very close in mass, requires special consideration. For the mass splitting Δm_ν such that $\Delta m_\nu/m_\nu << 1$ the photon energy will be $\sim E_\nu \Delta m_\nu/m_\nu$, i.e., it could be below the detection threshold of many detectors. Bouchez et al. (88) searched near a nuclear reactor for decay photons in the visible energy range. The experiment restricts the radiative lifetime for $\Delta m_\nu/m_\nu$ in the interval 10^{-7}—10^{-3}.

At neutrino masses $m_{\nu_H} > 1$ MeV the dominant decay mode is e^+e^- pair production as discussed above. As the expected lifetimes decrease rapidly with increasing neutrino mass, the experimental tests for large masses are within the range, or better, than the calculated lifetime for full mixing. These experiments, therefore, have been interpreted as providing limits on the mixing parameters $|U_{lH}|^2$ as discussed in the preceding Section.

4.2.2 *Lepton number violation*

Neutrino decays, discussed in this Section, are intimately related to the neutrino flavor oscillations described in Chapter 5; both are examples of processes in which the individual lepton numbers N_e, N_μ, N_τ, ... are not conserved, although the total lepton number $N = N_e + N_\mu + N_\tau + \ ...$ might be a conserved quantity.

There are other processes, not involving real neutrinos, which would violate individual lepton numbers. A typical example is the decay

$$\mu \rightarrow e + \gamma \, . \tag{4.41}$$

Large experimental efforts have been devoted to finding evidence for this decay; none has been found and the present limit, expressed as the branching ratio with respect to the standard muon decay, is 4.9×10^{-11} (Bolton et al. 88). Small limits have also been established for branching ratios of other lepton number violating processes, such as

$$\mu^+ \rightarrow e^+e^-e^+ \, , \ \mu^- + zA \rightarrow e^- + zA \, , \ K_L \rightarrow \mu^+e^- \text{ or } \mu^-e^+ \, . \tag{4.42}$$

(For the first of these processes a limit of 1.0×10^{-12} has been established (Bellgardt et al. 88).)

The decays (4.41) or (4.42) can take place as a result of various causes not directly related to the neutrino mass, such as the flavor changing neutral currents. Indeed, considerable care was required, when different models were developed, to avoid such currents. The decay (4.41) could also proceed if weak interaction involving right-handed currents is present (the rate ~ $1/M_{W_R}^4$), and the decay $\mu \rightarrow 3e$ (4.42) could be mediated by a new "horizontal" boson H. We shall not be concerned here with such possibilities.

On the other hand, if neutrinos are massive and mixed, processes (4.41) and (4.42) will necessarily exist at some level. We shall briefly discuss the relation between the phenomenon of neutrino mixing and the rate of the lepton number violating decays, using (4.41) as an example.

The decay (4.41) is described by the Feynman graphs in Figure 4.17, analogous to the graphs in Figure 4.15, describing neutrino radiative decay. (Graphs containing virtual Higgs particles must also be considered, care must be taken to preserve gauge invariance, or ensure cancellation of gauge noninvariant parts.) The analysis of the $\nu_H \rightarrow \nu_i\gamma$ as well as $\mu \rightarrow e\gamma$ decays in the framework of the general local gauge theories has been carried out by Heil (83). Here, we shall describe a simpler evaluation for the case of two mixed neutrinos (Petcov et al. 77). (The same simplified mixing scheme is extensively used in Chapter 5.) We assume that the ν_e and ν_μ neutrinos are the only ones mixed; the mixing matrix (4.1) is then characterized by a single parameter θ, $U_{e1} = U_{\mu 2} = \cos\theta$, $U_{e2} = -U_{\mu 1} = \sin\theta$. The rate of the $\mu \rightarrow e\gamma$ decay, evaluated in the extended standard Weinberg-Salam model

Figure 4.17. Feynman graphs describing the decay $\mu^+ \rightarrow e^+ + \gamma$. Note the analogy to Figure 4.15.

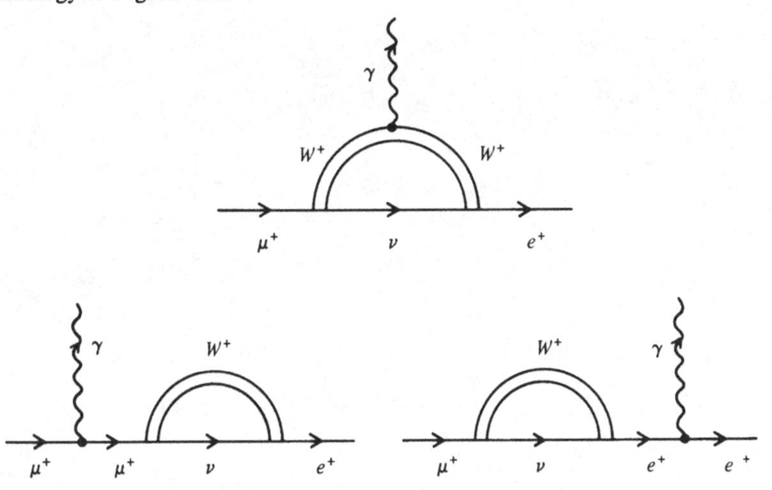

with left-handed currents only and with Dirac neutrinos, is then

$$\Gamma(\mu \to e + \gamma) = \frac{1}{16} \frac{G_F^2 m_\mu^5}{128\pi^3} \frac{\alpha}{\pi} \left(\frac{m_1^2 - m_2^2}{M_W^2} \right)^2 \sin^2\theta \cos^2\theta \,. \tag{4.43}$$

The branching ratio with respect to the standard muon decay is, therefore,

$$R_\mu = \frac{3}{32} \frac{\alpha}{\pi} \left(\frac{m_1^2 - m_2^2}{M_W^2} \right) \sin^2\theta \cos^2\theta \,. \tag{4.44}$$

The similarity between formulae (4.35) and (4.43) is now obvious. One can recognize the mixing factor $\sin^2\theta \cos^2\theta = |\sum_i U_{\mu i} U_{ei}|^2$; in (4.43) the contributions of neutrinos ν_1 and ν_2 enter with opposite sign, and therefore the "GIM cancellation" factor m_i^2/M_W^2 in (4.35) is replaced by $(m_1^2 - m_2^2)/M_W^2$ in (4.40).

The branching ratio R_μ (4.44) is exceedingly small for small values of $\Delta m^2 = |m_1^2 - m_2^2|$;

$$R_\mu = 5 \times 10^{-48} \left[\Delta m^2 \, (\text{eV}^2) \right]^2 \sin^2\theta \cos^2\theta \,.$$

Thus, neutrino mixing with Δm^2 in the eV^2 range leads only to an unobservably small rate of $\mu \to e\gamma$ and similar decays. The rate would be substantially larger, however, if heavy neutral leptons with masses in the GeV range exist. It should be stressed that various generalizations of the Standard Model leading to much faster $\mu \to e\gamma$ and $\nu_H \to \nu_i \gamma$ decays have been proposed, and experimental limits on R_μ are, therefore, important as tests of these ideas.

5

Neutrino Oscillations

The concept of neutrino oscillations has already been introduced in Section 1.3. There, we have shown that the neutrino mass matrix may be nondiagonal when expressed in terms of the weak eigenstates ν_l or, equivalently, that the weak lepton current may be nondiagonal when expressed in terms of the mass eigenstates ν_i. In the latter case the wave function describing neutrinos propagating freely will have several components, with phases depending on time or distance. As a consequence, the count rate in a detector, placed at varying distances from the neutrino source, will vary periodically with distance, hence the name "neutrino oscillations". Thus, the primary motivation for exploring neutrino oscillations is the determination of the parameters of the neutrino mass matrix.

Only massive neutrinos can oscillate and, consequently, the observation of oscillations furnishes information on the neutrino masses (or more precisely, on the mass differences). To put this in context, we have seen (in Chapter 2) that kinematic tests for the mass of the electron neutrino are sensitive to masses down to tens of eV, while being, however, considerably less sensitive for muon and tau neutrino masses. Searching for heavy neutrino admixtures (as discussed in Chapter 4) provides a mass sensitivity larger than a few keV, while double beta decay (described in Chapter 6) is capable of exploring neutrino mass down to a few eV. Neutrino oscillations, on the other hand, open up a mass range well below 1 eV and, therefore, offer by far the most sensitive method for exploring small masses. In fact, the results of these experiment may not be too far from the values typically predicted by GUTs. (Here, for the sake of this qualitative argument, we have assumed that $\Delta m^2 = |m_1^2 - m_2^2| \approx m^2$, where m is the larger of the masses m_1 and m_2.)

As we shall see, neutrino oscillations depend on the ratio (distance L)/(energy E). The typical sensitivity of an experiment is Δm^2 (eV^2) $\approx$ E (MeV)/L (m), which for low energy laboratory experiments may be as small as 10^{-2} eV^2. To gain a perspective of the range of sensitivity of various oscillation experiments, we show in Figure 5.1 an overview of the work discussed in this Chapter.

In closing this introduction, we mention the following review articles on oscillations: Bilenky & Pontecorvo (78), Frampton & Vogel (82), Boehm & Vogel (84), Vuilleumier (86), Bilenki & Petcov (87), Kayser et al. (89), and Moscoso (91).

5.1 Introduction and Phenomenology of Neutrino Oscillations

As we have seen, neutrinos ν_l participating in weak interactions (l stands for various lepton flavors, electron, muon, tau, etc.) may be superpositions of states ν_i of definite mass m_i

$$\nu_l = \sum_{i=1}^{N} U_{li}\nu_i \ . \tag{5.1}$$

When the matrix U is not exactly diagonal we are led to the concept of neutrino oscillations.* To explain the main features of this phenomenon we proceed as follows (using for this Section the system of units $\hbar = c = 1$). Consider a beam of neutrinos moving along the x axis, all having a common, fixed momentum p_ν. Assume that all neutrino masses m_i are much smaller than p_ν and, consequently,

$$E_i = (p_\nu^2 + m_i^2)^{1/2} \approx p_\nu + \frac{m_i^2}{2p_\nu} \ . \tag{5.2}$$

Now, suppose that the beam was created at the time t=0 at x=0 in a weak process in which the charged lepton l was absorbed (or the antilepton $\bar{l}$ emitted). The beam is then described by the wave function

$$\Psi(x,t) = \sum_i U_{li} e^{ip_\nu x} e^{-iE_i t} \nu_i$$

$$\approx e^{ip_\nu(x-t)} \sum_i U_{li} e^{-i\frac{m_i^2}{2p_\nu}t} \nu_i \ . \tag{5.3}$$

The detector, located at a point $x = L$, is capable of detecting the weak eigenstate neutrino $\nu_{l'}$. The count rate in the detector is then proportional to the probability of finding $\nu_{l'}$ at x=L ($t \approx L$ in our units and for light neu-

---

*The lines of the unitary mixing matrix U are labeled by the weak eigenstates and the columns are labeled by the mass eigenstates. It is customary to order the ν_is in such a way that U_{li} is as nearly diagonal as possible, and one can then use the approximate term "electron neutrino mass", meaning the mass of the dominant component of electron neutrino, etc.

trinos), i.e.,

$$P_{ll'}(p_\nu,L) = \left|\sum_i U_{li} U^\dagger_{il'}\, e^{-i\frac{m_i^2}{2p_\nu}t}\right|^2 \tag{5.4}$$

$$= \sum_i \left|U_{li} U^\dagger_{il'}\right|^2 + \mathrm{Re}\sum_i\sum_{j\neq i} U_{li} U^\dagger_{il'} U_{lj} U^\dagger_{jl'}\, e^{i\frac{|m_i^2-m_j^2|L}{2p_\nu}} \; .$$

We see that the signal depends periodically on the distance L between the neutrino source and the detector. For each pair of masses m_i, m_j, the pattern repeats itself, as the distance is increased by an integral multiple of the oscillator length

$$L_{osc} = 2\pi\frac{2p_\nu}{|m_i^2-m_j^2|} \approx 2\pi\frac{2E_\nu}{|m_i^2-m_j^2|} \; . \tag{5.5}$$

If we measure the length in m, the neutrino beam energy in MeV, and the neutrino masses in eV, we obtain

$$L_{osc}(\mathrm{m}) = \frac{2.48\times E_\nu\,(\mathrm{MeV})}{|m_i^2-m_j^2|(\mathrm{eV}^2)} \; . \tag{5.5'}$$

The periodic behavior seen in (5.4) is known as neutrino oscillations. This is a typical interference effect of the different components in the wave function $\Psi(x,t)$; it requires that the different components are coherent. From the above discussion it follows that neutrino oscillations are observable only if the neutrino masses are nonvanishing and nondegenerate, and if the mixing matrix U has some nonvanishing nondiagonal matrix elements.

There are two principal ways to study neutrino oscillations. In experiments referred to as "appearance" or "exclusive" experiments, one starts with the ν_l beam and observes neutrinos $\nu_{l'}$ with a *different* flavor at a distance L from the neutrino source. If oscillations are present, the probability $P_{ll'}\,(p_\nu,L)$ for $l\neq l'$ is nonvanishing. Here, one needs neutrinos of energy E_ν larger than the rest mass of the charged lepton l' to be created in the detection reaction. A practical requirement is that the neutrino beam should not contain the $\nu_{l'}$ neutrinos to start with (or that the number of $\nu_{l'}$s is small and

Figure 5.1. Illustration of regions of L/E accessible in various experiments.

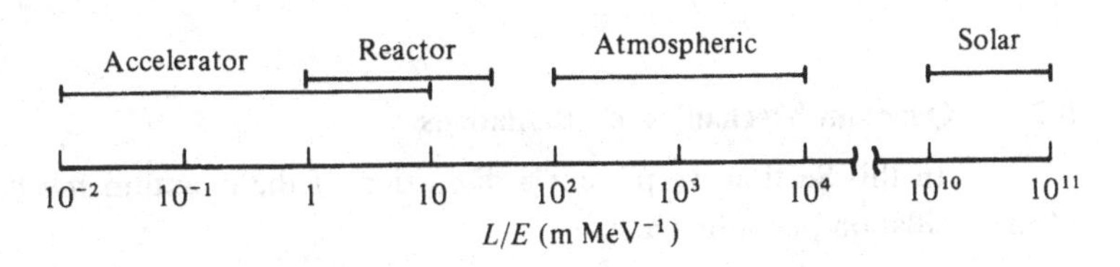

known). These experiments explore primarily the coupling of ν_l to $i \approx l'$ neutrinos ($\nu_l \to \nu_{l'}$ channel).

In "disappearance" or "inclusive" experiments, the neutrino beam, which started as a ν_l beam, is intercepted by a detector sensitive to the *same* flavor of neutrinos at varying distances L. If oscillations are present, the probability $P_{ll}(p_\nu,L)$ is smaller than unity. In these experiments for which the neutrino beam can have any energy E_ν, the coupling of ν_l to all ν_i, is explored, and we refer to this mode as the $\nu_l \to x$ channel.

Generally, the study of oscillations does not furnish the neutrino masses themselves, but rather the quantity $\Delta m_{ij}^2 = |m_i^2 - m_j^2|$, obtained from the oscillation length L_{osc} (see (5.5)). Information on the mixing coefficients U_{li} is obtained from the oscillation amplitudes. The oscillation pattern depends on L/E, the ratio of the distance to the neutrino energy. In an experiment, both the L dependence and the E dependence can be used to explore oscillations. In Figure 5.1 we show the regions of L/E accessible to experiments using different neutrino sources. In each experiment, the maximum sensitivity to the neutrino mass splitting Δm^2 (in eV^2) of the order of E/L (in MeV m^{-1}) is indicated.

So far, no experimental evidence for neutrino oscillations has been found. With this in mind we may simplify the analysis of the experimental data by considering just two neutrino states. This description includes only two parameters, a mixing angle θ ($\sin\theta = U_{12} = -U_{21}$; $\cos\theta = U_{11} = U_{22}$) and a mass parameter $\Delta m^2 = |m_2^2 - m_1^2|$. The probability that a neutrino ν_l will change into $\nu_{l'}$ ($l \neq l'$), after traveling a distance L, is then (in practical units)

$$P_{ll'} = \sin^2 2\theta \sin^2 \frac{1.27 \times \Delta m^2(\mathrm{eV}^2) \times L(\mathrm{m})}{E_\nu(\mathrm{MeV})} \quad l \neq l' . \tag{5.6}$$

The probability that ν_l will remain ν_l is given by

$$P(E_\nu, L, \Delta m^2, \theta) = 1 - \sin^2 2\theta \sin^2 \frac{1.27 \times \Delta m^2(\mathrm{eV}^2) \times L(\mathrm{m})}{E_\nu(\mathrm{MeV})} . \tag{5.7}$$

The results of experiments are presented in the form of "exclusion plots" exhibiting the allowed and forbidden regions in the ($\sin^2 2\theta$, Δm^2) parameter space.

5.2 Quantum Mechanics of Oscillations

In this Section we present a discussion of the quantum mechanics of the oscillation phenomena.

Phase relations. To appreciate the subtleties of the quantum mechanics of oscillations, and at the same time to justify our simplified discussion in the previous Section, let us consider a typical oscillation experiment based on the pion decay $\pi^+ \to \mu^+ \nu_\mu$. To perform the $\nu_\mu \to \nu_e$ appearance experiment, for example, one places a detector, capable of detecting the electron neutrinos, at varying distances L from the pion decay region which represents the neutrino source. Other devices may be present to measure the pion and muon momenta and energies. Using kinematics, it is possible, at least in principle, to determine the energy and momentum (and therefore also the mass) of the emerging neutrino in each event.

Assume now that the muon neutrino created in the decay is a superposition of two mass eigenstates m_1, m_2. Its wave function then has two components, each with its own energy and momentum and

$$E_1 - E_2 = \frac{m_1^2 - m_2^2}{2E_\pi} , \tag{5.8}$$

$$p_1 - p_2 = \frac{m_1^2 - m_2^2}{p_1 + p_2} \frac{2E_\pi - E_1 - E_2}{2E_\pi} , \tag{5.9}$$

where E_π is the energy of the decaying pion. The phase difference of the two components of the neutrino wave function is

$$\Phi(x,t) = (E_1 - E_2)t - (p_1 - p_2)x$$

$$= \frac{m_1^2 - m_2^2}{E_1 + E_2} t + (p_1 - p_2) \left[\frac{p_1 + p_2}{E_1 + E_2} t - x \right] . \tag{5.10}$$

The first term in (5.10) is the standard one, leading to the familiar definition of the oscillation length (see (5.5)).

$$L_{osc} = 2\pi \frac{E_1 + E_2}{|m_1^2 - m_2^2|} \approx 2\pi \frac{2\bar{E}}{|m_1^2 - m_2^2|} . \tag{5.5''}$$

The second term is usually not considered (see (5.3)). It vanishes for the points related by

$$x = v_o t = \frac{p_1 + p_2}{E_1 + E_2} t , \tag{5.11}$$

where v_o is the average velocity of the two components. The two components of the neutrino wave function can retain their coherence, and one can observe neutrino oscillations only for space-time points obeying (5.11). For these points the standard approach is justified.

Compatibility of oscillation and "double peak" experiments. One can determine the neutrino mass by measuring precisely the energy and momentum of the initial pion and final muon. Then it will be known, for each event, whether ν_1 or ν_2 is involved. (We have discussed experiments of this type in Chapter 4.) In the case under consideration, the muon spectrum consists of two peaks separated by $p_1 - p_2$. However, the neutrino mass determination is possible only if the two peaks are experimentally resolved, i.e., if the momentum resolution is better than $\Delta p = |p_1 - p_2|$. Noting that

$$\Delta p L_{osc} \approx \frac{2E_\pi - E_1 - E_2}{2E_\pi} \leqslant 1 , \tag{5.12}$$

and using the uncertainty relation $\Delta p \Delta x \approx 1$, we see that the "double peak" experiment implies delocalization of the neutrino source by an amount larger than L_{osc}. The "double peak" and neutrino oscillation experiments are, therefore, incompatible; we can perform one or the other, but not both of them simultaneously. (A wave packet treatment of this problem has been discussed by Kayser (81).)

Coherence length. We should also consider the question of the coherence length. How far away from the source will coherence and, therefore, oscillations survive? Let us assume that initially the source is localized within an interval $\delta x \approx 1/\delta p$. The uncertainty in the phase of the neutrino wave function will increase with distance as

$$\delta\Phi(x) \approx \frac{\Delta m^2}{2p} \cdot \frac{\delta p}{p} x . \tag{5.13}$$

Coherence can be maintained only as long as $\delta\Phi(x) \leqslant 1$. Therefore, a natural definition of the coherence length is the distance where $\delta\Phi(x = L_{coh}) = 1$. The maximal number of observable oscillations is then

$$N = L_{coh}/L_{osc} \approx p/\delta p \approx p\delta x . \tag{5.14}$$

Equivalently, the coherence length L_{coh} can be defined as the distance where the two wave packets, which are moving with different velocities $v_1 = p_1/E_1$ and $v_2 = p_2/E_2$, spatially separate by δx, the localization uncertainty of the source.

In usual experimental arrangements L_{coh} is an enormous distance. In practice, however, the oscillation pattern may disappear before the coherence limit is reached, because the probabilities (5.7) must be averaged (non-coherently) over the physical size of the neutrino source and detector, and over the energy resolution of the detector. This averaging has nothing to do with coherence of the neutrino beam; it represents deviations of the actual experiment from the ideal one.

For an isotropic neutrino source of size D, the averaged oscillation function (5.7) is of the form

$$\bar{P}(E_\nu,\bar{L},\Delta m^2,\theta) = 1 - \frac{1}{2}\sin^2 2\theta \left(1 - \frac{\sin\frac{\pi D}{L_{osc}}}{\frac{\pi D}{L_{osc}}} \cos\frac{2\pi L}{L_{osc}} \right). \qquad (5.7')$$

Thus the amplitude of the oscillating part is damped for $L_{osc} < \pi D$. In other words, for

$$\Delta m^2(\text{eV}^2) >> 2.48 E_\nu\,(\text{MeV})\,/\pi D(\text{m})\,,$$

the probability (5.7') is a constant, independent of distance.

Similarly, for a detector with energy resolution ΔE, the averaged probability (5.7) is of the form

$$\bar{P}(\bar{E}_\nu,L,\Delta m^2,\theta) \approx 1 - \frac{1}{2}\sin^2 2\theta \left(1 - \frac{\sin\frac{\delta\pi L}{L_{osc}}}{\frac{\delta\pi L}{L_{osc}}} \cos\frac{2\pi L}{L_{osc}} \right), \qquad (5.7'')$$

where $\delta = \Delta E/E_\nu$ and $L > L_{osc}$ was assumed. Thus, at distances $L >> L_{osc}/\pi\delta$, the probability is again a constant, independent of distance.

5.3 General Formalism

To describe neutrino physics completely, one has to begin with the most general neutrino Lagrangian, consisting of the kinetic energy term, the mass term, and the weak interaction term which couples neutrinos and charged leptons to the gauge bosons. We assume that this list exhausts all interactions in which neutrinos participate.

In Chapter 1 we have described how neutrino mixing arises if the mass term is not diagonal, when expressed in terms of the weak interaction eigenstates. At the present time there is no fundamental theory allowing the determination of the parameters in the mass term from first principles. Instead, one must resort to neutrino oscillation experiments to determine, or at least constrain, the values of these parameters. In this Section we discuss the general features of such a program.

Diagonalization of the mass term. Neutrino oscillations described so far are "flavor" oscillations; the individual flavor lepton numbers (electron, muon, tau, etc.) are no longer conserved but their sum, the total lepton number N_{lepton}, is conserved. Neutral current weak interactions are not affected by these "flavor" oscillations, and no distinction between Dirac and Majorana neutrinos can be made.

However, the "flavor" oscillations are not the most general ones. The analysis of the general case begins with the mass Lagrangian containing both the Dirac and Majorana mass terms,

$$-\mathbf{L}_M = \sum_{lk}\left[M^D_{lk}\bar{\nu}_{lL}\nu_{kR} + M^L_{lk}(\bar{\nu}^c_l)_R\nu_{kL} + M^R_{lk}(\bar{\nu}^c_l)_L\nu_{kR} \right] + h.c. \quad (5.15)$$

Here, $\nu_{l,L(R)}$ describe physical neutrinos of definite chirality, $(\nu^c)_{L(R)}$ is the charge conjugated field of definite chirality, and the indices l,k label the different neutrino flavors (N flavors altogether). The matrix M^D of Dirac masses is a complex $N\times N$ matrix with $2N^2$ real parameters, the matrices M^L,M^R of Majorana masses are symmetric complex $N\times N$ matrices with $N(N+1)$ parameters each. Altogether, $\mathbf{L}_M$ contains $4N^2 + 2N$ real parameters; if CP conservation is required, the mass matrices are real and the number of parameters is reduced to $2N^2 + N$.

The operator $\mathbf{L}_M$ and the corresponding kinetic energy operator can be simultaneously diagonalized by a unitary $2N\times 2N$ matrix* V leading to $2N$ neutrino mass eigenstates $\Phi_A = \Phi_{AL} + \Phi_{AR}$

$$\Phi_{AL} = \sum_l\left[V_{Al}\nu_{lL} + V_{A(l+N)}\nu^c_{lL}\right],$$

$$\Phi_{AR} = \sum_l\left[V^*_{Al}\nu^c_{lR} + V^*_{A(l+N)}\nu_{lR}\right]. \quad (5.16)$$

The operator Φ_A describes a Majorana-like particle of mass m_A (Kobzarev et al. 81). The operators Φ_{AL}, Φ_{AR} are projections of a definite chirality; they represent auxiliary quantities, the eigenstates of $\mathbf{L}_M$ are Φ_A.

Left-handed weak interactions. To proceed further and see how one could determine the relevant parameters (and how many parameters we should determine), we must consider the weak interactions. In the standard theory, only the left-handed neutrinos participate in the charged current weak interactions via the Lagrangian

$$-\mathbf{L}_L = \frac{g_L}{\sqrt{2}}\sum_l \bar{\nu}_{lL}\gamma^\mu l_{lL} W^+_{L\mu} + h.c. \equiv \frac{g_L}{\sqrt{2}}\sum_{A,l}\bar{\Phi}_{AL}V_{Al}\gamma^\mu l_{lL} W^+_{L\mu} + h.c. \quad (5.17)$$

*We denote the general mixing matrix by V to distinguish it from the special case of flavor oscillations with mixing matrix U. In reality U represents a submatrix of V.

Here l_l is the charged lepton partner of the neutrino ν_{lL} and W_L is the intermediate vector boson which couples to the left-handed current. It is seen that only half of the mixing matrix V (matrix elements V_{Al}, A=1, ... , $2N$, $l \leqslant N$) appears in $\mathbf{L}_L$ and, therefore, can be determined by experiments based on the charged current weak interactions. This part of V contains $3N^2$ real parameters, of which N are unphysical phases.

To determine the mixing parameters, one can, first of all, perform a series of "flavor" oscillation experiments as described above. Here the neutrino beam contains at time t=0 neutrinos of flavor ν_l (that is, an antilepton l_l^+ was created at t=0 in a weak decay). The amplitude for the creation at time t of lepton $l_{l'}$ through the weak charged current (5.17) is proportional to

$$A(l \rightarrow l') = \sum_{A=1}^{2N} e^{\left(\frac{-im_A^2 t}{2E}\right)} V_{lA}^{+} V_{Al'} \quad , \tag{5.18}$$

where m_A are mass eigenvalues and E is the common energy of all components (we assume $E >> m_A$). The probability of finding the lepton $l_{l'}$ at time t (or distance x=t) is proportional to $|A(l \rightarrow l')|^2$ and has the characteristic oscillatory pattern. By performing all N disappearance experiments $l \rightarrow l'$ (for varying distances x) it is possible to determine the $2N-1$ mass differences $|m_A^2 - m_B^2|$ and the $2N^2-N$ independent absolute values $|V_{Al}|^2$ of the mixing matrix V. In addition, by performing all appearance experiments $A(l \rightarrow l')$, $l \neq l'$, it is possible to determine the $(N-1)(2N-1)$ independent phase differences of V_{Al} and $V_{Al'}$ for A=2,...,$2N$.

Flavor oscillations do not violate the total lepton number N_{lepton}; the oscillation takes place between neutrino states of the same helicity. When the Majorana mass terms M^L or M^R are nonvanishing, oscillations $l \rightarrow \bar{l}'$ with violation of the total lepton number $|\Delta L| = 2$ are possible. Thus, the same beam of neutrinos ν_l can also create, at distance x, a charged antilepton l_l^+. The amplitude of such a process $A(l \rightarrow \bar{l}')$ contains, however, a helicity suppression factor m_A/E (Bahcall & Primakoff 78). Thus

$$A(l \rightarrow \bar{l}') = \sum_{A=1}^{2N} \frac{m_A}{E} e^{\left(\frac{-im_A^2 t}{2E}\right)} V_{Al}^{*} V_{Al'}^{*} \quad . \tag{5.19}$$

The helicity suppression makes oscillation experiments of this type impractical. We shall encounter, however, exactly this kind of term in Chapter 6 when discussing the mass mechanism of neutrinoless double beta decay. If one could somehow perform the whole set of $l \rightarrow \bar{l}$ experiments, the additional $2N-1$ phases of the matrix elements V_{A1}(A=2,...,$2N$) could be determined. Thus, we see that oscillation experiments allow us, in principle, to determine the $2N^2-N$ absolute values and $2N^2-N$ phases of the matrix elements V_{Al}. This constitutes an overdetermination of the matrix $V_{Al}(A=1,...,2N,\ l=1,...,N)$.

Right-handed currents. The other half of the matrix V ($V_{A(i+N)}$) appears in amplitudes of observable processes only if right-handed currents participate in weak interactions. In that case the $|\Delta N_{lepton}| = 2$ processes are no longer suppressed by the helicity factor m_A/E; instead, the amplitude contains terms proportional to the strength of the right-handed current weak interaction. Again, we shall discuss the effect of right-handed currents on the $|\Delta N_{lepton}| = 2$ processes in Chapter 6 in connection with the neutrinoless double beta decay.

CP or T noninvariance. If *CP* or *T* invariance is violated, the mixing matrix V will contain some matrix elements with imaginary parts which cannot be removed by a suitable transformation of the base states. In that case the oscillation patterns in the channels $l \rightarrow l'$ and $l' \rightarrow l$ ($l \neq l'$) will not be the same and, as pointed out by Cabibbo (78), large differences in the oscillation amplitudes might exist. (Note that *CP* or *T* noninvariance has no observable effects in disappearance, $\nu_l \rightarrow x$, experiments.)

The number of *CP* violating phases in the mixing matrix V depends on the form of the mass Lagrangian $\mathbf{L}_M$. If $\mathbf{L}_M$ has only Dirac mass terms $M^D \neq 0$, $M^L = M^R = 0$, there are $(N-1)(N-2)/2$ possible *CP* violating phases in V as in the Kobyashi-Maskawa case. If one of the Majorana mass terms (M^L for example) is nonvanishing, while $M^D=0$ and $M^R=0$, there are $N(N-1)/2$ *CP* violating phases. The number of phases which could be determined from flavor oscillation experiments is $(N-1)(N-2)/2$. The remaining $N-1$ phases must be determined from the $|\Delta N_{lepton}| = 2$ neutrino-antineutrino oscillations. Thus, in agreement with our previous claim, the study of "flavor" oscillations ($\Delta N_{lepton} = 0$) cannot distinguish between the Dirac and Majorana mass terms, even in the presence of *CP* noninvariance.

Second class oscillations. When both Dirac and Majorana mass terms are nonvanishing simultaneously, "second class" oscillations may exist. One expects the existence of states ν^c_{lL}, which are singlets under weak interactions (and therefore "sterile"). The oscillation amplitudes (5.18) are now no longer normalized,

$$\sum_{l=1}^{N} |A(l \rightarrow l')|^2 < 1 \ .$$

These oscillations of the "second class", i.e., oscillations involving noninteracting weak singlets, affect both the charged and neutral current weak interactions. (Barger et al. 80).

5.4 Low Energy Experiments

Low energy studies of neutrino oscillations have been carried out with the help of nuclear reactors. These experiments test the disappearance channel $\overline{\nu}_e \rightarrow x$. Appearance channels cannot be detected at low energy as discussed above.

Reactors are abundant sources of electron antineutrinos with energies up to about 8 MeV. They are created in nuclear beta decay of neutron rich fission products. Nuclear power reactors, such as the reactors at Gosgen and Bugey which have a thermal power around 2800 MW, emit about $5\times10^{20}\overline{\nu}_e\ s^{-1}$ (corresponding to a source strength of about 10^4 megacuries). In contrast, the emission rate for ν_e associated with positron decay or electron capture in the decay of the much rarer neutron poor isotopes is lower by a factor of 10^{-5}—10^{-8} (Schreckenbach 84) depending on energy.

If oscillations are present, the energy spectrum of neutrinos as observed in a detector at distance L from the source (the reactor core) will be modified from the no-oscillation spectrum in accordance with (5.7). The integrated yield will also be affected by oscillations. The neutrino energy E_ν and the distance L are thus two independent variables in an experimental study of oscillations.

In order to evaluate the observed neutrino spectrum in terms of the oscillation parameters Δm^2 and $\sin^2 2\theta$, one needs to know the fission yields, as well as the composite neutrino spectrum of all the fission products. Calculations of the neutrino spectrum emitted by the reactor have been carried out by several researchers. These calculations as well as the experimental studies of the electron spectra accompanying fission are discussed in the following Subsection. The uncertainties inherent in these calculations are bound to limit the accuracy of the analysis of the oscillation parameters. An obvious way to reduce these uncertainties in the neutrino source spectrum greatly is to compare the experimental yields at two or more distances L. This procedure, which is virtually independent of the reactor source spectrum, has been adopted in the Gosgen experiments (Vuilleumier et al. 82; Gabathuler et al. 84; Zacek et al. 85; Zacek et al. 86) in which spectra were recorded for three distances, L = 38 m, 46 m, and 65 m, and also in the Bugey experiment (Cavaignac et al. 84) with data taken at two distances, L = 13.6 m and 18.3 m. An earlier experiment was performed by Kwon et al. (81) at the ILL research reactor in Grenoble at a distance of 8.8 m.

The reactor neutrino spectrum. The Gosgen and Bugey reactors are pressurized water reactors. The Gosgen reactor (2800 MW thermal) operates for an eleven month period, followed by a scheduled shut down of about one month to allow the replacement of one third of the fuel elements. At the beginning of a power cycle, the fuel composition expressed in per cent fission contribution was as follows: 69% from ^{235}U, 7% from ^{238}U, 21% from ^{239}Pu, and 3% from ^{241}Pu. Other fissioning isotopes such as ^{236}U, ^{240}Pu, ^{242}Pu, etc. contribute less than 0.1% to the neutrino spectrum and can, therefore, be neglected.

During a burning cycle, the U isotopes are breeding the fissionable Pu isotopes. As a function of time the relative contribution of fissioning isotopes to the total number of fissions varies as depicted in Figure 5.2. Since the $\bar{\nu}_e$ spectra of the mentioned isotopes are all different from each other, one has to evaluate the composite $\bar{\nu}_e$ spectrum as a function of time and average over the measuring period. The relative contributions of the fissioning isotopes to the reactor power, averaged over the measuring time for the three Gosgen experiments, (Zacek et al. 86), the Bugey experiment (Cavaignac et al. 84), and the ILL experiment (Kwon et al. 81) are summarized in Table 5.1.

Table 5.1 *Mean contributions (in %) to the reactor power from different isotopes.*

Fissioning Isotope Energy per Fission (MeV)	^{235}U 201.7	^{239}Pu 205.0	^{238}U 210.0	^{241}Pu 212.4
Gosgen 37.9 m	61.9	27.2	6.7	4.2
Gosgen 45.9 m	58.4	29.8	6.8	5.0
Gosgen 64.7 m	54.3	32.9	7.0	5.8
Bugey 13.6 m	62.1	26.4	7.6	3.9
Bugey 18.3 m	47.9	36.9	8.2	7.0
ILL 8.8 m	>98.5	<1	<0.5	0

The neutrino spectra emitted by the fissioning isotopes (the $L = 0$ spectra) were obtained experimentally, and also by calculation. For ^{235}U and ^{239}Pu, the dominant isotopes, and for ^{241}Pu, the electron spectra following fission were measured at the ILL reactor in Grenoble by Schreckenbach et al. (81, 85), by Feilitzsch et al. (82), and by Hahn et al. (89). For the isotope ^{238}U no experimental data exist, and one has to resort to the calculated neutrino spectra.

In the experimental work at Grenoble, targets with enriched ^{235}U, ^{239}Pu, and ^{241}Pu in the form of thin foils sandwiched between two nickel foils were exposed to the high thermal neutron flux from the ILL reactor. This flux is comparable to the neutron flux in the reactor core. The electrons emitted from the target traversed a 13 m vacuum pipe before entering a double focusing, iron core beta spectrometer. The spectrometer had a momentum resolution of $\approx 10^{-3}$; it employed a multiwire detector at its focal plane. The energy range in these experiments was 1.5—10.5 MeV. The spectrometer's momentum setting and transmission were determined accurately with the help of internal conversion lines emitted by several neutron capture sources placed at the target location. The absolute yield for the electron spectra could be determined with an accuracy of 3%.

The electron spectra were then transformed to obtain the neutrino spectra. The conversion procedure is based on the fact that in each beta decay the electron and antineutrino share the available energy. Thus, for each branch of beta decay with the endpoint E_o, the probabilities of emitting an electron of energy E_e and an antineutrino of energy $E_o - E_e$ are equal. For many branches the electron spectrum is

$$N_\beta(E_e) = \sum a_i P_\beta^{(i)}(E_e, E_o^{(i)}, Z) \, , \qquad (5.20)$$

where a_i are the branching ratios, P_β is the spectrum shape of a single branch, and the summation is over all branches with endpoints larger than

Figure 5.2. Relative contribution from the most important fissioning isotopes to the reactor power as a function of reactor operating time.

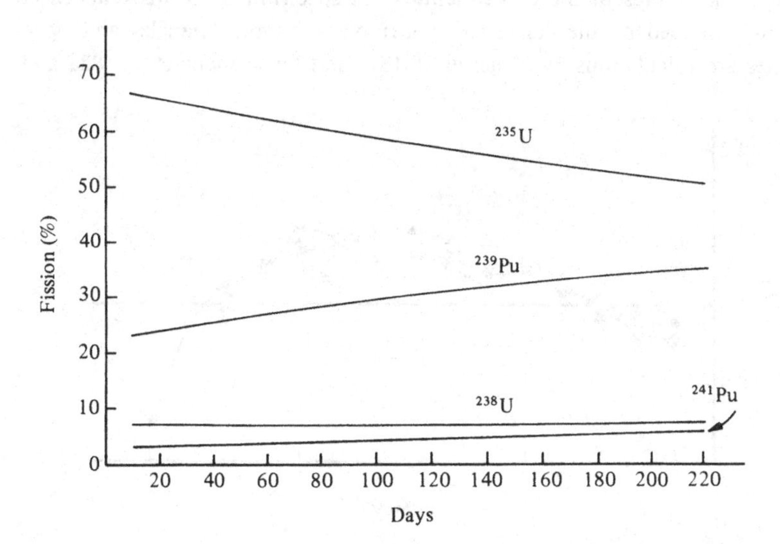

E_e. The antineutrino spectrum is determined by the same set of branching ratios a_i

$$N_\nu(E_\nu) = \sum a_i P_\beta^{(i)}(E_0^{(i)} - E_\nu, E_0^{(i)}, Z). \tag{5.21}$$

The conversion procedure works, therefore, as follows: The measured electron spectrum is used and a set of branching ratios a_i with endpoints $E_0^{(i)}$ is determined by a fit. Once these quantities are known the antineutrino spectrum is calculated. It was shown by Davis et al. (79) and Schreckenbach et al. (81, 85) that this procedure introduces very little additional uncertainty into the neutrino spectrum beyond the inherent experimental uncertainty of the measured electron spectrum.

To illustrate this point, we show in Figure 5.3 the ratio $N_{e^-}/N_{\bar{\nu}}$ of the electron to neutrino spectrum (the electron spectrum is evaluated in terms of the full electron energy). The function $K(E)$ is close to unity over the whole energy interval of interest and there is very good agreement between the $K(E)$ as determined experimentally by Schreckenbach et al. (85) and the calculated values.

In order to minimize the role of the uncertainties in the reactor neutrino spectrum, the experimental positron spectra measured at different distances (and thus at different times in the reactor cycle) were analyzed simultaneously. In this analysis the largest remaining uncertainty stems from the spread of the calculated spectra of ^{238}U and ^{241}Pu by Vogel et al. (81) and

Figure 5.3. Ratio of the observed beta spectrum to the neutrino spectrum. Full circles represent the experimental beta spectrum by Schreckenbach et al. (85) divided by the converted neutrino spectrum. Triangles and open circles are calculations by Vogel et al. (81) and by Klapdor et al. (82a and b).

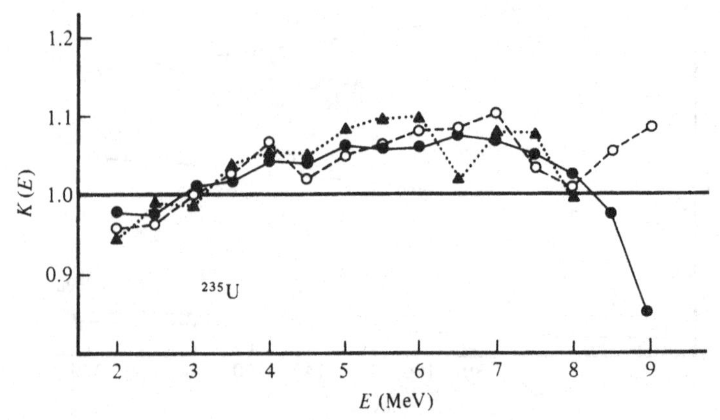

by Klapdor et al. (82a and b), and amounted to 1.3%. The reactor neutrino spectra in the three Gosgen experiments differed from each other by less than 4% for any relevant energy bin. This difference was taken into account by the calculation, with an estimated uncertainty of less than 1%. (In the Bugey work, the reactor power varied considerably over several running periods, resulting in an uncertainty of 4.8% in the normalization of the spectra at the two distances.)

The detector reaction. For most neutrino oscillation experiments at reactors, the neutrinos are detected by means of the reaction

$$\bar{\nu}_e + p \to e^+ + n \, . \tag{5.22}$$

This reaction which has a relatively large cross section ($\approx 10^{-42}$ cm^2 for reactor energies) provides a convenient time correlated pair of positron and neutron signatures.

The antineutrino capture on protons (5.22) is one of the neutrino induced reactions on nucleons discussed in Chapter 3. Here, we are concerned only with its cross section near threshold, i.e., at energies relevant to the neutrino oscillation studies at nuclear reactors.

Positron and antineutrino energies are connected by the relation

$$E_{\bar{\nu}} = E_e + (M_n - M_p) + O(E_{\bar{\nu}}/M_n) \, , \tag{5.23}$$

where the small last term reflects the recoil of the final neutron. Thus, a measurement of the positron energy allows an accurate determination of the energy of the incoming antineutrino. The cross section of the detector reaction near threshold is determined by the rate of the neutron decay, and for a given energy E_e of the outgoing positron we obtain

$$\sigma(E_e) = \frac{2\pi^2 \hbar^3}{m_e^5 c^7 f \tau_n} p_e E_e \, , \tag{5.24}$$

where τ_n is the neutron lifetime, and f is the usual statistical function for neutron beta decay. Wilkinson (82) calculates $f = 1.71465(15)$; this value includes a number of corrections besides the usual Coulomb interaction. The cross section formula above is derived in the lowest approximation, in which the nucleons are assumed to be infinitely heavy and the radiative effects are neglected. Inclusion of the effects of nucleon recoil, weak magnetism, and radiative corrections (Vogel 84b; Zacek et al. 86; Fayans 85) results in a few per cent modification of the cross section formula.

For the neutron lifetime τ_n, the recommended value of $\tau_n = 898 \pm 18$ s (Wilkinson 82) has been assumed. A more recent value for

τ_n of 888.6 ± 3.5 (Review of Particle Properties 90) is consistent but much more accurate.

The reaction has a threshold $E_T = M_n + m_e - M_p \approx 1.805$ MeV; only antineutrinos with $E_{\bar{\nu}} > E_T$ can create positrons. From threshold, the cross section increases approximately quadratically with energy. This steep increase is modified at higher energies, where one can no longer neglect the nucleon recoil (see Chapter 3).

Neutrino detector system. Neutrinos identified through reaction (5.22) can be detected by measuring the emerging, time correlated positrons and neutrons in appropriate detector subunits. An efficient neutrino detector system with good energy resolution was built by Kwon et al. (81) for the oscillation experiment at the ILL reactor in Grenoble. It makes use of liquid scintillation counters for the detection and energy measurement of the positrons, and ^{3}He wire chambers for the identification of neutrons as schematically illustrated in Figure 5.4. This system was later improved to be used at the Gosgen reactor (Zacek et al. 86). At the same time, a similar detector system was built for experiments at the Bugey reactor (Cavaignac et al. 84). Below, we shall describe the Gosgen detector in some detail as an example of a low energy neutrino detector.

In the Gosgen detector, shown in Figure 5.5, a composite liquid scintillation counter serves as a proton target, a positron counter, as well as a

Figure 5.4. Schematic drawing of the neutrino detector system for reactor experiments.

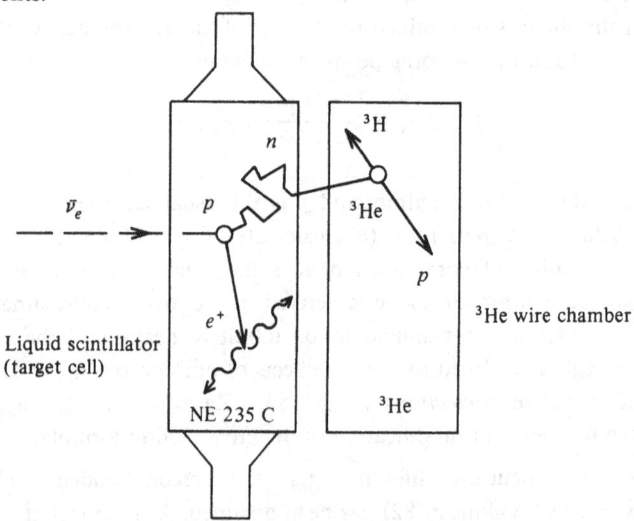

neutron moderator. It consists of 30 rectangular boxes (cells) made from Lucite in such a way as to give maximum light output, and is filled with proton rich, liquid scintillator (of the type NE 235 C). Each cell is viewed by two photomultiplier tubes at each end. Six cells are stacked vertically to form a plane, (1.2×0.9) m^2 in size. There are five planes holding a total of 377 l of liquid scintillator. An event can be localized within the detector by identifying the cell that fired, and by using the time-of-flight technique within each cell. Good energy resolution in the target cells is of importance to permit a sensitive search for oscillation effects in the positron spectrum. At Gosgen the energy resolution was studied with a ^{65}Zn source (1.1 MeV gamma ray) and the final resolution function, which is required in the deconvolution of the observed spectra, was obtained by including, with the help of Monte Carlo calculations, various secondary effects, such as annihilation in flight and bremsstrahlung. The resolution functions versus energy are shown in Figure 5.6. Effective energy resolutions at 1 MeV and 5 MeV were 0.3 MeV and 0.7 MeV, respectively.

Because of the special property of the scintillation liquid (NE 235 C), identification of heavy, charged particles by means of pulse shape discrimination is possible (Kwon et al. 81). A neutron may transfer its energy by

Figure 5.5. Experimental arrangement for the neutrino detector at Gosgen. (From Boehm & Vogel 84.)

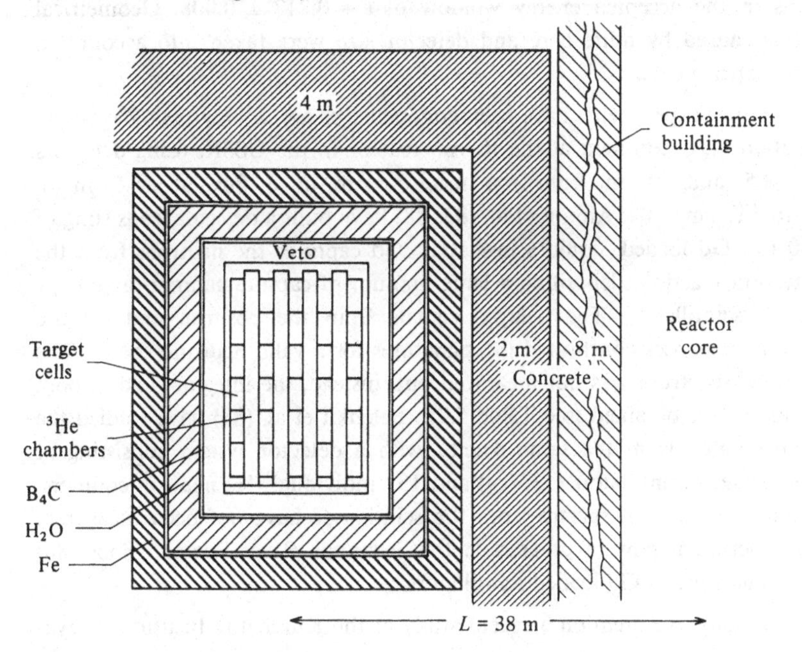

recoil to a proton in the scintillator giving rise to a light pulse with a decay time longer than that from a positron or gamma ray. The neutron subsequently thermalizes and is captured in the adjacent neutron counter, simulating a time correlated event. The purpose of the pulse shape discriminator is to label these unwanted events.

The second element of the detector system is a set of neutron counters. These are ^{3}He filled, multiwire, proportional chambers ($(1.2\times0.9\times0.08)$ m^3 in size), sandwiched between the stacks of target cells as illustrated in Figure 5.5. The chambers are filled with 95% ^{3}He and 5% CO_2 at atmospheric pressure. All materials utilized for the fabrication of the chambers were selected for low alpha radioactivity. The neutron capture events in the ^{3}He counter can be localized in space by virtue of the multiwire structure.

A true event is characterized by a pulse in the scintillator (above a hardware threshold of 0.7 MeV), followed within 250 μs by a neutron capture pulse in the ^{3}He wire chamber. The efficiency of the detector for neutrons from the detection reaction having energies up to 50 keV was measured with a Sb-Be neutron source ($\approx$ 25 keV neutrons). A special target cell was built which allowed the Sb-Be source to be moved by remote control along the coordinate axes. This cell could be substituted for any of the target cells and a complete spatial distribution of the efficiency could be obtained. The calibration procedure yielded a weighted efficiency for neutrons in the accepted energy window of $\varepsilon = 0.217 \pm 0.008$. Geometrical effects caused by finite core and detector size were taken into account in these experiments.

Other experiments have been conducted or are in progress. We briefly mention an experiment at the Rovno reactor in the USSR, using detectors at 18.5 and 25 m from the reactor core at a flux at 18.5 m of $6\times10^{12}\ \bar{\nu}_e\ \mathrm{cm}^{-2}\ \mathrm{s}^{-1}$. Afonin et al. (83, 88) have built a detector, consisting of 240 l of Gd loaded, liquid scintillator. Gd captures the neutrons from the detection reaction, depositing a large amount of capture gamma ray energy in the scintillator. The delayed signals from the positron and capture gamma responses constituted the signatures for a valid neutrino event. The positron spectrum was not measured but, instead, the integral yield at both positions was obtained (see Table 5.2). Belenkii et al. (83) have studied the neutrino flux from the same reactor with a detector system consisting of polyethylene scintillators (136 kg) and 132 individual ^{3}He neutron counters. Again, the positron spectra from the scintillator have not been measured. An experiment now in progress at the Savannah River Reactor (Sobel 86) also makes use of Gd loaded scintillators.

Finally, we mention a recent study at the Kurchatov Institute (Vidyakin et al. 90a) with two detectors at 57 m and 231 m, respectively, from the Institute's reactors. Only the neutrons were detected in these experiments, the detectors being ^{3}He proportional counters embedded in polyethylene moderators.

Shielding. Among the potential sources of background are the gamma rays and neutrons from the reactor, the alpha, beta, and gamma rays from radioactive nuclei in components of the detector system, such as photomultipliers, chamber walls, and wires, and the cosmic ray particles. The reactor associated background can be eliminated entirely by adequate shielding. In the case of the Gosgen experiment, the reactor containment building provided 8 m of concrete shielding, and the bunker housing the detector contributed an additional 2 m lateral shielding. The detector was surrounded with water (20 cm), steel (15 cm), and sheets of boron carbide. The water and the boron carbide served as moderator and absorber, respectively, for neutrons associated with cosmic rays. In the Gosgen experiment there was no difference in singles count rates for reactor-on or reactor-off.

The nucleonic component of the cosmic rays can be reduced

Figure 5.6. Response functions for monoenergetic positrons of 1, 3, and 5 MeV in a scintillation counter, including secondary processes. (From Zacek et al. 86.)

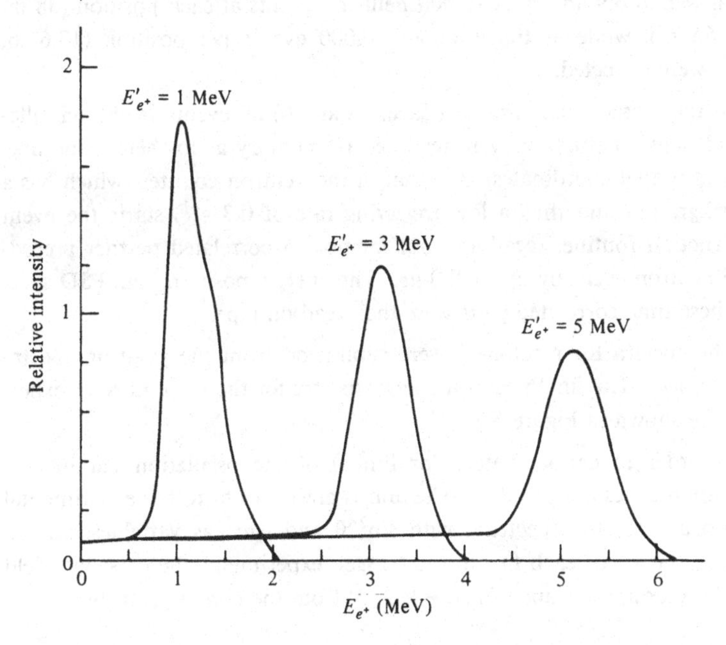

significantly with concrete shielding. The experimental bunker at Gosgen had 4 m of concrete overhead, reducing the flux of the nucleonic component by a factor of about 10^4 (to about 10^{-8} cm^{-2} s^{-1}). The muon component, on the other, hand was reduced by only a factor of 2, to about 7×10^{-3} cm^{-2} s^{-1}. This residual muon component can be effectively tagged with the help of a veto counter. At Gosgen the neutrino detector was completely surrounded by a liquid scintillation counter in the form of a box $(1.5\times1.5\times1.5)$ m^3, made from six scintillation tanks (Figure 5.5). The effectiveness of the veto was 99.8%. Some muons, however, may stop in the material surrounding the veto, liberating neutrons as a result of nuclear excitation and capture.

One of the most significant radioactive backgrounds in the detector stems from ^{40}K. This isotope occurs in concrete and particularly in the glass of the photomultiplier tubes and gave rise to a singles background rate (above 0.7 MeV), summed over all target cells, of 300 s^{-1}. The background in the ^{3}He counter was much lower (0.3 s^{-1}). (The numbers quoted are for Gosgen, but were similar for Bugey.)

Measurements and results. Owing to the small neutrino cross section, the event rates in the detector system are as low as a few counts per hour. Therefore, the data acquisition time required to obtain a positron spectrum with good statistical accuracy is several months. In the Gosgen experiments the goal was to obtain about 10,000 neutrino events at each position (38 m, 48 m, 65 m), while at Bugey about 30,000 events per position (13.6 m, 18.3 m) were collected.

In the Gosgen experiments (Zacek et al. 86) all events in the scintillation and neutron counters were recorded in memory as to their amplitude, time, and spatial coordinates. A signal in the neutron counter (which has a low background and thus a low triggering rate of 0.3 s^{-1}) starts the event reconstruction routine, recalling from memory a correlated partner preceding the neutron event by up to 320 μs. The energy, position, and PSD spectra of these time correlated pairs were then read on tape.

The spectra for reactor-off were subtracted from the positron spectra thus obtained. The final positron energy spectra for the three Gosgen experiments are shown in Figure 5.7.

In order to extract values (or limits) of the oscillation parameters, several approaches are possible. The aim is always to fit to the experimental spectrum a predicted spectrum with $\sin^2 2\theta$ and Δm^2 as variables, and to determine the χ^2 for each fit. In the Gosgen experiments the expected yield in the detector at a distance L_j (j = 1, 2, 3) from the core is given by

$$Y(E_{e^+},L_j,\Delta m^2,\theta) = Y_{no-osc}(E_{e^+},L_j)P(E_{e^+},L_j,\Delta m^2,\theta)\ , \qquad (5.25)$$

where

$$Y_{no-osc}(E_{e^+},L_j) = \frac{1}{4\pi L_j^2}a_{PSD}(E_{e^+})\varepsilon n_p \sigma(E_{e^+})S(E_{e^+})\ . \qquad (5.26)$$

Here $a_{PSD}(E_{e^+})$ is the PSD acceptance, ε is the detector efficiency, n_p is the number of protons in the target, $\sigma(E_{e^+})$ represents the detection cross section (it includes the neutron lifetime τ_n), and $S(E_{e^+})$ is the reactor neutrino spectrum. $P(E_{e^+},L_j,\Delta m^2,\theta)$ is the oscillation function given in (5.7) (where the

Figure 5.7. Positron spectra from the reaction $\bar{\nu}_e p \to e^+ n$ as obtained with the Gosgen detector, for several distances from the reactor core. The solid lines are the fitted spectra. The dashed lines represent the spectra predicted for no oscillations, based on the measured electron spectra accompanying fission and on calculations.

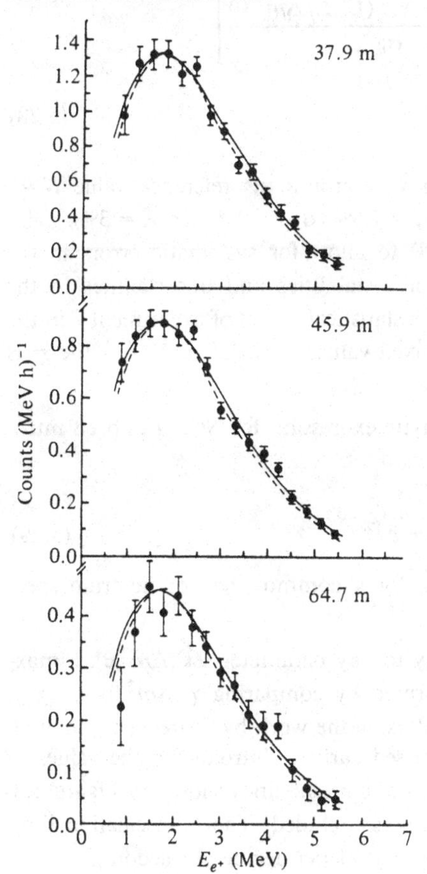

positron energy E_{e^+} must be replaced by the corresponding neutrino energy $E_{\bar{\nu}}$). Expression (5.25) has to be folded with the detector energy response and also with the finite size of reactor core and detector. After performing this folding integral over E and L, the yield Y becomes an averaged yield denoted by $\bar{Y}$. The 16 discrete energy bins of the Gosgen spectra are denoted with the subscript k. It is convenient to factor out the terms that are different in the three experiments by writing the no-oscillation yield as follows,

$$Y_{no-osc} = \frac{1}{4\pi L_j^2}(1+\eta_{kj})X_k \tag{5.27}$$

where η_{kl} is a small term reflecting the difference in reactor fuel composition for the three experiments and X_k describes the position independent no-oscillation yield for the energy bin k. Following Zacek et al. (86), we define a χ^2 as follows,

$$\chi^2(\Delta m^2,\theta) = \sum_{jk}\left[\frac{Y_{exp}(E_k,L_j)-N_j\bar{Y}(E_k,L_j,\Delta m^2,\theta)}{\sigma_{jk}}\right]^2$$
$$+\sum_{j=1}^{3}\left(\frac{N_j-1}{\sigma_{Nj}}\right)^2 . \tag{5.28}$$

N_j is the normalization which can vary around its reference value N_j=1 within the systematic errors $\sigma_{Nj}(\sigma_{N1} = 1.5\%$, $\sigma_{N2} = 1.5\%$, $\sigma_{N3} = 3\%)$. The last three terms are added in (5.28) to allow for systematic errors in the three measurements due to detector instabilities and uncertainties in the knowledge of the reactor power. The statistical errors of experiment j in the energy bin k are denoted σ_{jk}. For fixed values of $\sin^2 2\theta$ and Δm^2, the χ^2 is minimized by varying N_j and X_k.

Instead of varying X_k, an analytic expression for $S(E)$ has been introduced (Zacek et al. 86) as follows,

$$S(E) = e^{(\sum_{l=0}^{2} A_l E^l)} \tag{5.29}$$

with simultaneous fit parameters A_l for a common reactor neutrino spectrum. Now, N_j and A_l are varied.

To assign a relative probability to any parameter set $(\Delta m^2,\theta)$, a maximum likelihood ratio test is performed by comparing $\chi^2(\Delta m^2,\theta)$ to χ^2_{Min}. The spectrum $S(E)$ can be evaluated using the work by Schreckenbach et al. (81, 85) and Vogel et al. (81) discussed earlier. Introducing the values of $\sigma(E_{e^+})$ and ε, Zacek et al. (86) find the contour lines shown in Figure 5.8. The region to the right of the curves is excluded. The no-oscillation limit $(\theta = 0)$ represents a good solution ($\chi^2 = 0.91$ per degree of freedom).

The Bugey data (Bouchez 89), while originally suggesting evidence for oscillations (Cavaignac et al. 84), also exclude neutrino oscillations at, however, a somewhat less restrictive parameter range ($\Delta m^2 < 0.07\text{eV}^2$ for full mixing).

Finally, we summarize in Table 5.2 the ratios of experimental to predicted positron yields, integrated over the spectrum, assuming no oscillations.

We conclude this Section by summarizing the low energy neutrino results. Detailed experimental studies of energy spectra of reactor neutrinos at several distances so far have not revealed any evidence for neutrino oscillations, with limits for oscillation parameters given in Figure 5.8. For maximum mixing the quantity Δm^2 cannot exceed its present sensitivity limit of 2×10^{-2} eV^2.

As to the future of these experiments, it seems possible to improve the

Figure 5.8. Limits at 68% and 90% CL for the oscillation parameters Δm^2 and $\sin^2 2\theta$ from the three Gosgen experiments, and from these experiments combined with the no-oscillation spectrum. The excluded area is to the right of the curves.

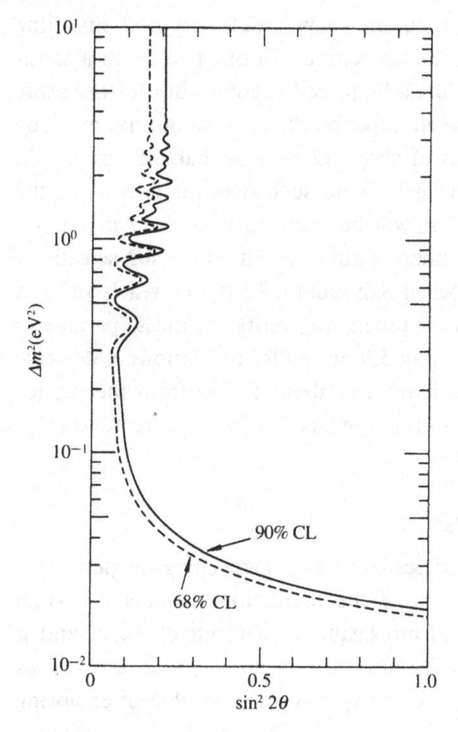

Table 5.2. *Ratios of integral experimental to predicted (no-oscillation) positron yields and ratios of integral experimental yields at different distances.*

	$\int Y_{exp} / \int Y_{th.no-osc.}$	$\int Y^{(i)}_{exp} / \int Y^{(1)}_{exp}$
Gosgen 38 m (1)	1.018 ± 0.065	-
Gosgen 46 m (2)	1.045 ± 0.065	1.027 ± 0.034
Gosgen 65 m (3)	0.975 ± 0.072	0.958 ± 0.053
ILL 8.8 m	0.955 ± 0.110	-
Bugey 13.6 m (1)	-	-
Bugey 18.3 m (2)	-	1.007 ± 0.03
Rovno 18.5 m (1)	0.997 ± 0.06	-
Rovno 25 m (2)	-	0.986 ± 0.047

sensitivity in a significant way by building larger detectors and installing them at larger distances from the reactor station. In order to reach a sensitivity of, say, 10^{-3} eV^2, a detector must be placed twenty times further away from the reactor than in the Gosgen experiment, or at about 1 km. The detector must have a fiducial mass of about 12 tons so that the results will have comparable statistical significance. One such detector is now in the planning stage (Boehm et al. 91). It will be installed near the San Onofre Nuclear Generating Station in Southern California. Its projected sensitivity is shown in Figure 5.11 (curve labelled San Onofre 12 t). To reach an even higher sensitivity, such as that required to verify a mass parameter $\Delta m^2 \approx 10^{-4}$ eV^2, as discussed in Section 5.9 on matter oscillations, a detector with a mass of 1000 tons or more located at about 13 km from the reactor station will be needed (see Figure 5.11, curve labelled San Onofre 1000 t).

5.5 High Energy Experiments

In this Section we shall describe several high energy experiments which have contributed in determining the oscillation parameters. High energy accelerators provide ν_μ neutrino beams as a result of K, π, and μ decay, allowing the study of the channels $\nu_\mu \rightarrow \nu_e$, $\nu_\mu \rightarrow \nu_\tau$, as well as $\nu_\mu \rightarrow x$. Appearance experiments, such as $\nu_\mu \rightarrow \nu_e$, are capable of exploring very small mixing angles, the sensitivity being limited only by the purity of the ν_μ beam, while the ν_μ beam serves as a monitor. We shall organize the discussion below according to the particular oscillation channel, starting with a discussion of those experiments which involve coupling to ν_e.

The $\nu_\mu \to \nu_e$ channel. If ν_μ were to oscillate into ν_e, there should be an excess of ν_e in a ν_μ beam. The beam lines producing the incident ν_μ and $\bar{\nu}_\mu$ neutrinos are designed in such a way as to minimize the impurities of ν_e and $\bar{\nu}_e$ neutrinos. Some of these neutrinos are always present, primarily as a result of the decays $K^+ \to \pi^o e^+ \nu_e$ ($K^- \to \pi^o e^- \bar{\nu}_e$). Unlike in a disappearance experiment, such as those described in Section 5.4, the ν_μ spectrum does not need to be known with great accuracy.

A search for the $\bar{\nu}_\mu \to \bar{\nu}_e$ channel has been undertaken by Wang et al. (84), at LAMPF (Los Alamos Meson Physics Facility) as a by-product of a $\nu_e e \to \nu_e e$ elastic scattering study. The proton beam (800 MeV, 600 μA) is stopped in the beam dump and produces predominantly positively charged pions which decay through the reactions $\pi^+ \to \mu^+ \nu_\mu$, and $\mu^+ \to e^+ \nu_e \bar{\nu}_\mu$. Thus, the neutrino beam consists mostly of ν_e, ν_μ and $\bar{\nu}_\mu$. However, there is a contamination of $\bar{\nu}_e$ as a result of the decay chains $\pi^- \to \mu^- \bar{\nu}_\mu$ and $\mu^- \to e^- \bar{\nu}_e \nu_\mu$. Fortunately, this contamination is small (2×10^{-3}), partly because the proton beam produces less π^- than π^+, but mainly because the π^- are captured and absorbed by nuclei in the beam stop. With this in mind, the number, or upper limit, of $\bar{\nu}_e$ seen in the experiment provides a measure of the $\bar{\nu}_\mu \to \bar{\nu}_e$ oscillations. A spectrum of the three neutrino types is shown in Figure 5.9. In the experiment one looks for $\bar{\nu}_e$ which may have been produced from oscillation of $\bar{\nu}_\mu$. The detection reaction is $\bar{\nu}_e + p \to e^+ + n$. The detector consists of a stack of plastic scintillators and flashtubes with a total mass of 14 tons. It is located 10 m from the beam dump, with 6.3 m of steel in between to shield against neutrons. The signature of a true event is a single electron track, required to have an energy of $35 < E_e < 53$ MeV. If all $\bar{\nu}_\mu$ were to oscillate into $\bar{\nu}_e$ one would expect to see 930 ± 163 events. Instead, only −1.25 ± 7.9 events were observed, leading to the following limits for the oscillation parameters: $\Delta m^2 < 0.49$ eV^2 for full mixing, and $\sin^2 2\theta < 0.028$ for $\Delta m^2 > 2$ eV^2. A somewhat less stringent limit was obtained earlier at LAMPF by Willis et al. (80) and Nemethy et al. (81).

The oscillation experiment at BNL (Brookhaven National Laboratory) described by Ahrens et al. (85) (BNL 734) is also a by-product of a $\nu_\mu e^- \to \nu_\mu e^-$ scattering study. It is based on the comparison of the observed and calculated ν_e fluxes. What is measured are the reactions $\nu_e n \to e^- p$

and $\nu_\mu n \to \mu^- p$. The target neutrons are bound in nuclei; however, at high energy, to good approximation, these neutrons can be considered free.

The neutrino detector, located at 96 m from the neutrino source, consists of 112 planes of liquid scintillator, each with dimensions $(4 \times 4 \times 0.08)$ m^3, and 224 planes of proportional drift chambers $(4.2 \times 4.2 \times 0.038)$ m^3. These detectors were subdivided into a large number of cells to allow the determination of a fine topology of the events. Downstream from the detectors were two shower counters. The $e^- p$ and $\mu^- p$ final states were measured, the first by recording the electron angles and energies, the second by identifying the μ^- and recording the proton kinematics. Owing to the larger acceptance for the $e^- p$ events, the ratio of ν_e/ν_μ fluxes could be measured down to 10^{-3}. There were 418 $e^- p$ events in the energy region $0.9 < E_\nu < 5.1$ GeV, while the number of $\mu^- p$ events was 1370. The neutrino fluxes $\Phi(E_{\nu_e})$ and $\Phi(E_{\nu_\mu})$ were obtained from the experimental yields of the reactions $\nu_e n \to e^- p$ and $\nu_\mu n \to \mu^- p$, using the respective acceptance functions for the two reactions and the known magnitude and energy dependence of the quasi-elastic cross sections. The observed flux ratio, $\Phi(E_{\nu_e})/\Phi(E_{\nu_\mu})$ is shown in Figure 5.10(a), and compared with the results of neutrino beam calculations. Subtracting the actually observed ν_e flux from that predicted by calculations based on K and μ decay (see Figure 5.10(b)), an upper limit for the number of ν_e produced by oscil-

Figure 5.9. Neutrino spectra from stopped π^+. Spectra of this form are emitted from the beam stop at medium energy facilities, such as LAMPF.

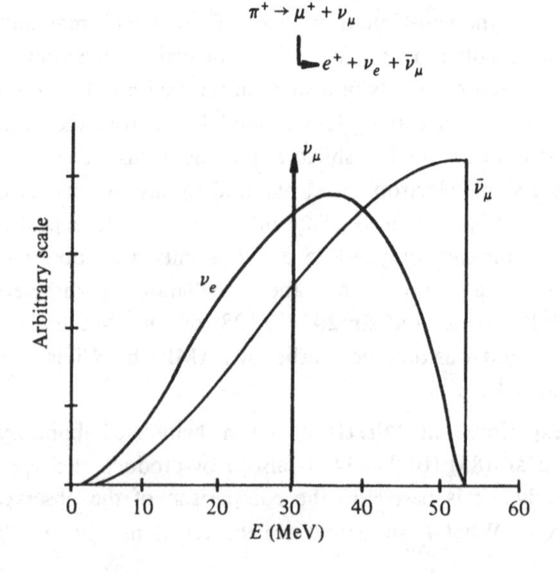

lations has been obtained. From this limit the values of the Δm^2 and $\sin^2 2\theta$ excluded at 90% CL have been calculated and are shown in Figure 5.11.

The $\nu_\mu \rightarrow \nu_e$ oscillation channel was also explored at CERN in experiments carried out by the European Bubble Chamber (BEBC) collaboration (Athens, Padova, Pisa, Wisconsin collaboration; Angelini et al. 86), and also by the CHARM collaboration (CERN, Hamburg, Amsterdam, Rome, Moscow; Bergsma et al. 84). In these experiments use was made of the long baseline facility installed at the CERN PS (proton synchrotron) with the specific purpose of studying neutrino oscillations.

The proton beam of 19.2 GeV from the PS strikes a beryllium target. The pions and kaons are allowed to decay in a 50 m space. Following the

Figure 5.10. (a) Ratio of the neutrino fluxes, $\Phi(E_{\nu_e})/\Phi(E_{\nu_\mu})$, for the Brookhaven experiment. The solid lines are calculations from a neutrino beam transport code. (b) Difference between the measured and calculated ratios. The expected difference in the presence of oscillations is denoted by the dashed line. (From Ahrens et al. 85)

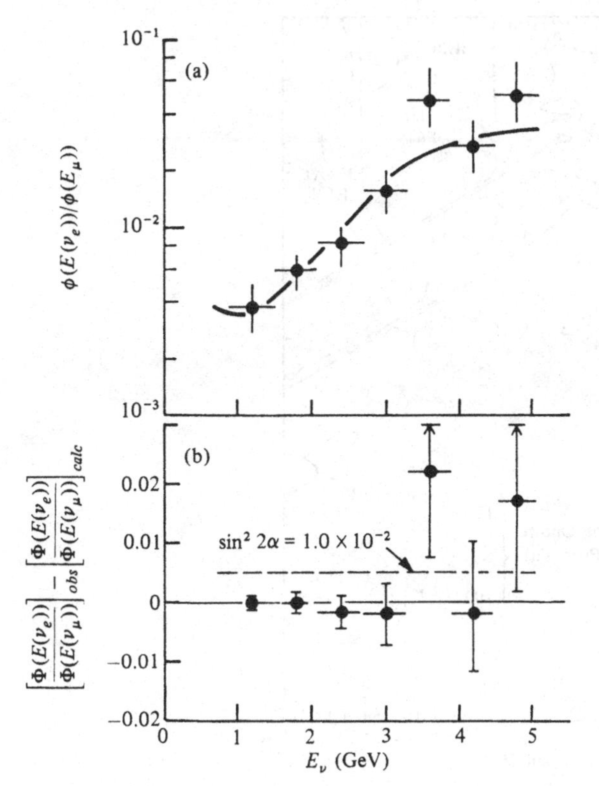

decay section there is 70 m of iron and earth shielding. The detectors of the BEBC and CHARM collaborations were about 900 m away from the neutrino production target, as illustrated in Figure 5.12. For the BEBC experiment the neutrino flux was enhanced by a factor of 7 by focusing the pions with the help of a magnetic horn.

The BEBC bubble chamber was filled with 70% Ne and 30% H_2 and had a fiducial mass of 14 tons. The ν_e and ν_μ fluxes, whose average energies were 2.5 GeV, were evaluated by Monte Carlo calculations and are shown in Figure 5.13. The goal of the experiment was to look for an excess of ν_e over the calculated rate. A total of 794,000 pictures was taken of which 228,000 have been analyzed in the quoted publication. A total of 150

Figure 5.11. Limits for the oscillation parameters Δm^2 and $\sin^2 2\theta$ that involve coupling to ν_e from high energy accelerator experiments and low energy reactor experiments. The excluded region is to the right of the curves. Sensitivities for proposed experiments are shown by dotted lines. (For references see text.)

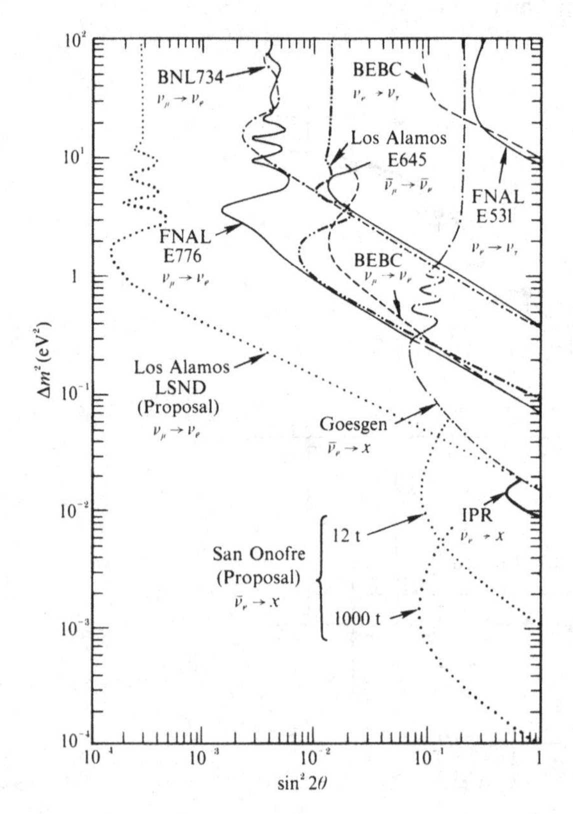

charged current events involving a $\mu^- X$ in the final state were seen. In contrast, only one candidate for a ν_e induced event was found leading to a $e^{mdash}\pi^+ pp$ final state. This roughly corresponds to the expected rate based on the known beam composition. From this finding, the limits for the oscillation parameters shown in Figure 5.11 were derived.

The CERN CHARM collaboration has carried out a $\nu_\mu \rightarrow \nu_e$ search, using two detectors running simultaneously at different distances from the source. The advantage of using two detectors is that an absolute normalization of the neutrino flux is not necessary. The neutrino target and arrangement described above is depicted in Figure 5.12. The $\nu_e(\bar{\nu}_e)$ contamination of the ν_μ beam was $< 0.5\%$, and the ν_μ spectrum was calculated by the Monte Carlo method. The CHARM fine grain calorimeter consists of 78 subunits, each consisting of 20 scintillation counters, a layer of 256 streamer tubes, a marble plate, $(3\times3\times0.08)\ \mathrm{m}^3$ in size, and a layer of 128 proportional drift tubes. For the neutrino search, the first 18 subunits of the calorimeter (fiducial mass of 36 tons) were detached from the main detector

Figure 5.12. Layout of experimental area at CERN. (From Angelini et al. 86.)

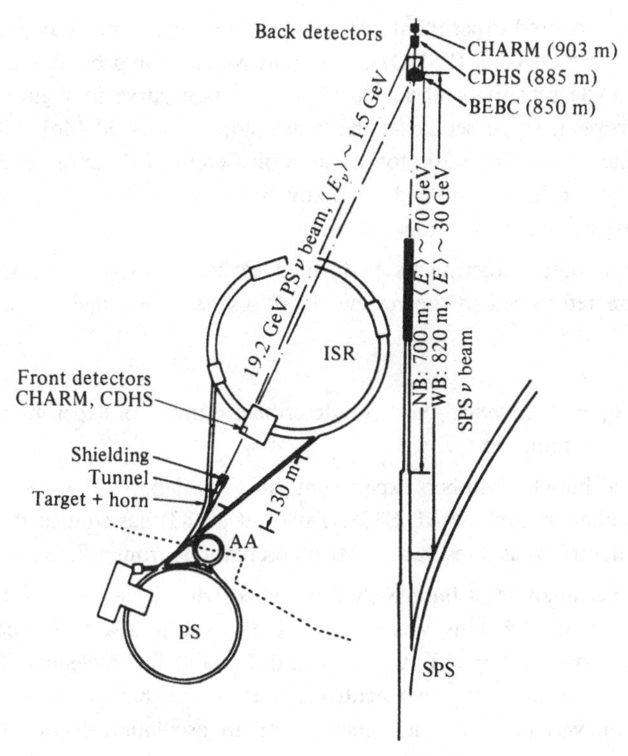

and installed at 123 m from the target. The remaining 60 subunits (fiducial mass of 120 tons) were at 903 m (see Figure 5.12). The average ratio of solid angle subtended by a subunit of each detector was (123 m / 903 m)^2 = 0.0186. The detector could identify electron neutrino events with an average energy of $E_{\nu_e} \approx 2.2$ GeV. From the original preselected 2043 candidates for electron events in the close detector and 220 in the far detector, only 66 events survived stringent conditions in the close detector and 19 in the far detector. The ratios of the ν_e and ν_μ induced charged current events, $(\nu_e/\nu_\mu)_c$, were found to be $(9.2 \pm 1.1 \pm 1.1)\%$ for the close detector and $(8.9 \pm 2.0 \pm 1.0)\%$ for the far detector. From the difference of these ratios it was found that at most 2.7% of the ν_μ have oscillated into ν_e at 90% confidence level. The resulting exclusion plot is shown in Figure 5.11.

An experiment at the beam stop of the Los Alamos Meson Facility (LAMPF) (LAMPF 645; Durkin et al. 88) has provided limits similar to those of the BEBC experiment. The exclusion curve of this experiment is also shown in Figure 5.11.

Most recently, the Fermilab (FNAL) collaboration (FNAL E776; Blumenfeld et al. 89, 91) have obtained more stringent limits for the mass parameters as shown in the exclusion curve depicted in Figure 5.11.

As to proposed experiment, we mention the Los Alamos Large Scintillation Neutrino Detector (LSND) experiment which should be able to reach a mixing angle sensitivity of $\sin^2 2\theta$ of 3×10^{-4} (see curve in Figure 5.11). This experiment, to be set up at the beam stop of the 800 MeV LAMPF proton linac, consists of a 200 ton mineral oil Cerenkov detector containing a wave length shifter. Combined Cerenkov and scintillation light will permit good timing as well as directional resolution.

Several other experiments performed earlier to explore the $\nu_\mu \rightarrow \nu_e$ channel are referenced in the reviews by Wotschack (84) and by Shaevitz (83).

The $\nu_\mu \rightarrow \nu_\tau$ *channel.* Next, we describe a number of experiments that explore the couplings to ν_μ.

In an bubble-chamber experiment at Fermilab, a Hawaii-Berkely-Fermilab collaboration (FNAL E338; Taylor et al. 83) has studied the creation of $\bar{\nu}_\tau$ neutrinos as a result of neutrino oscillations from a $\bar{\nu}_\mu$ beam.

The Fermilab 15 ft bubble chamber filled with a Ne - H_2 mixture of fiducial mass of 11.8 tons was used as a detector to search for charged current reactions $\bar{\nu}_\tau N \rightarrow \tau^+ X$, where N and X stand for nucleons. The τ^+ decays into a positron and two neutrinos with a 17% branching ratio, and this positron was taken as the signature for an oscillation event. The $\bar{\nu}_\mu$

beam had an energy between 20 and 200 GeV. As in the previously described experiments, the contamination of $\bar{\nu}_e$ had to be minimized and taken into account. This contamination in the beam arises mainly from K decay, but some of it is also due to the decay of charmed particles. A total of 427 $\bar{\nu}_\mu$ induced charged current events were identified, and 4 $\bar{\nu}_e$ induced events were seen. The latter value should be compared to the expected $\bar{\nu}_e$ background of 3.8 ± 0.5. Thus, there is no evidence that oscillations produced $\bar{\nu}_\tau$ and the mentioned limit allows the construction of the exclusion curve. It should be noted that this experiment also gives limits on the $\bar{\nu}_\mu \rightarrow \bar{\nu}_e$ oscillations

The limits quoted were somewhat improved by a 1986 FNAL bubble chamber experiment (FNAL E564; Brucker et al. 86), excluding areas shown in Figure 5.14.

Also at Fermilab, a new emulsion experiment was carried out (FNAL

Figure 5.13. Expected ν_μ and ν_e fluxes at the BEBC detector at CERN. (From Angelini et al. 86.)

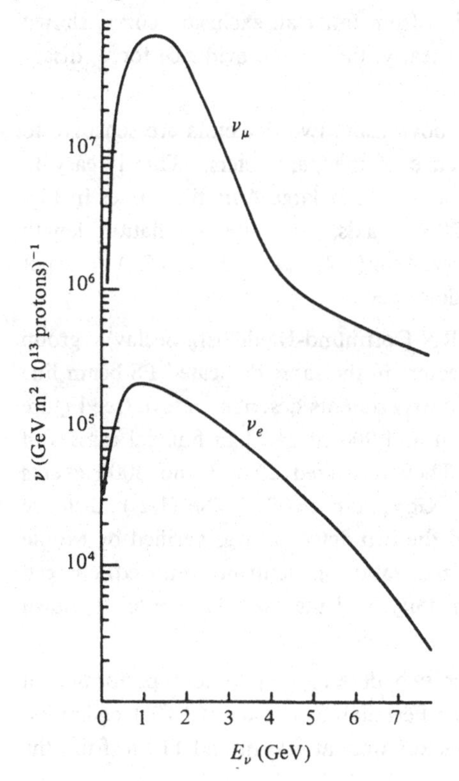

E531; Ushida et al. 86). It has provided the best limits for oscillations in the $\nu_\mu \rightarrow \nu_\tau$ channel. The limits from this work are also show in Figure 5.14

To study the coupling to ν_μ further, several experiments are now in the proposal state. Figure 5.14 shows the proposed sensitivity limits for two projects, a CERN SPS Proposal (CPS-P254; Armenise et al. 1990), and a Fermilab Proposal (FNAL P803; 1990). In both experiments, the aim is to observe $\nu_\mu \rightarrow \nu_\tau$ oscillations by observing the decay of the τ.

The $\nu_\mu \rightarrow x$ *channel.* In searches for ν_μ disappearance it is most convenient to use two detectors at two different distances from the neutrino source. This technique has the advantage over a single detector experiment that the appreciable uncertainties in the neutrino flux, reaction yield, and detector efficiency drop out in the ratio. The normalized ν_μ induced rates in the two detectors are expected to differ if there are neutrino oscillations.

The CHARM experiment (Bergsma et al. 84) described above utilizes two detectors at 920 m and 120 m. In this experiment there were 2000 and 270 ν_μ induced charged current events in the close and far detector, respectively. The observed-to-expected rates were found to be consistent with unity. From the maximum deviation from unity an exclusion curve, shown in Figure 5.14, has been derived. Clearly, there is no evidence for ν_μ disappearance.

Studies like those described above using two detectors are sensitive to oscillations only in a restricted range of the parameters. This is easy to understand in terms of oscillation length. For large Δm^2 the curves in Figure 5.14 turn back to the $\sin^2 2\theta = 1$ axis, since the oscillation length $L = 2.5\, E_\nu / \Delta m^2$ is large and the term $\sin^2(1.27\, \Delta m^2 L/E)$ in (5.7) does not vary appreciably between the two detectors.

The CERN CDHS (CERN-Dortmund-Heidelberg-Seclay) group (Dydak et al. 84) installed two detectors in the same dedicated PS beam line that served the BEBC and CHARM experiments described above (see Figure 5.12). The detectors were at 130 m and 900 m and had fiducial masses of 100 and 600 tons, respectively. They registered 20,000 and 3000 events with a mean neutrino energy of 3 GeV, respectively. The $(1/L)^2$ distance dependence of the neutrino flux in the two detectors was verified by Monte Carlo calculations. The normalized ratio of neutrino induced charged current events was consistent with unity, and the excluded region is shown in Figure 5.14.

Finally, we mention another two detector experiment performed at Fermilab by the Chicago-Columbia-Fermilab-Rochester (CCFR) collaboration (Stockdale et al. 85). The detectors were at 700 m and 110 m from the neutrino source. The observed normalized ratio of ν_μ events was found to be consistent with unity. The corresponding exclusion curve is shown in Figure 5.14.

A proposed search for $\nu_\mu \rightarrow x$ at Los Alamos LSND (Mann et al. 1991) is also included in Figure 5.14.

Beam dump experiments. Neutrinos produced in high energy accelerators have energies up to 200 GeV. The question whether these very high energy neutrinos oscillate can be explored with beam dump experiments. Here the high energy (400 GeV) proton beams at CERN or Fermilab are completely stopped in a thick target (the beam dump). The proton-nucleon collisions give rise to high energy hadrons, among them charmed particles, some of which decay promptly into charged leptons and high energy neutrinos. At high energies, the phase space for this semileptonic decay into muons or electrons is essentially the same. Thus, one expects that the number of muon neutrinos emitted is equal to the number of electron neutrinos, as required by universality.

In addition to these fast decaying or "prompt" hadrons, there will also

Figure 5.14. Limits for neutrino oscillation parameters summarizing the results of experiments that explore coupling to ν_μ. The excluded regions are to the right of the curves. Sensitivities for proposed experiments are shown by dotted lines. (For references see text.)

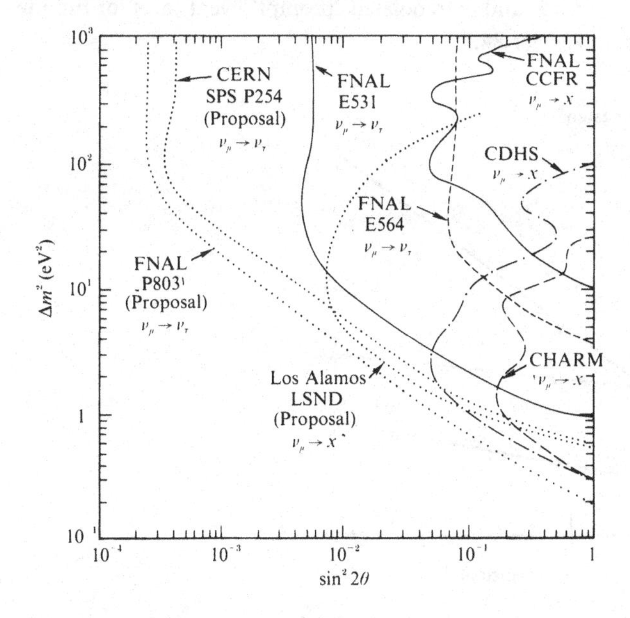

be the "conventional", long lived, light hadrons such as the π and the K. These energetic hadrons also decay into ν_μ and ν_e, but not before having lost most of their energy through interaction with target nucleons. Thus, their decay neutrinos (ν_μ , ν_e) will have relatively low energy (< 1 GeV) and can be separated from the high energy, prompt neutrinos. In eliminating them from the data of a beam dump experiment, one avoids having to deal with phase space related unequal yields in the neutrino production from the decay of low energy K,π,μ, etc., complicating the mentioned universality argument.

Thus, the goal of a beam dump experiment is to study the relative strength of prompt ν_e and ν_μ neutrino beams in a detector at distance L from the production region, assuming this relative strength was unity at creation. If more ν_μ are detected than ν_e, for example, one would have to conclude that neutrinos oscillate.

The method of determining the prompt fraction of the neutrinos emerging from the dump is to vary the density of the dump target. This has been accomplished by splitting the target into various slices displaced from each other. In a split target (low average density), some π or K hadrons may decay in the split region and produce high energy neutrinos. For an infinitely dense target, all emerging high energy neutrinos are prompt.

Figure 5.15. The rates of neutrino events for two tungsten targets differing in density by a factor of three and extrapolated "prompt" event rates for infinite target density. (From Reeder 84.)

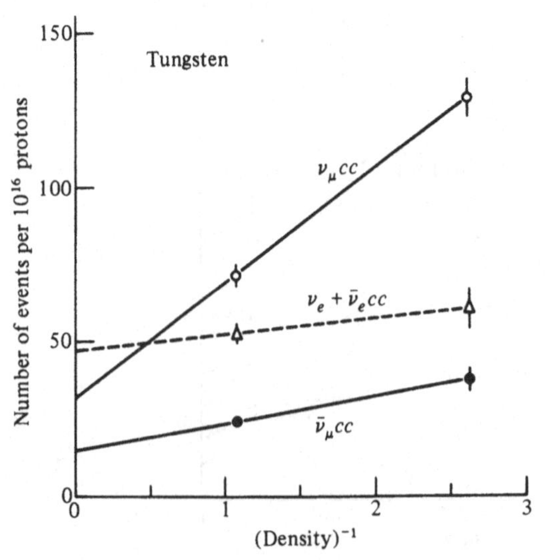

Thus, the technique used is to extrapolate the neutrino yield vs. target density to infinite target density, as depicted in Figure 5.15.

The ν_μ and ν_e beams emerging from the dump are measured with a neutrino detector by studying the muons and electrons produced in charged current reactions. At CERN the detectors of the BEBC, CHARM, and CDHS collaborations mentioned previously in this Section have been used. These detectors, installed at about 450 m from the dump, are capable of measuring the muons and electrons and their energies. Both ν_μ and $\bar{\nu}_\mu$ as well as ν_e and $\bar{\nu}_e$ events could be identified through charged current reactions. We refer to a review by Hulth (84) for a description of these experiments. The extrapolated prompt ratio of electron neutrinos and muon neutrinos is shown in Figure 5.16. This Figure also includes an experimental result from Fermilab (Reeder 84) (which is also based on the extrapolation technique for the target density). With a detector at 56 m from the target, it carefully explores the energy dependence of the ratio N_{ν_e}/N_{ν_μ} and finds that ratio to be consistent with unity for all energies from 20—200 GeV.

The conclusion from these studies is that three of the four results are consistent with $R = N_{\nu_e}/N_{\nu_\mu} = 1$, that is no oscillations, while the fourth

Figure 5.16. Display of prompt ratios $R = N_{\nu_e}/N_{\nu_\mu}$ from beam dump experiments from CERN (BEBC, CDHS, CHARM) and Fermilab (FMOW). The neutrino rates N are averages over the energy range of 20—200 GeV.

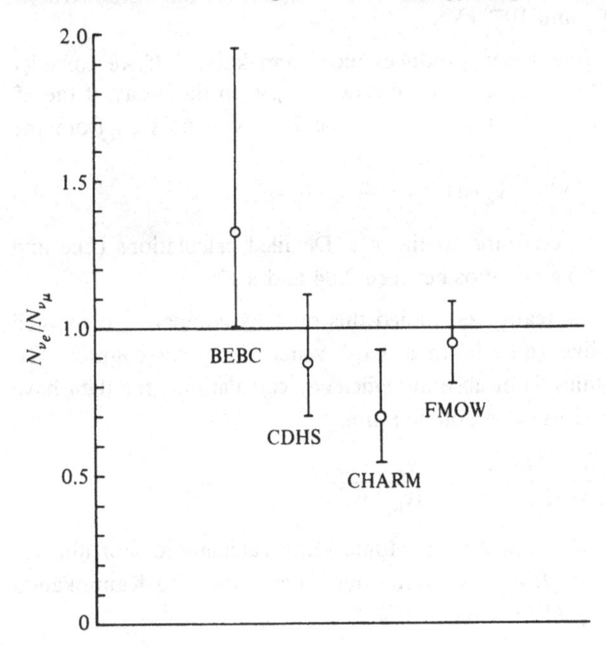

(the CHARM experiment: Bergsma 86) deviates from $R = 1$ by about 1.5 σ. The CERN experiments taken at the same distance from the source should give consistent results with or without oscillations. The fact that they disagree with each other may point to a problem in understanding the data.

In concluding this Section, we summarize that none of the experiments carried out at high energy accelerators has revealed evidence for neutrino oscillations. Among the oscillation channels explored were $\nu_\mu \to \nu_e$, $\nu_\mu \to \nu_\tau$, and $\nu_\mu \to x$. One of four beam dump experiments gives a puzzling result, however, which requires clarification. The parameter space excluded is summarized in Figures 5.11 and 5.14.

5.6 Atmospheric Neutrinos

The interaction of energetic hadrons in cosmic rays with nuclei in the upper atmosphere gives rise to what is known as atmospheric neutrinos. The flight path of these neutrinos from production to detector extends from $\approx$ 20 km (thickness of the atmosphere) to 12,000 km (diameter of the Earth) and thus represents a very long baseline experiment. The spectrum of the neutrinos is a decreasing function of energy. The usable energy range in experiments with water Cerenkov detectors is between around 100 MeV (threshold for track recognition) and 2 GeV (diminishing intensity of neutrino flux). Experiments with atmospheric neutrinos, therefore, are well suited for exploring neutrino disappearance through oscillations in the range of Δm^2 between 10^{-2} and 10^{-4} eV^2.

The hadronic interaction produces pions and kaons. If we consider the pion decay processes, we expect the ratio of ν_μ/ν_e in the decay of the π^+ produced in energetic nuclear reactions to be 2, as can be seen from the decay chain,

$$\pi^+ \to \nu_\mu + \mu^+; \quad \mu^+ \to \nu_e + \bar{\nu}_\mu + e^+,$$

with a similar result pertaining to the π^-. Detailed calculations (Lee and Koh 91; Barr et al. 89) give ratios between 2.04 and 2.17.

The Kamiokande team has studied this ratio by comparing contained μ-like rings and e-like rings in their large water Cerenkov counter. To reduce the uncertainties from absolute efficiency calculations, the data have been presented (Kajita 91) as a double ratio,

$$\frac{[\mu\text{-}like/e\text{-}like]_{data}}{[\mu\text{-}like/e\text{-}like]_{MC}} = \frac{[\nu_\mu / \nu_e]_{data}}{[\nu_\mu / \nu_e]_{MC}} = 0.61 \pm 0.10,$$

where the subscript MC stands for a Monte Carlo calculation. For illustration, Figure 5.17 taken from the quoted reference shows the Kamiokande results for e-like and μ-like events.

Recently, Casper et al. (91) have published new results from the IMB water Cerenkov detector. The data is presented as a double ratio R,

$$R = \frac{[single\ \mu\text{-}rings/all\ single\ rings]_{data}}{[single\ \mu\text{-}rings/all\ single\ rings]_{MC}} = \frac{[\nu_\mu/(\nu_\mu+\nu_e)]_{data}}{[\nu_\mu/(\nu_\mu+\nu_e)]_{MC}}.$$

It is interesting to compare the values of R for the Kamiokande data, the IMB 3 data, the earlier IMB 1 data obtained from a reanalysis (see Casper et al. 91), as well as the data from the Frejus (Berger et al. 90) and NUSEX (Aglietta et al. 90) detectors. To help the comparison, we display the ratio R for all experiments in Figure 5.18. The Kamiokande and IMB ratios R are around 0.8, implying a deficiency of ν_μ or an enhancement of ν_e. An independent ratio was derived from the data by considering delayed coincidence events attributable to muon decay. This ratio is also around 0.8 and thus confirms the prompt data. The Frejus and Nusex detectors, both fine grain iron calorimeters, find a ratio of μ-like events to total events of around 1.0, with errors of about 0.1.

Notwithstanding a slight disagreement between the results of the

Figure 5.17. Observed e-like and μ-like rings from the Kamiokande detector (Kajita 91) compared to Monte Carlo simulations.

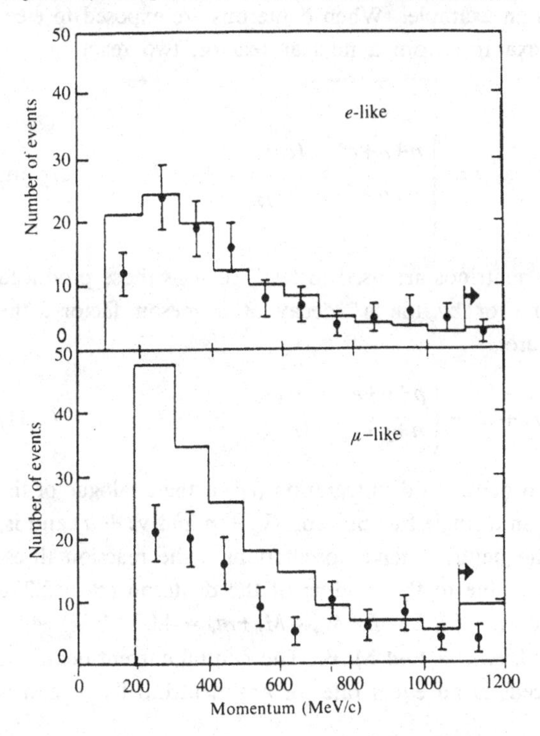

Kamiokande and IMB water Cerenkov detectors and those of the Frejus and Nusex detectors, we may argue that all results are consistent with a muon to total ratio of about 0.8. Assigning a conservative error of 0.1, we can now interpret this as a deficiency of ν_μ (ν_μ disappearance) at large distance,

$$R = 1 - (1/2)\sin^2 2\theta \approx 0.8 \pm 0.1,$$

associated with a mixing angle of $\sin^2 2\theta \approx 0.4 \pm 0.2$. Assuming an average energy of 1 GeV and a distance of 10,000 km, we have $\Delta m^2 > 10^{-4}$ eV^2. In their work, Casper et al. (91) point out, however, that the energy spectra and angular dependence of the events do not support oscillations. Clearly, more studies are needed to confirm this remarkable result.

5.7 Neutrino Induced Nuclear Reactions

The oscillation searches described in Section 5.4 are based on the antineutrino capture on protons $\bar{\nu}_e + p \to n + e^+$. Several other targets containing bound nucleons have been also used for this purpose. At low and moderate energies, the nuclear binding causes a major modification of the cross section formulae, as we have seen in Chapter 3.

The deuteron represents the simplest of bound nuclear targets and will be discussed in detail as an example. When deuterons are exposed to electron antineutrinos, for example from a nuclear reactor, two reactions are possible

$$\bar{\nu}_e + d \to \begin{cases} n + n + e^+ & (cc)_{\bar{\nu}} \\ n + p + \bar{\nu}_e & (nc)_{\bar{\nu}} \end{cases} . \tag{5.30}$$

Similarly when electron neutrinos are used instead, such as those produced in ^{8}B decay in the Sun, or by the μ^+ decay at a meson factory, the corresponding reactions are

$$\nu_e + d \to \begin{cases} p + p + e^- & (cc)_{\nu} \\ n + p + \nu_e & (nc)_{\nu} \end{cases} . \tag{5.31}$$

The charged current deuteron disintegration (cc) is the analogue of the charged current capture on a single free nucleon ($\bar{\nu}_e + p$ and $\nu_e + n$) and is, therefore, sensitive to the neutrino flavor oscillations. The reaction threshold, however, is different due to the binding of the deuteron (B = 2.226 MeV) and we obtain for $(cc)_{\bar{\nu}}$; $E_T = B + M_n - M_p + m_e$ = 4.03 MeV, while for $(cc)_\nu$; $E_T = B + M_p - M_n + m_e$ = 1.44 MeV. The neutral current deuteron disintegration (nc) proceeds at an equal rate for any neutrino flavor and is insensitive to the flavor oscillations. The threshold E_T of the *nc* reaction* is just the deuteron binding energy B.

Figure 5.18. Ratio of atmospheric-neutrino induced muon events to total events, divided by their respective Monte Carlo simulation value. KAMIOKANDE (Kajita 91) is a water Cerenkov counter and so is IMB (Casper et al. 91). Frejus (Berger et al. 90) and NUSEX (Aglietta et al. 90) are iron calorimeters. "μ-ring" designates a single muon-like Cerenkov ring corresponding to an incident neutrino with 0.1—1.5 GeV energy; "μ-decay" refers to a delayed muon event; "μ-like" refers to a neutrino induced event with a muon in the final state.

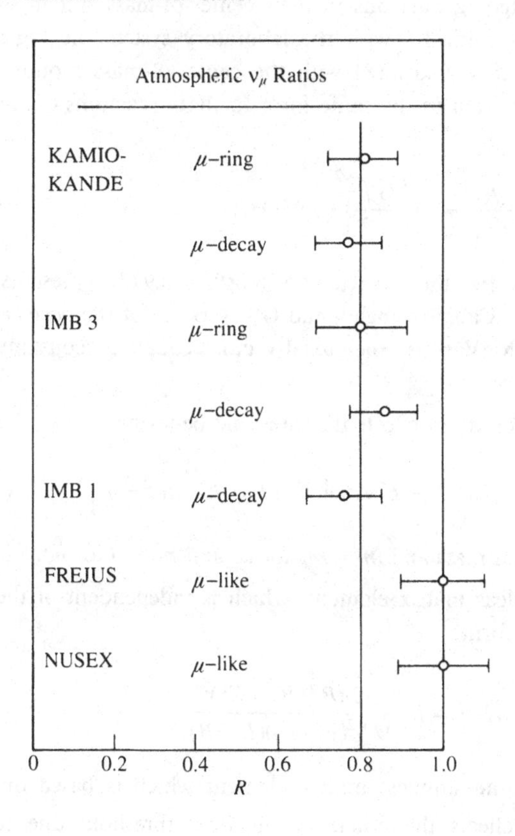

*Here, we are considering the inelastic channel of the neutral current deuteron disintegration. The elastic reaction, in which the deuteron is left in its bound state, would be very difficult to observe. Besides, while the inelastic deuteron disintegration near threshold is dominated by the isovector ($T = 0 \rightarrow T = 1$) part of the interaction, the elastic neutrino scattering is entirely isoscalar ($T = 0 \rightarrow T = 0$).

Remembering that the ground state of the deuteron has spin $S = 1$, isospin $T = 0$, and is predominantly an s-state of the relative motion of the two nucleons, we see that near threshold the final state for both reactions must be dominantly $S = 0$, $T = 1$ s-state, and thus only the axial-vector isovector part of the weak nucleon current contributes. In particular, this means that in the standard Weinberg-Salam model the (nc) cross section near threshold does not depend on the Weinberg angle $\sin^2\theta_W$. In evaluating the cross section, one has to take into account the final state strong interaction of the outgoing nucleons. This is done usually by using the experience from the deuteron photodisintegration (Bethe & Longmire 50).

For monoenergetic neutrinos of low energy E_ν ($E_\nu \ll M_p$) we express the differential cross section in the "reduced nucleon energy E_r "; E_r is the energy of the two outgoing nucleons in their center-of-mass system, which for practical purposes coincides with the laboratory system, i.e., one can neglect the kinetic energy associated with the center-of-mass motion. The cross section can be written in the same form for both reactions (Ahrens & Lang 71)

$$\frac{d\sigma}{dE_r} = 2\pi \left(\frac{G_A^{eff}}{\sqrt{2}} \right)^2 (\hbar c)^2 |ME|^2 \rho \ , \tag{5.32}$$

where $G_A^{eff} = G_A \cos\theta_c$ for the cc reaction ($\cos\theta_c = 0.973$ represents the quark mixing, θ_c is the Cabbibo angle), and $G_A^{eff} = G_A/2$ for the nc reaction. ($G_A = -1.42 \times 10^{-11}$ MeV^{-2} is the axial-vector coupling constant for $G_A/G_V = -1.25$.)

The final state density ρ reflects the threshold behavior

$$\rho = \frac{M^{3/2}}{8\pi^4} E_r^{\frac{1}{2}} (E_\nu - E_T - E_r + m) \left[(E_\nu - E_T - E_r + m)^2 - m^2 \right]^{\frac{1}{2}} , \tag{5.33}$$

where M is the nucleon mass and $m = m_e$ for cc and $m = 0$ for nc.

Finally, the nuclear matrix element, which is independent of the lepton energies, is of the form

$$|ME|^2 = b \, \frac{32\pi B^{\frac{1}{2}} (B^{\frac{1}{2}} + E_s^{\frac{1}{2}})^2}{M^{3/2} (E_r + E_s)(E_r + B)^2} \ . \tag{5.34}$$

Eq. (5.34) represents the simplest matrix element which is based on zero effective range. It reflects the enhancement near threshold due to the existence of the "virtual singlet level" E_s. The coefficient b above is $b = 1$ for cc and $b = 2$ for nc. The quantity E_s is the energy of the virtual singlet state and is related to the scattering length by the relation $E_s = 1/Ma_s^2$. In evaluating the cross section, one has to take into account the apparent isospin symmetry breaking of nuclear force as revealed by the experimentally determined scattering lengths $a_{np} = -23.7$ fm and $a_{nn} = -18.5$ fm (Gabioud et al. 79, 84). In detailed analysis, the finite range of the nuclear force as well as the ($\approx$ 5%) effect of the pion exchange current should be included. For the *cc* reaction with ν_e, the Coulomb interaction of the final state protons also plays a role (Kelly & Uberall 66). At higher energies the interference between the axial-vector and the weak-magnetism parts of the hadronic weak current contributes to the separation of the *cc* reaction cross sections induced by the electron neutrinos and electron antineutrinos, respectively. The behavior of the cross sections is illustrated for energies from threshold to 50 MeV in Figure 5.19. (See O'Connel 72 and Ali & Dominguez 75.)

To judge the difficulty of various experiments, it is of interest to fold the cross section with the corresponding neutrino spectra from various sources and evaluate the averaged cross sections. For reactor neutrinos one obtains

$$\bar{\sigma} = 1.2 \times 10^{-44} \text{ cm}^2 \text{ per fission} \quad \text{for } (cc)_{\bar{\nu}} ,$$

$$\bar{\sigma} = 2.9 \times 10^{-44} \text{ cm}^2 \text{ per fission} \quad \text{for } (nc)_{\bar{\nu}} .$$

Figure 5.19. Cross sections for neutrino interactions with deuterons at low and intermediate energies. The charged current reactions are shown by the full curves, and the neutral current reactions are shown by the dashed curves.

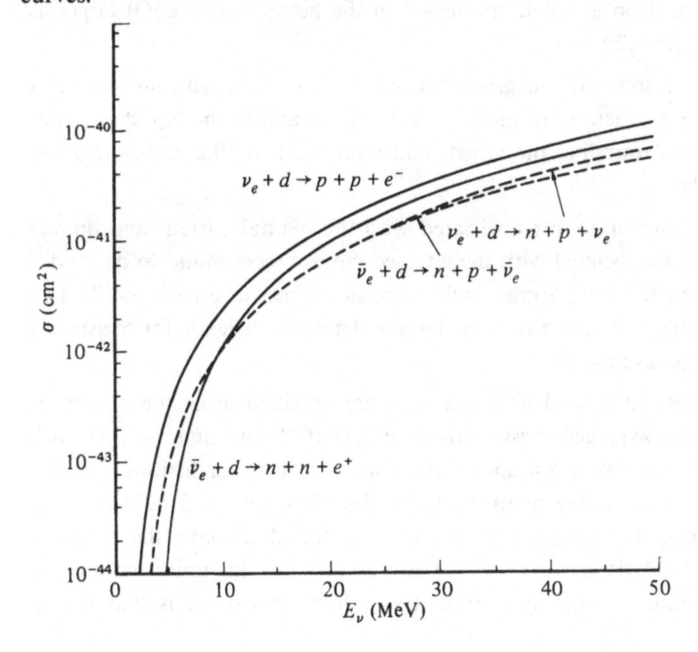

These cross sections are about 20 times smaller than the cross section for protons as targets. The difference, which makes the deuteron target experiments correspondingly harder, can be attributed to the higher threshold and to the relatively small overlap between the initial bound state and the final continuum state.

Similar averages can be evaluated for the ^8B beta decay spectrum yielding

$$\bar{\sigma} = 1.2 \times 10^{-42}\ \text{cm}^2 \text{ per } ^8\text{B decay} \quad \text{for } (cc)_\nu ,$$

$$\bar{\sigma} = 0.6 \times 10^{-42}\ \text{cm}^2 \text{ per } ^8\text{B decay} \quad \text{for } (nc)_\nu ,$$

and for the spectrum of ν_e from the μ^+ decay at rest at the meson factory beam stop (see Figure 5.9 for the spectrum),

$$\bar{\sigma} = 4.4 \times 10^{-41}\ \text{cm}^2 \text{ per } \mu^+ \text{ decay} \quad \text{for } (cc)_\nu ,$$

$$\bar{\sigma} = 2.1 \times 10^{-41}\ \text{cm}^2 \text{ per } \mu^+ \text{ decay} \quad \text{for } (nc)_\nu .$$

We now turn to the experimental verification of the antineutrino-deuteron reaction (5.30) and its cross sections, and to the application of the $\bar{\nu} + d$ reaction in shedding light on neutrino mass and neutrino oscillations.

The $\bar{\nu}_e + d$ reactions (5.30) were studied by Pasierb et al. (79) and by Reines et al. (80) at the Savannah River reactor. The detector, situated at 11.2 m from the core in a neutrino flux of 2.5×10^{13} cm^{-2} s^{-1} consisted of several ^{3}He neutron counters immersed in the heavy water (D_2O) target as shown in Figure 5.20.

Reactor associated backgrounds came from reactor neutrons, neutrons from the $\bar{\nu}_e + p$ reaction on protons in the target and in the liquid scintillation counter surrounding the target, and amounted to 10.2 ± 0.7 neutron counts per day.

Single neutron counts associated with the neutral current and double neutron counts associated with the charged current were found to be 70 d^{-1} and 4 d^{-1}, respectively. Some single neutron counts have been ascribed to the *cc* reactions and are caused by the low detector efficiency for registering double neutron events.

The expected *cc* and *nc* event rates are obtained using the known $\bar{\nu}_e$ flux, the energy averaged cross sections of 2.1×10^{-45} cm^2 and 5×10^{-45} cm^2, respectively (see Section 5.4 above, there are $\approx 6.1\ \bar{\nu}_e$ per fission), and the known single and double neutron efficiencies. The observed ratio R of *cc* and *nc* events, divided by that expected, is found to have the value of $R = 0.74 \pm 0.23$ (Reines 83), essentially consistent with unity within one standard deviation. The conclusion drawn from this result is that the *cc* event rate is not significantly attenuated by neutrino oscillations of the type $\bar{\nu}_e \to x$.

The *cc* and *nc* reactions on deuterons were observed also by Vidyakin et al. (90b) at a nuclear reactor. Their statistical accuracy is, however, poorer than the results of Pasierb et al. (79) and Reines et al. (80).

The virtue of studying the ratio of *cc* and *nc* events is that the *nc* process which is not altered by oscillations plays the role of a monitor of the flux of neutrinos of all flavors. Also, this ratio is relatively insensitive to the details of the neutrino spectrum from the reactor.

Next, we shall discuss the neutrino-deuteron reactions (5.31). Among these the *cc* reactions are of particular interest since their inverse reactions, $pp \to de^+\nu_e$ and $ppe^- \to d\nu_e$, are believed to be responsible for the primary energy release in the Sun. This *cc* reaction has been studied by Willis et al. (80) and by Nemethy et al. (81) at LAMPF (Los Alamos Meson Physics Facility). Neutrinos from the beam stop, whose energy spectra are shown in Figure 5.9, traversed 6 m of iron shielding and impinged on a 6000 l D_2O

Figure 5.20. Detector used for the study of the antineutrino interaction with deuterons (Pasierb et al. 79).

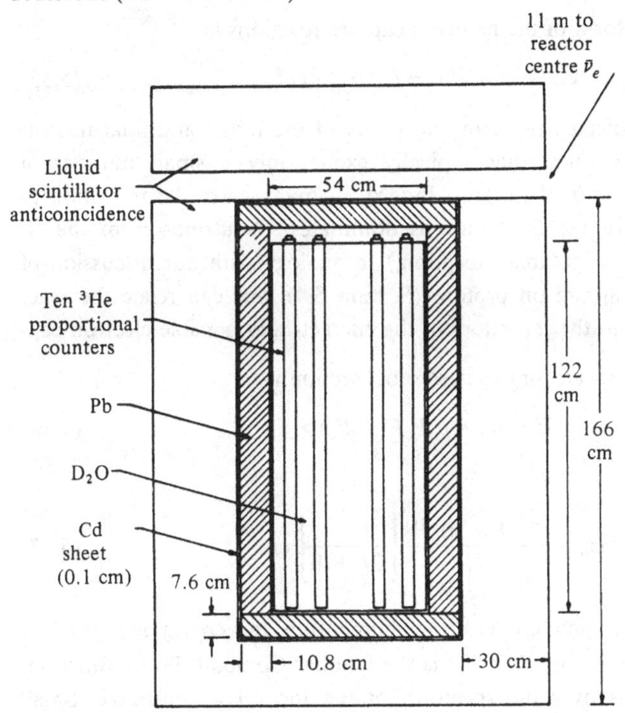

Čerenkov total energy calorimeter placed at 9 m from the beam stop. The ν_e flux from the chain $\pi^+ \rightarrow \mu^+\nu_\mu$, $\mu^+ \rightarrow e^+\nu_e\bar{\nu}_\mu$ in the beam stop was 2×10^7 cm^{-2} s^{-1}. The D_2O counter was surrounded with a scintillation veto counter and shielding. Above 25 MeV the $\nu_e d$ event rate was about 20 d^{-1}, with a cosmic ray background of 120 d^{-1}. The cross section for this *cc* reaction was found to be $0.52 \pm 0.18 \times 10^{-40}$ cm^2, in agreement with theory.

As we have shown in Section 5.4, protons are the most convenient targets in studies of charged current reactions with low energy electron antineutrinos. If one were to use ^{3}He nuclei instead of protons, the reaction threshold would be even lower, $E_T = 1.04$ MeV (as compared to $E_T = 1.8$ MeV for proton target), and the cross section for reactor antineutrinos would be larger, $<\sigma> \approx 1.3\times10^{-42}$ cm^2 per fission (as compared to $\approx 0.6\times10^{-42}$ cm^2 per fission for proton target). However, various practical considerations (e.g., the availability and cost of ^{3}He) make this possibility somewhat academic.

The situation is very different for the charged current reactions induced by electron neutrinos, in particular, the solar neutrinos originating in the $p+p \rightarrow d+e^++\nu_e$ and $p+p+e^- \rightarrow d+\nu_e$ reactions, in the electron capture on ^{7}Be, and in the positron decay of ^{8}B. A variety of nuclei has been proposed as detectors of these neutrinos.

The general form of the neutrino capture reactions is

$$\nu_e + {}^{Z-1}A(I^\pi) \rightarrow e^- + {}^{Z}A(I'^\pi) , \tag{5.35}$$

where we have indicated the spin and parity of the initial and final nuclear states. Low energy neutrinos typically excite only a small number of discrete states in the final nucleus. Indeed, in many cases the transition to the ground state of the final nucleus dominates. (Neutrinos from the ^{8}B positron decay are a notable exception.) In analogy with our discussion of the antineutrino capture on protons (Section 5.4), one can relate the reaction rate of (5.35) to the $ft_{\frac{1}{2}}$ values of the energetically possible electron capture $I'^\pi \rightarrow I^\pi$. The spectrum averaged cross section is

$$\bar{\sigma} = \sigma_o <p_e E_e F(Z,E_e)> , \tag{5.36}$$

where

$$\sigma_o = \frac{2.64 \times 10^{-41}}{ft_{\frac{1}{2}}} \left(\frac{2I' + 1}{2I + 1} \right) \text{cm}^2 . \tag{5.37}$$

Here, p_e, E_e are the outgoing electron momentum and energy measured in units of electron mass m_e, $F(Z,E_e)$ is the familiar Coulomb Fermi function, and the average is over the spectrum of the incoming neutrinos. Small

Table 5.3. *Average electron neutrino capture cross sections for several sources of solar neutrinos and a variety of detectors. Contributions from excited states are included.*

Source		$p+p$	$p+p+e^-$	^{7}Be + e^-	^{8}B [b]
Source characteristics		Continuum $E_\nu^{Max}=$ 420 keV	Discrete $E_\nu=$ 1.442 MeV	Discrete[c] -	Continuum $E_\nu^{Max}=$ 14.02 MeV
Detector	Threshold	$\bar{\sigma}$ (10^{-46} cm^2)			
^{2}H	1.44 MeV	-	-	-	1.2×10^4
^{7}Li	862 keV	-	655	9.6	3.9×10^4
^{37}Cl	813 keV	-	16	2.4	1.1×10^4
^{71}Ga	236 keV	11	215	73.2	2.4×10^4
^{81}Br	322 keV	-	75	18.3	2.7×10^4
^{115}In	-495 keV[a]	78	576	248.0	2.5×10^3
^{127}I	789 keV[d]	-	-	20.0	2.0×10^4

[a]Ground state transition is 4th forbidden, the allowed transition to $7/2^+$ state has a threshold of 118 keV.

[b]Final 2^+ state in ^{8}Be is a $\approx$ 2 MeV broad resonance.

[c]Two branches, E_ν = 862 keV (90%), E_ν = 384 keV (10%).

[d]To the first excited $3/2^+$ state.

corrections to (5.36) arise from overlap and exchange contributions. The former is related to the atomic excitations of the final state caused by the sudden change in the nuclear charge, and the latter is related to the indistinguishability of the outgoing electron from the initially present atomic electrons. The averaged cross sections for nuclei used or proposed as detectors of solar neutrinos are listed in Table 5.3 adapted from Bahcall (89). The entries for ^{127}I are from Engel et al. (91).

The ^{8}B neutrinos have high enough energy so that transitions to the excited states of the daughter nucleus dominate the reaction yield. For such transitions Eq. (5.37) is not applicable. Instead, one has to rely either on a theoretical evaluation of the cross section (or, equivalently, of the so-called beta strength), or on indirect experimental data on related processes, such as the (p,n) forward angle cross section which is proportional to the corresponding beta strength. For the case of ^{37}Cl an ingenious method based on the analogy with the ^{37}Ca decay (isobar analogue) can be used (Garcia 91).

Besides the nuclei listed there, several others (^{97}Mo, ^{98}Mo and ^{205}Tl) have low thresholds for neutrino capture, and the final nuclei are relatively long lived (10^6—10^7 years). Geochemical methods of neutrino detection are considered in these cases. For these three nuclei, one cannot calculate accurately, with the available experimental information, the neutrino capture cross section for the various excited states of importance. Finally, neutrino capture on deuterons, discussed previously, and neutrino electron scattering can also serve as detectors of solar neutrinos.

5.8 Neutrinos from the Sun

Exothermic nuclear reactions in the interior of the Sun are the source of the energy reaching the surface of the Earth. The main reactions are those of the proton-proton (*pp*) chain. The net effect of this chain of reactions is the conversion of four protons and two electrons into an α particle,

$$2e^- + 4p \rightarrow {}^4\text{He} + 2\nu_e + 26.7 \text{ MeV} . \tag{5.38}$$

Approximately 97% of the energy released in the *pp* chain resulting in (5.38) is in the form of charged particle and photon energy, which is the source of the Sun's thermal energy. The remaining approximately 3% is in the form of the kinetic energy of the neutrinos. Thus, for each $\approx$13 MeV of the generated thermal energy we expect one neutrino. Assuming that the Sun is in equilibrium, and has been in stable equilibrium for a sufficiently long time, the rate with which the thermal energy is generated is equal to the rate at which the energy is radiated from the surface. The solar constant, the energy of solar flux reaching the Earth's atmosphere, is $S = 1.4\times10^6$ erg cm^{-2} s^{-1} = 8.5×10^{11} MeV cm^{-2} s^{-1}. Thus, the neutrino flux is $\Phi_\nu = S/13 = 6\times10^{10}\ \nu_e$ cm^{-2} s^{-1}. Experimental verification of this neutrino flux (or of one of its components) would constitute an important test of our understanding of the energy production in the Sun. At the same time, as the path of solar neutrinos to a detector on Earth is much longer than the neutrino path length for any terrestrial neutrino detector, observation of solar neutrinos constitutes a very stringent test for the inclusive oscillations of electron neutrinos.

5.8.1 *The predicted neutrino spectrum from the Sun*

In order to test the neutrino production in the Sun in a quantitative way, we have to evaluate the neutrino flux in greater detail than the crude estimate above. We need to know not only the total flux, but also its energy distribution. In particular, the various possible detector reactions listed in Table 5.3 have different thresholds and therefore are sensitive to different components of the solar neutrino flux. The *pp* chain of reactions, schematically indicated in (5.38), is shown in detail in Figure 5.21. We see that the *pp* chain results in the neutrino spectrum having two continuous components (*pp* and ^{8}B decay), and three discrete lines (*pep* and the two branches of ^{7}Be electron capture). In addition, one has to include the contributions of the solar CNO cycle (decays of ^{13}N and ^{15}O). The resulting neutrino spectrum is depicted in Figure 5.22.

The branching ratios in Figure 5.21, and the neutrino spectrum in Figure 5.22 are obtained from calculations based on the "standard solar model". The equation of state of the model relates the pressure $P(r)$, the density $\rho(r)$, the temperature $T(r)$, the energy production rate per unit mass $\varepsilon(r)$, the luminosity $L(r)$, and the opacity $\kappa(r)$. The equation also depends on the

Figure 5.21. Nuclear reactions in the *pp* chain. The branching ratios are deduced from standard solar model calculations.

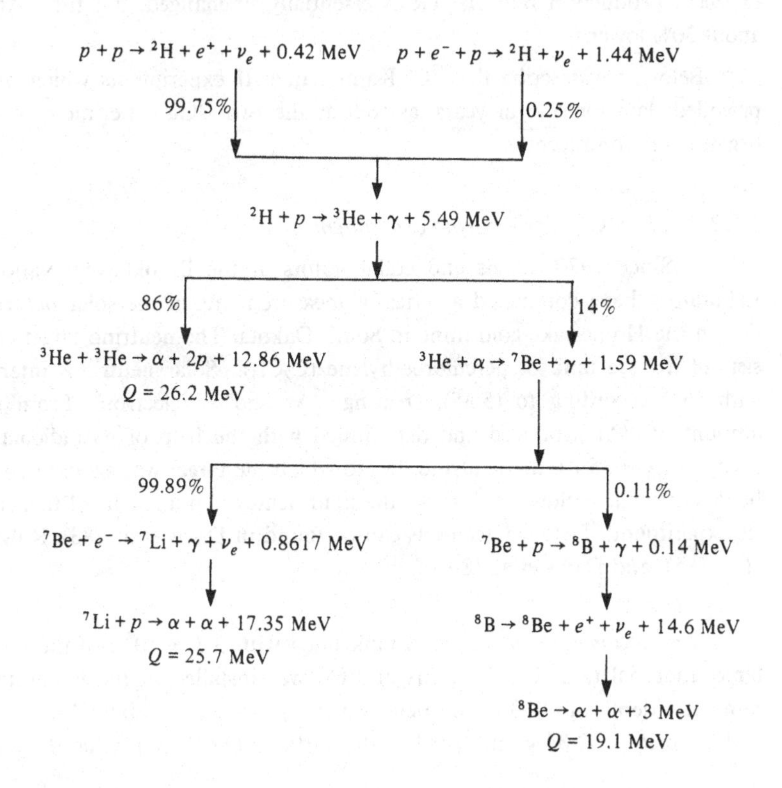

fractional abundances of hydrogen, helium, and the heavier elements. The solution of the equation of state is complicated because the opacity κ is a complex function of density, temperature, and composition, which can be estimated only from detailed knowledge of the atomic transitions involved. The energy production rate ε depends on the detailed knowledge of cross sections for the reactions shown in Figure 5.21. These cross sections have to be extrapolated from laboratory energies to the much lower solar energies.

The equations of the standard solar model have been carefully solved in a number of papers and the uncertainties in the solar neutrino flux have been estimated (see Bahcall 89 for details). We present in Table 5.4 the neutrino fluxes of individual components. Multiplying the flux by the averaged cross sections given earlier in Table 5.3, we arrive at the predicted reaction rates for the ^{37}Cl and ^{71}Ga detectors also shown in Table 5.4.

A completely independent calculation of the solar neutrino flux has been carried out by Turck-Chieze et al. (88). The final solar model is similar to the model used for entries in Table 5.4 but uses different nuclear parameters and opacities. There is a complete agreement in the *pp* flux but the ^{8}B flux is 35% lower and the ^{7}Be flux is 11% lower. Correspondingly, the expected production rate of ^{71}Ge is essentially unchanged, but the ^{37}Ar is about 30% lower.

Below, we describe the ^{37}Cl Kamiokande II experiments which have provided data for several years, as well as the two ^{71}Ga experiment which began operation recently.

5.8.2 *The ^{37}Cl solar neutrino experiment*

Since 1970, Davis and collaborators at the Brookhaven National Laboratory have conducted a series of measurements of the solar neutrino flux in the Homestake gold mine in South Dakota. The neutrino target consists of a large tank of perchloroethylene (C_2Cl_4). Solar neutrinos interact with ^{37}Cl according to (5.35), creating ^{37}Ar and an electron. From the amount of ^{37}Ar produced and determined with the help of its radioactive decay ($T_{1/2}$ = 35 d), the neutrino flux to which the target was exposed could be determined. Below we describe the main features and results of this classic experiment. Detailed accounts can be found in Davis et al. (83), Rowley et al. (85), and Davis et al. (89).

Description of experiment. A tank filled with 3.8×10^5 l of the C_2Cl_4 target material (2.2×10^{30} atoms of ^{37}Cl) was installed in the Homestake mine at a depth of 4000 mwe (meter water equivalent), or about 1400 m of rock (density of 2.84 g cm^{-3}) below the surface. The ^{37}Ar produced by the

Table 5.4. *Neutrino fluxes and reaction rates in SNU (1 SNU = 1 capture per second per 10^{36} target atoms) for chlorine and gallium detectors based on the standard solar model. Numbers of ^{37}Ar and ^{71}Ge atoms produced per day in the respective detectors, having* 2.2×10^{30} *atoms of* ^{37}*Cl, and* 1.0×10^{29} *atoms of* ^{71}*Ga, are given in columns 4 and 6.*

	Flux 10^{10} cm^{-2} s^{-1}	^{37}A SNU	^{37}A d^{-1}	^{71}Ge SNU	^{71}Ge d^{-1}
pp	6.0	0	0	70.8	0.61
pep	0.014	0.2	0.04	3.0	0.03
^{7}Be	0.47	1.1	0.21	34.3	0.30
^{8}B	0.00058	6.1	1.16	14.0	0.12
^{13}N	0.06	0.1	0.02	2.8	0.02
^{15}O	0.05	0.3	0.06	6.1	0.05
Total	6.6	7.9	1.5	132	1.14

Figure 5.22. Solar neutrino energy spectrum according to the standard solar model. The ordinate represents the solar neutrino flux at the surface of the Earth. The monoenergetic neutrinos from the *pep* and ^{7}Be processes are given in cm^{-2} s^{-1}. The continuum neutrinos are given in cm^{-2} s^{-1} MeV^{-1}.

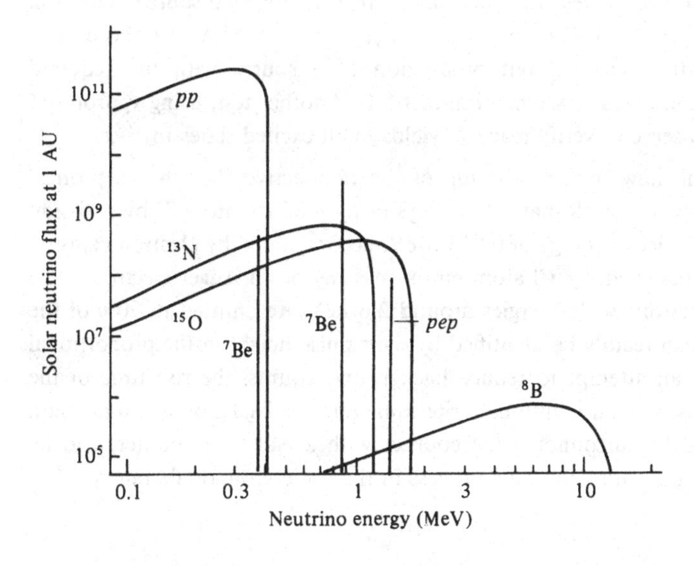

neutrino interaction on ^{37}Cl was periodically separated and counted. At the onset of an experimental run, each lasting from 35 to 50 days, an amount of 0.2 cm^3 (STP) of ^{36}Ar or ^{38}Ar was added to the target volume to serve as the carrier for the ensuing argon extraction. At the end of each exposure, the argon was removed by circulating 6×10^5 l of helium through the liquid and gas in the tank. The helium, carrying with it the argon, was cleaned with the help of a condenser (−32°C) and a molecular sieve, before entering a liquid nitrogen cooled (−196°C) charcoal trap. Here the argon was frozen (freezing point of argon −189°C) and absorbed. The trap was then warmed up and the argon transferred to an apparatus, which allowed the determination of the volume of argon and the identification of the atomic mass. It was established that 95% of the argon introduced as the carrier could be removed in this separation. The argon was then transferred to a proportional counter with a volume of 0.3—0.5 cm^3 and, after adding some methane quenching gas, serves as the fill gas for the counter. Each sample was counted for about eight months.

Although the extraction efficiency of neutral argon atoms was found to be as high as 95%, it still needed to be demonstrated that the ^{37}Ar which, following neutrino capture, is formed as a positive ion, behaves the same way as a neutral atom. This was shown with the help of a C_2Cl_4 sample labeled with radioactive ^{36}Cl. The $^{36}Ar^+$ ions, resulting from the decay $^{36}Cl \rightarrow {}^{36}Ar^+ + e^- + \bar{\nu}_e$, could be removed with the helium circulation technique using ^{38}Ar as a carrier. The ^{36}Ar was quantitatively identified with activation analysis.

An ideal test of the extraction technique as well as of the reaction yield would be the use of neutrinos emitted form a radioactive source. The beta decay $^{65}Zn \rightarrow {}^{65}Cu + e^+ + \nu_e$, with $E_{\nu Max} = 1.35$ MeV, has been considered for this purpose, but production of a source with the required strength of 1 megacurie seemed impractical. Another test, using ν_e from μ^+ decay, could serve to verify reaction yields to all excited states in ^{37}Ar.

Turning now to the counting of the radioactive ^{37}Ar in the proportional counter, we recall that ^{37}Ar decays ($T_{1/2} = 35$ d) into ^{37}Cl by emission of a ν_e with a decay energy of 0.813 MeV, accompanied by electron capture, after which the excited ^{37}Cl atom emits an x-ray or an Auger electron. The K Auger electrons with energies around 2.6 keV are emitted in 90% of the decays and can readily be identified by their pulse height in the proportional counter. In an attempt to reduce background counts, the rise time of the pulses was used as an additional criterion. Further background suppression was achieved by surrounding the counter with a NaI veto counter, and by carrying out the entire counting process in the underground laboratory.

Besides the solar neutrinos, there are two principal nonsolar sources that may contribute to the production of ^{37}Ar: muons from cosmic rays, and fast neutrons from fission or (α,n) reaction of ^{238}U in the rock. The cosmic ray muon flux in the Homestake Laboratory has been measured and found to be 1.5×10^3 m^{-2} y^{-1} $ster^{-1}$. These muons give rise to a cascade containing pions, protons, and neutrons as well as their evaporation protons producing ^{37}Ar by ^{37}Cl (p,n) ^{37}Ar. The magnitude of the background has been estimated by measuring the ^{37}Ar production rate by cosmic rays in three smaller tanks of C_2Cl_4 at depths of 254, 327, and 819 mwe. Scaled to the neutrino detector at 4000 mwe this background is found to be 0.08 ± 0.03 d^{-1}. (An illustration of muon flux vs. depth is contained in Figure 6.11.) Fast neutrons can be shielded effectively with water. In comparing the ^{37}Ar rate with and without a water shield, it was concluded that the background contribution from fast neutrons is negligible.

Results. Altogether, 90 completed runs were carried out between 1970 and 1990 and the experiment is continuing. In the first 61 runs the total number of events in the proportional counter with proper pulse height and rise time was 774. By fitting these events to an exponential decay of ^{37}Ar superimposed on a constant background, 339 events were attributed to ^{37}Ar decay. The average background count rate in the region of the ^{37}Ar Auger electron peak was 0.033 d^{-1} over 61 runs. (For the more recent runs this background was 0.01 d^{-1}.) The average production rate in 1970—88 for ^{37}Ar was 0.518 ± 0.037 atoms d^{-1}. After subtraction of the cosmic muon production rate, a solar production rate of 0.438 ± 0.050 atoms d^{-1} was derived. This translates into 2.33 ± 0.25 SNU. As can be seen from Table 5.4, this neutrino flux is about three times lower than predicted. We show in Figure 5.23 a plot of the observed neutrino flux, respectively the observed ^{37}Ar production rate. The predicted ^{37}Ar production rate, corrected for the small cosmic rate background is $7.9(1 \pm 0.33)\text{SNU}/5.35 + 0.08 \pm 0.03 = 1.56 \pm 0.49$, where the "theoretical uncertainty" corresponds to 3σ.

This significant discrepancy is not understood at the present time. The recent results of the Kamiokande experiment (see Section 5.8.3 below) confirm the basic conclusion of the ^{37}Cl experiment, i.e., the reduction of the higher energy solar neutrino flux when compared with the standard solar model expectations. The various suggestions made to explain the discrepancy can be divided into two broad classes: speculations that the temperature in the center of the Sun, and therefore also the decay rate of ^8B, is different from that predicted by the standard solar model, as well as more radical ideas relating the reduction of the neutrino flux to neutrino oscillations, either in the interior of the Sun (see Section 5.9), or during the passage to the Earth.

A separate, although possibly related problem, is the apparent variability of the neutrino flux as measured by the ^{37}Ar production rate. The data in Figure 5.23 have uncertainties too large to reveal any time variations. However, if one plots instead the yearly average ^{37}Ar production rate, as in Figure 5.24, a periodicity of 10—11 years seems to emerge. The frequency, and phase of this effect seems to be anticorrelated with solar activity, characterized by the number of sunspots, which follow a well-known 11 year cycle. Since one believes that the neutrino luminosity of the solar core is constant over long periods of time, and the solar activity is related to processes in the solar conductive zone, far away from the core, it is difficult to see how these two processes could be related.

Interest in this effect has been strengthened following the suggestion of Voloshin et al. (86), who proposed a physical scenario for this time dependence involving a relatively large (10^{-11}—10^{-10} μ_{Bohr}) neutrino static or transitional magnetic moment precessing in the magnetic field of the solar convective zone. Various generalizations of this idea have been proposed, among them Lim & Marciano (88) and Akhmedov (88) proposed the

Figure 5.23. Production rate of ^{37}Ar as a function of time. The Standard Solar Model prediction of Bahcall & Ulrich (88) is 1.56 ± 0.49 per day, after correction for the cosmic ray background. (From Bahcall & Press 91.)

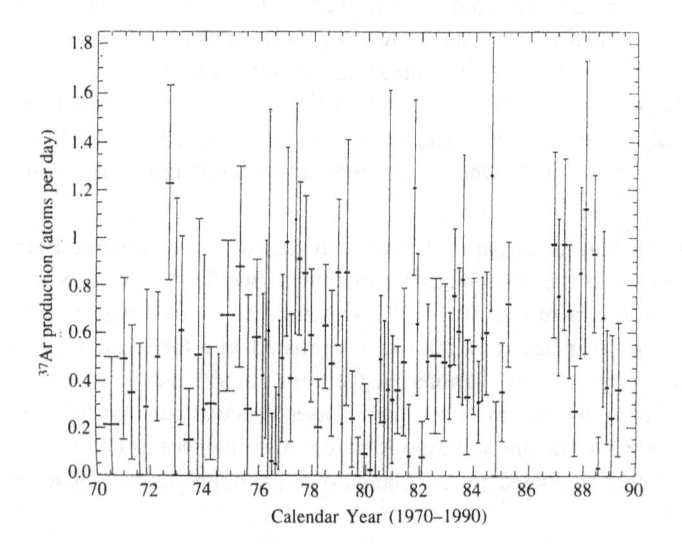

resonant enhancement of the neutrino helicity flip by the coherent interaction with the solar matter (see Section 5.9).

Massless neutrinos have vanishing magnetic moment. Moreover, the neutrino magnetic moment of required magnitude would exceed the expectations (see Section 1.2.3) by many orders of magnitude. It would not violate, however, limits set by terrestrial experiments. Astrophysical considerations (see, e.g., Raffelt 89) constrain the magnetic moment more severely, and new experiments to tighten the limits on μ_ν have been proposed. Quite apart from that, there is a continuing debate concerning the statistical significance of the apparent correlation between the neutrino flux and the solar cycle (Filippone & Vogel 90; Bieber et al. 90; Bahcall & Press 91).

5.8.3 *The Kamiokande II experiment*

When neutrinos scatter on electrons by the reaction $\nu_e + e \rightarrow \nu_e + e$, the recoil electrons are scattered in the forward direction. By measuring the recoil energy and direction one can identify the neutrino source, and determine its spectrum. Also, if the direction to the source is known (as is the case of solar neutrinos) one can measure the background by looking in the direction away from the neutrino source. These are the most important ingredients that made it possible to measure on-line the upper energy part of the solar neutrino spectrum in the Kamiokande II detector.

Kamiokande, a water Čerenkov detector, was built in 1983 to search for proton decay. In 1985—6 the detector was upgraded. The threshold energy was lowered and the background was reduced substantially. Since the

Figure 5.24. Yearly averages of (a) the ^{37}Ar production rate, and (b) the sun-spot numbers. (From Bieber et al. 90.)

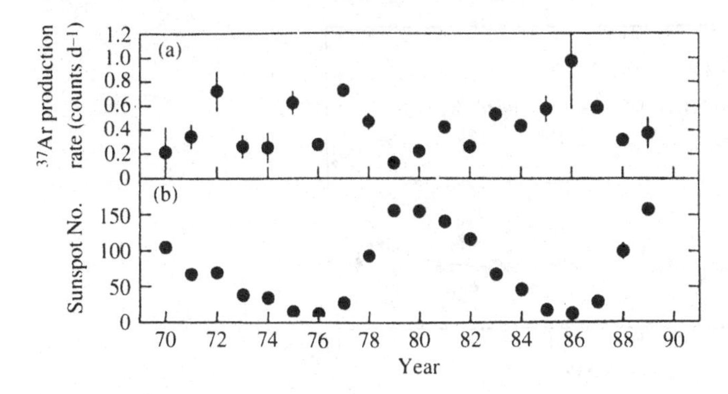

end of 1986 the detector was used to measure the solar neutrino flux for electron recoil energies above $E_e \geqslant 9.3$ MeV, and from June 1988 for $E_e \geqslant 7.5$ MeV. The whole detector contains 4,500 t of purified water, of which the central 680 t form the fiducial volume for solar neutrino measurement. (The same detector was used to observed the neutrino from SN1987A, see Section 7.2.) The detector is equipped with 948 photomultipliers (d = 50 cm), that are placed along the cylindrical walls. The number of Čerenkov photons is measured by counting the number of lit PMTs. The strong directionality of Čerenkov light is used to determine the direction of the scattered electrons (angular resolution for electrons is 10^0 at 10 MeV). The experiment, background conditions, and data analysis are described in Hirata et al. (90, 91).

In Figure 5.25 the data are plotted as a function of the angle with respect to the Sun. An excess in the direction $\cos(\theta_{sun})$ = 1.0 is evident. The height of the experimental peak is lower than the expected signal based on the standard solar model; the reduction factor is $R_K = 0.46 \pm 0.05 \pm 0.06$ where, as usual, the first error is statistical and the second one is systematic. The energy spectrum of recoil electrons is shown in Figure 5.26. It agrees in shape, but not in magnitude, with expectation. Finally, unlike the chlorine data described above, within the limited statistics the rate appears to be constant, without time variations.

The Kamiokande result thus confirms qualitatively the "solar neutrino puzzle", i.e., the reduction of the solar neutrino flux when compared to

Figure 5.25. Angular distribution of the Kamiokande data sample after the reduction procedure for the threshold and live timed. The histogram is the calculated signal distribution based on the full SSM value and includes multiple scattering and angular distribution of the detector. (From Hirata et al. 90.)

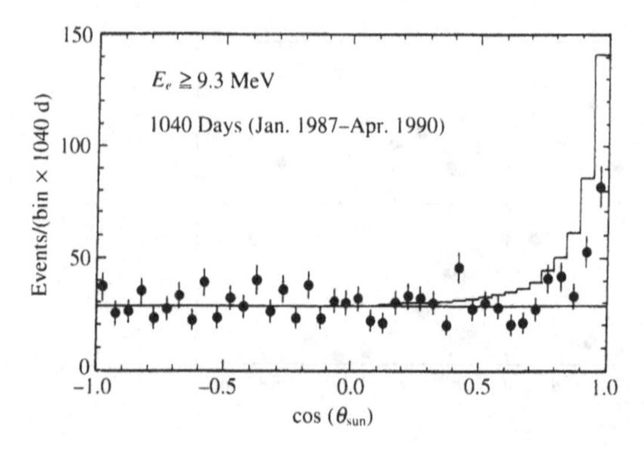

expectation based on the standard solar model. In Section 5.10 we shall analyze these data in terms of neutrino oscillations, and discuss the consistency (or lack of it) with the ^{37}Cl data.

5.8.4 *The* ^{71}Ga *experiment*

Recently, two gallium experiments have been devised and installed in underground laboratories: they are GALLEX (European Gallium Collaboration) and SAGE (Soviet American Gallium Experiment). The gallium detectors are based on the neutrino capture reaction ^{71}Ga(39.6%) (ν_e, e^-) ^{71}Ge which has a neutrino energy threshold of $E_\nu = 236$ keV. The decay of ^{71}Ge by electron capture has a half life of 11.43 d. The production rate for ^{71}Ge predicted by the standard solar model (see Table 5.4) is 132 ± 20 SNU (3σ), or 1.2 atoms/d for 30 t of Ga, of which 74 SNU are due to the *pp* and *pep* neutrinos. Clearly, gallium detectors, are well suited to probe the abundant primary *pp* reactions.

GALLEX. The GALLEX detector (Kirsten 86, 91) consists of 30.3 t of gallium in form of an aqueous gallium chloride solution with a mass of 101 t (53.5 m^3). Standard solar model neutrinos will induce 1.1 captures per day. After each 20 d exposure, approximately 13 ^{71}Ge atoms will be present; they are extracted chemically after addition of about 1 mg Ge carrier. The volatile $GeCl_4$ is purged with nitrogen and then converted to GeH_4. The latter is mixed with Xe and used to fill tiny (1 cm^3 volume) proportional counters which in turn are enclosed in a NaI anticoincidence

Figure 5.26. Energy distribution of the recoil electrons from the 1040 d Kamiokande sample. The histogram is the best fit of the expected Monte Carlo calculated distribution of the data. (From Hirata et al. 90.)

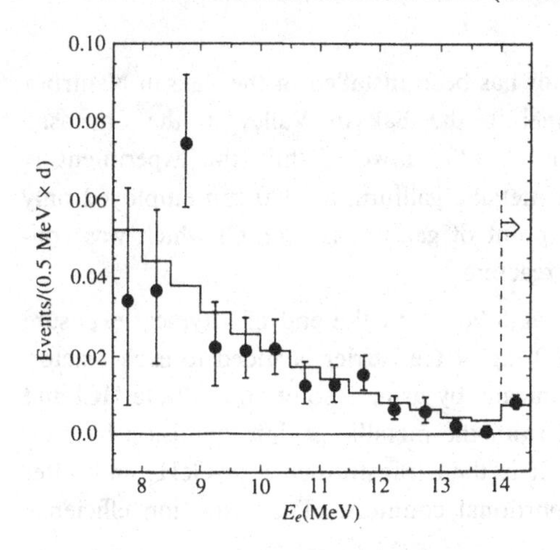

well counter surrounded by lead shielding. The K and L Auger electrons from ^{71}Ga with energies of 10.4 and 1.2 keV, respectively, are counted in the underground laboratory. Just as in the ^{37}Cl experiment, the signal pulses from Auger electrons are separated from γ-ray background with the help of pulse shape discrimination utilizing the long decay time associated with an electron pulse in the proportional counter. Each sample is counted for a period of 90 d.

The experiment has been installed in Hall A of the Gran Sasso Underground Laboratory in Italy at a depth of 3500 mwe. During industrial preparation of the gallium solution and storage above ground, cosmogenically produced radioisotopes ^{68,69,71}Ge build up leading to a significant radioactive Ge background. These Ge isotopes were purged repeatedly before and after installation of the detector underground. After removal of a resilient ^{68}Ge radioactivity by means of heating of the solution, the experiment is now operational and data taking has commenced in Spring 1991.

Inasmuch as the isotope ^{71}Ge can also be produced by other sources, notably from protons via the reaction ^{71}Ga(p,n)^{71}Ge, great care must be taken to account for these backgrounds. Such protons may arise from reactions triggered by external neutrons, or as products of interactions of residual cosmic ray muons. Estimates show that the production rate of ^{71}Ge from background sources is small, and may be responsible for at most 1.5% of the solar neutrino signal.

To help verify reaction yield and separation efficiency, plans exist to prepare an artificial ^{51}Cr neutrino source (electron capture, $E_\nu = 431$ and 751 keV) with a strength of about 1 megacurie. This source will be inserted into a well in the gallium chloride tank provided for that purpose. The source will be made by neutron activation of enriched ^{50}Cr.

SAGE. The SAGE detector has been installed in the Baksan Neutrino Observatory located in a tunnel in the Baksan Valley in the Caucasus Mountains, USSR, at a depth of 4700 mwe. While the experiment is capable of handling 60 tons of metallic gallium, a 1990 run employed only 30 tons of the metal (melting point of gallium is 29.8 C) which was contained in four heated chemical reactors.

The experiment proceeds as follows: At the end of a typical exposure time of 3 – 4 weeks, about 160 mg of Ge carrier is added to each reactor. The ^{71}Ge is then extracted chemically by using a solution of dilute HCl and H_2O_2. The $GeCl_4$ is separated from the metallic gallium emulsion by bubbling the extracts with argon. It is then transformed into GeH_4 and, after mixing with Xe, filled in proportional counters. The extraction efficiency was determined to be about 80%. The *K* Auger electrons are counted for a period of 2 – 3 months in an arrangement similar to that described above for GALLEX. The counting efficiency is around 37%.

The background problems arising from cosmogenically produced Ge isotopes such as ^{68}Ge are similar to those discussed for the GALLEX experiment and were resolved by multiple purging. Additional background problems associated with a radon contamination have been overcome.

The results of the January – July 1990 runs (five extractions) have been reported by Abazov et al. (91) and Elliott (91). At the end of a four-week exposure period, the mean number of detected *K* Auger electrons from decay of ^{71}Ge falling within the proper energy and rise time window, expected in each run, is about 4.0. In comparison, about four counts were observed in this period. However, from the time dependence of the counts in all five runs, it was concluded that almost all the counts can be attributed to a background constant in time. The best fit for the production rate is

$$\text{Neutrino Capture Rate} = 20^{+25}_{-20}\,(stat) \pm 32\,(syst)\ \text{SNU}$$

with upper limits of 55 SNU (68% CL) and 79 SNU (90% CL). From this fit, the best value of observed ^{71}Ge atoms is 2.6, as compared to 17 atoms predicted for the runs reported.

This early result qualitatively supports the suppression scenario of the chlorine and Kamiokande experiments, and extends it to the low energy *pp* neutrinos which are insensitive to the underlying solar neutrino model. However, we consider it too early to draw definite conclusions about the nature of the suppression and, in particular, about its relation to neutrino mass and mixing.

The experiment is continuing during 1991 and beyond with the final amount of 60 tons of gallium.

5.9 Neutrino Oscillations in Matter

The neutrino oscillations discussed so far are vacuum oscillations, caused by the propagation of a highly relativistic neutrino of mass m through vacuum,

$$\nu(t) = \nu(0)e^{i(px-Et)} \approx \nu(0)e^{-it\frac{m^2}{2p}} . \tag{5.39}$$

However, when neutrinos propagate through matter, the phase factor above is changed, $ipx \rightarrow ipnx$, where n is the index of refraction. The oscillations of a neutrino beam through matter are then modified, as first recognized by Wolfenstein (78). Under favorable conditions, a resonant amplification of oscillations (Mikheyev & Smirnov 85,86,87) may occur. (This process is often referred to by the acronym MSW.)

The index of refraction deviates from unity owing to the weak interaction of the neutrinos. Thus, for neutrinos of weak interaction flavor l and of momentum p the index of refraction is

$$n_l = 1 + \frac{2\pi N}{p^2} f_l(0) , \tag{5.40}$$

where N is the number density of the scatterers and $f_l(0)$ is the forward scattering amplitude. For practical purposes, we can neglect the absorption of neutrinos by matter and, therefore, consider only the real part of $f_l(0)$.

Matter is composed of nucleons (or quarks) and electrons. The contribution of nucleons (or quarks) to the forward scattering amplitude is described by the neutral current (Z^o exchange), as illustrated in Figure 5.27(a); it is identical for all neutrino flavors. (We assume that there are no flavor changing neutral currents.) For electrons the situation is different; the electron neutrinos interact with electrons via both the neutral current (Figure 5.27(a)) and the charged current (W^+ exchange), shown in Figure 5.27(b). All other neutrino flavors interact only via the neutral current.

That part of the refractive index (or forward scattering amplitude) which is common to all neutrino flavors is of no interest to us, because it modifies the phase of all components of the neutrino beam in the same way. However, the fact that ν_e ($\bar{\nu}_e$) interact differently has important consequences. Evaluating the graphs in Figure 5.27 one finds (Wolfenstein 78) that

$$\Delta f(0) = f_e(0) - f_\alpha(0) = -\sqrt{2}\,\frac{G_F\, p}{2\pi} , \tag{5.41}$$

where α denotes all neutrino flavors, except e, ($\alpha = \mu, \tau, \ldots$), and G_F is the Fermi constant $G_F \approx 10^{-5} M_p^{-2}$. For $\bar{\nu}_e$ the sign of $\Delta f(0)$ is reversed. (Eq. (5.41) differs from the original Wolfenstein (78) formula by the factor $\sqrt{2}$

Figure 5.27. Feynman graphs describing neutrino interaction in matter.

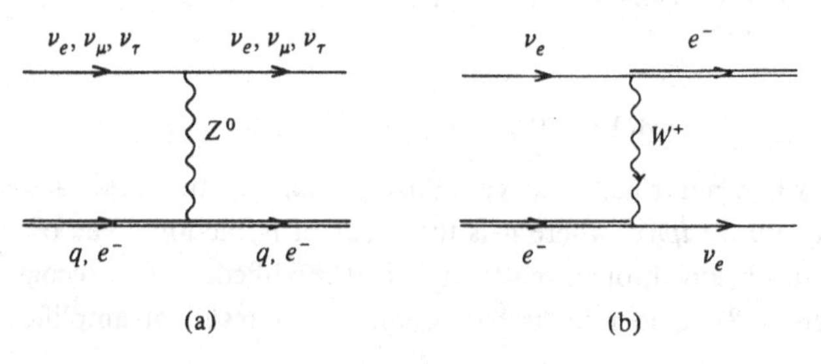

first obtained by Lewis (80), and by the sign change, derived by Langacker et al. (83).)

It is, perhaps, instructive to outline the derivation of Eq. (5.41). For a neutrino of mass m and momentum $p >> m$ propagating in matter, the effective energy is

$$E_{eff} \cong p + \frac{m^2}{2p} + <e\nu|H_{eff}|e\nu> \quad ,$$

where H_{eff} represents the W exchange in Figure 5.27(b).

$$\begin{aligned} H_{eff} &= \frac{G_F}{\sqrt{2}} J^{\mu}_{W} J^{+}_{\mu W} \\ &= \frac{G_F}{\sqrt{2}} \bar{e}\gamma^{\mu}(1-\gamma_5)\nu\bar{\nu}\gamma_{\mu}(1-\gamma_5)e \\ &= \frac{G_F}{\sqrt{2}} \bar{e}\gamma^{\mu}(1-\gamma_5)e\bar{\nu}\gamma_{\mu}(1-\gamma_5)\nu \ . \end{aligned}$$

(In the last line we performed the Fierz transformation to the charge retaining form.) For electrons at rest, only the $\mu = 4$ term contributes and we obtain

$$<e|\bar{e}\gamma^{\mu}(1-\gamma_5)e|e> = N_e\delta_{\mu 4} \quad ,$$

$$<\bar{\nu}|\bar{\nu}\gamma_4(1-\gamma_5)\nu|\nu> = 2 \ .$$

Thus,

$$E_{eff} \cong p + \frac{m^2}{2p} + \sqrt{2}G_F N_e \quad ,$$

and the time dependence for a neutrino propagating through matter is

$$\nu(t) = \nu(0)\, e^{-it(\frac{m^2}{2p} + \sqrt{2}G_F N_e)} \quad ,$$

in agreement with (5.39)—(5.41).

Let us consider propagation through matter in the simplest case of two neutrino flavors e and α, which are superpositions of two mass eigenstates ν_1, ν_2. As before, we assume that

$$\nu_e = \nu_1 \cos\theta + \nu_2 \sin\theta \ ,$$

$$\nu_\alpha = -\nu_1 \sin\theta + \nu_2 \cos\theta \ .$$

The time development of a neutrino beam in a vacuum is described in a compact form by the differential equation

$$i\frac{d}{dt}\begin{pmatrix}\nu_1\\ \nu_2\end{pmatrix} = \begin{pmatrix} m_1^2/2p & 0 \\ 0 & m_2^2/2p \end{pmatrix} \begin{pmatrix}\nu_1\\ \nu_2\end{pmatrix} \quad , \tag{5.42}$$

leading to the familiar expression (5.5) for vacuum oscillations described by the oscillation length (2π divided by the difference of the diagonal matrix elements)

$$L_{osc} = 2\pi \frac{2p}{m_1^2 - m_2^2} \quad . \qquad (5.5a)$$

The amplitude of the oscillation is $\sin^2 2\theta$. The oscillation length L_{osc} in Eq. (5.5a) above may now have either sign, depending on the relative magnitude of the masses m_1, m_2. Previously, in Eq. (5.5) we used the absolute value of Δm^2, and thus L_{osc} was always positive.

In matter, we have an additional contribution to the time development of the electron neutrino beam

$$\nu_e(x) = \nu_e(0)e^{ipnx} = \nu_e(0)e^{-\sqrt{2}iG_F N_e x} \; , \qquad (5.43)$$

where in the last term the uninteresting overall phase is omitted. The phase factor in (5.43) leads to the definition of the "characteristic matter oscillation length" (Wolfenstein 78) that is, the distance over which the phase changes by 2π,

$$L_o = \frac{2\pi}{\sqrt{2}G_F N_e} \approx \frac{1.7 \times 10^7}{\rho\ (\text{g cm}^{-3})\frac{Z}{A}}\ \text{m}, \qquad (5.44)$$

where we used $N_e = \rho N_o Z/A$, Z/A being the average charge-to-mass ratio of the electrically neutral matter, and N_o is the Avogadro number. Unlike the vacuum oscillation length L_{osc}, the matter oscillation length L_o is independent of the neutrino energy. For ordinary matter (rock) $\rho \approx 3$ g cm^{-3} and $Z/A \approx 0.5$, while at the center of the Sun $\rho \approx 150$ g cm^{-3} and $Z/A \approx 2/3$, thus $L_o \approx 10^4$ km (rock) and $L_o \approx 200$ km (Sun).

Including the forward scattering in matter, we obtain the differential equation for the time development of the mass eigenstates ν_1, ν_2 in the form

$$i\frac{d}{dt}\begin{pmatrix}\nu_1\\ \nu_2\end{pmatrix} = \begin{pmatrix} \frac{m_1^2}{2p} + \sqrt{2}G_F N_e c^2 & +\sqrt{2}G_F N_e sc \\ +\sqrt{2}G_F N_e sc & \frac{m_2^2}{2p} + \sqrt{2}G_F N_e s^2 \end{pmatrix} \begin{pmatrix}\nu_1\\ \nu_2\end{pmatrix} , \qquad (5.42')$$

where $c = \cos\theta$, $s = \sin\theta$. The 2×2 matrix on the right-hand side of (5.42') can be brought to the diagonal form by the transformation

$$\nu_{1m} = \nu_e \cos\theta_m - \nu_\alpha \sin\theta_m = \nu_1 \cos(\theta_m - \theta) - \nu_2 \sin(\theta_m - \theta) \; ,$$

$$(5.45)$$

$$\nu_{2m} = \nu_e \sin\theta_m + \nu_\alpha \cos\theta_m = \nu_1 \sin(\theta_m - \theta) + \nu_2 \cos(\theta_m - \theta) \; .$$

The combinations ν_{1m} and ν_{2m} represent "particles" which propagate through matter as plane waves. The new mixing angle θ_m depends on the vacuum mixing angle, and on the vacuum and matter oscillation lengths L_{osc} and L_o through the relation

$$\tan 2\theta_m = \tan 2\theta \left(1 + \frac{L_{osc}}{L_o}\sec 2\theta\right)^{-1} . \tag{5.46}$$

(Note that L_{osc} through its dependence on Δm^2 can in this context have both the positive and negative signs.) The "effective" oscillation length in the presence of matter is obtained from the difference of eigenvalues of the 2×2 matrix in Eqs. (5.42') and is given by

$$L_m = L_{osc}\frac{\sin 2\theta_m}{\sin 2\theta} = L_{osc}\left[1 + \left(\frac{L_{osc}}{L_o}\right)^2 + \frac{2L_{osc}}{L_o}\cos 2\theta\right]^{-\frac{1}{2}}, \tag{5.47}$$

and the probability of detecting a ν_e at a distance x from the ν_e source is

$$P(E_\nu, x, \theta, \Delta m^2) = 1 - \sin^2\theta_m \sin^2\frac{\pi x}{L_m} , \tag{5.48}$$

where θ_m and L_m both depend on the vacuum oscillation angle θ and on the vacuum oscillation length L_{osc} (and through it on $\Delta m^2/E_\nu$).

We can now consider several special cases:

i) $|L_{osc}| \ll L_o$ (low density): matter has virtually no effect on oscillations. In experiments with terrestrial sources of neutrinos, one is at best able to study oscillations with $|L_{osc}| \leqslant$ Earth diameter, and thus this limiting case is relevant. Matter will have, therefore, only a small effect on experiments with neutrinos of terrestrial origin. (The effect could be observable as day-night or seasonal variation of the neutrino signal in a solar neutrino detector, see discussion below.)

ii) $|L_{osc}| \gg L_o$ (high density): the oscillation amplitude is suppressed by the factor $L_o/|L_{osc}|$, while the effective oscillation length $L_m \approx L_o$ is independent of the vacuum oscillation parameters.

iii) $|L_{osc}| \approx L_o$: in this case oscillation effects can be enhanced; in particular, for $L_{osc}/L_o = -\cos 2\theta$ one has $\sin^2 2\theta_m = 1$. However, the "effective" oscillation length is correspondingly longer, $L_m = L_{osc}/\sin 2\theta$.

From (5.46) it follows that when neutrinos of different energies propagate through matter, the resonance condition iii), $L_{osc}/L_o = -\cos 2\theta$, can be obeyed only for negative Δm^2 ($m_2 > m_1$), and for

$$\frac{E_\nu\,(\mathrm{MeV})}{|\Delta m^2|(\mathrm{eV}^2)} = \frac{6.8 \times 10^6}{\frac{Z}{A}\rho\,(\mathrm{g\,cm^{-3}})}\cos 2\theta . \tag{5.49}$$

Thus, for each density ρ the enhancement affects only certain values of $E_\nu/\Delta m^2$.

The behavior of the "effective" mixing angle θ_m as a function of L_{osc}/L_o for different values of the vacuum mixing angle θ is shown in Figure 5.28. The enhancement ($\sin^2 2\theta_m \approx 1$) is present over a finite region of values of the abscissa, i.e., over a finite range of L_0, and hence also over a finite range of matter densities.

The discussion so far has dealt with neutrino propagation in matter of a constant density. Eqs. (5.45)—(5.48) represent then an exact analytical solution of the coupled first order differential equations (5.42'). Before discussing further the obviously crucial problem of neutrino propagation when density is not constant, and, in particular, the situations when the resonance enhancement in point iii) above occurs, it is worthwhile stressing that neutrino oscillations in matter do not represent a new physical phenomenon; they are necessary consequences of vacuum oscillations.

When neutrinos propagate through a medium of variable density, such

Figure 5.28. Dependence of the effective mixing angle θ_m on the ratio L_{osc}/L_o of the vacuum oscillation length L_{osc} to the characteristic matter oscillation length L_o. The curves are labeled by the values of the vacuum oscillation mixing angle θ.

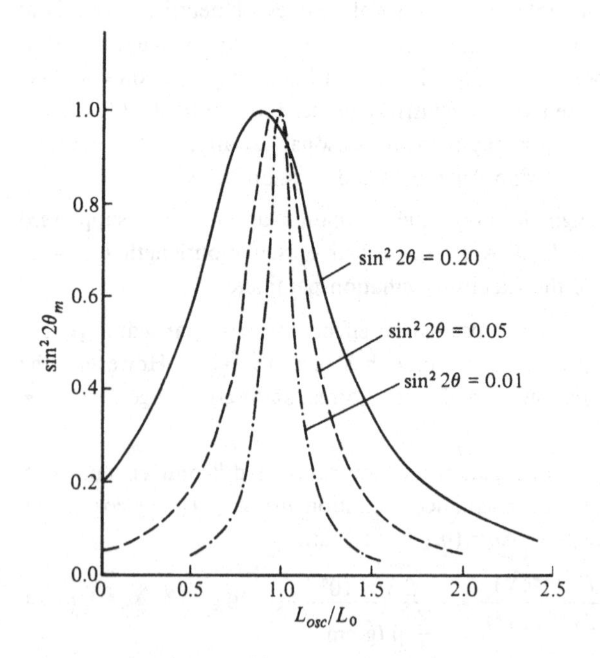

as from the center of the Sun, in a collapsing supernova, or through the center of the Earth, one has to solve the system of coupled differential equations (5.42'), where N_e depends on position, and through it on time. It is always possible to solve these equations numerically. For the case of the Sun this has been done in many papers, among the first ones were the original work of Mikheyev & Smirnov (85, 86), as well as the papers by Rosen & Gelb (86), Bouchez et al. (86), and Parke & Walker (86). We shall discuss the results and their relation to the possible solution of the solar neutrino problem in Section 5.10.

In Figure 5.29 we show what happens when neutrinos propagate from the central region of the Sun where $L_0 << |L_{osc}|$, through the region where the resonance condition is obeyed, and to the Sun surface where

Figure 5.29. Illustration of the propagation of neutrinos through the Sun. The abscissa represents the distance from the center of the Sun; it is schematically divided into three pieces: center, region of resonance, and surface. For the resonance region the density is shown in the upper part of the Figure. The effective mixing angle θ_m is also given there. The probability P that a neutrino emitted in the center as an electron neutrino remains an electron neutrino is shown in the middle part of the Figure. The bottom part shows the relative orientation of the relevant vectors.

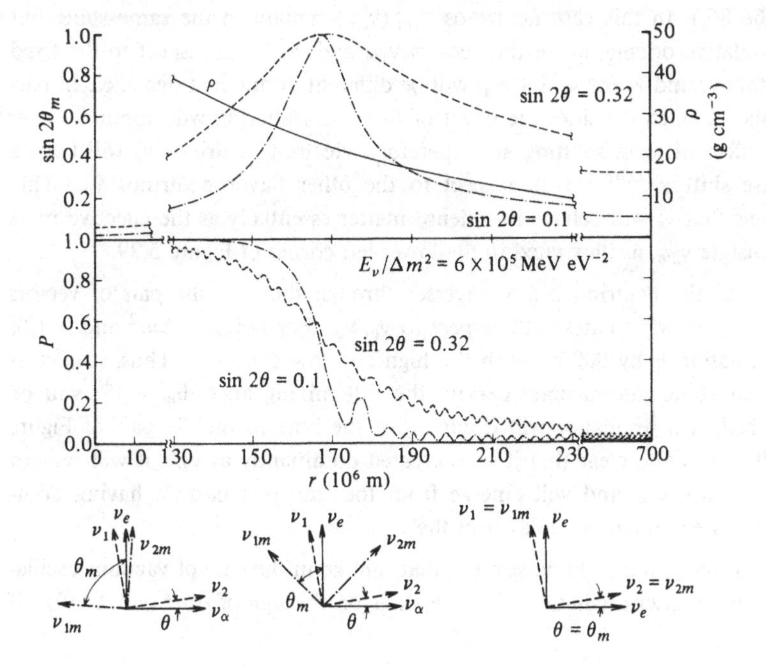

$L_0 >> |L_{osc}|$. The density ρ and the effective mixing angle θ_m are shown in the upper part, and the probability that an electron neutrino created in the center remains an electron neutrino is shown in the middle part of the Figure. One can see that, for this particular choice of $E_\nu/\Delta m^2$, the neutrinos leaving the Sun are no longer electron neutrinos; indeed, the probability P becomes very small at the solar edge. Numerical calculations show that if real neutrinos have $|\Delta m^2| \approx 10^{-7}$—$10^{-4}$ eV^2, matter oscillation enhancement would be able to explain the suppression of the neutrino flux as detected by the ^{37}Cl and Kamiokande detectors (see Section 5.8), even if the vacuum mixing angle is as small as $\sin^2 2\theta = 10^{-4}$.

Is it possible to understand intuitively this behavior of the probability P? In a vacuum (or at low density), the mass eigenstates ν_1, ν_2 are identical to the effective eigenstates ν_{1m}, ν_{2m}, and are rotated by the vacuum mixing angle θ (which is assumed to be small) with respect to the weak flavor eigenstates ν_e and ν_α. Furthermore, if $m_2 > m_1$ (that is, $\Delta m^2 < 0$ and thus the electron neutrino is dominantly the lighter mass eigenstate ν_1), the state ν_1 is shifted by a phase shift $e^{-i\Delta m^2 t/2p}$ with respect to ν_2. The situation is illustrated in the lower right corner of Figure 5.29.

Now, let us consider a beam of neutrinos passing through an object of slowly varying density. (Slow density variation means that the density can be regarded as constant over the effective oscillation length L_m. In the following we describe the so called adiabatic solution of the MSW effect, see Bethe 86.) In this case neutrinos ν_{1m} (ν_{2m}) remain in the same state, but the relative orientation of the vectors ν_{1m} and ν_{2m} with respect to the fixed vectors ν_1 and ν_2 (or ν_e and ν_α) will be different at different densities. In particular, at high densities the effect of forward scattering will dominate over the effect of mass splitting and, therefore, electron neutrinos ν_e will have a phase shift $e^{\sqrt{2} i G_F N_e x}$ with respect to the other flavor neutrinos ν_α. This means that ν_e will behave in a dense matter essentially as the effective mass eigenstate ν_{2m}, as illustrated in the lower left corner of Figure 5.29.

As the neutrino beam traverses through the star, the pair of vectors ν_{1m}, ν_{2m} slowly rotates with respect to ν_e, ν_α. For negative Δm^2 and $\Delta f(0)$, the rotation is by 90° between the high and low densities. Thus, it is clear that at some intermediate density the full mixing angle $\theta_m = 45^o$ will be reached, i.e., resonance will occur. (See the bottom middle part of Figure 5.29.) It is also clear that if ν_e is created dominantly as ν_{2m}, it will remain dominantly ν_{2m} and will emerge from the star as a particle having dominantly weak interaction flavor of the neutrino ν_α.

Finally, it should be stressed that, unlike in the case of vacuum oscillation, the matter oscillation effect depends on the sign of Δm^2 and $\Delta f_e(0)$. If either of these signs is reversed, by using antineutrinos instead of neutrinos (reversal of the sign of $\Delta f_e(0)$), or if in vacuum, electron neutrinos are the dominantly heavier neutrinos (reversal of the sign of Δm^2), there is no resonance and there is no attenuation of the corresponding flavor. In the case of the Sun where $\nu_1 \approx \nu_e$, an enhancement of matter oscillations will occur only if the admixed state $\nu_2 \approx \nu_\mu$ or ν_τ has higher mass than $\nu_1 \approx \nu_e$, a situation which is expected naturally in many schemes of grand unification. Also, the required values of the mass difference Δm^2 are quite close to the values expected in the see-saw mechanism of neutrino mass generation (Section 1.4).

For solar neutrinos, besides the numerical calculations mentioned above, and discussed in greater detail in Section 5.10, a valuable understanding can be obtained from approximate analytical solutions of the problem (Haxton 87; Kuo & Pantaleone 87; Pizzochero 87; Petcov 88; to name just a few). Briefly, if the central solar density is denoted ρ_{Max} one can write the probability that electron neutrino emerges from the Sun as electron neutrino in the form

$$P_{\nu_e \to \nu_e} = \frac{1}{2} + \left(\frac{1}{2} - P_x \right) \cos 2\theta_m(\rho_{Max}) \cos 2\theta \, , \qquad (5.50)$$

where P_x is the probability of transition $1 \leftrightarrow 2$ at the resonance ($P_x = 0$ in the adiabatic case). Usually $\cos 2\theta_m(\rho_{Max}) \simeq -1$ and thus

$$P_{\nu_e \to \nu_e} \simeq \sin^2\theta + P_x \cos 2\theta \, . \qquad (5.50')$$

To find conditions for the neutrino conversion we need to be sure that: a) the resonance occurs at all, and b) the quantity P_x is small.

Condition a) above is obeyed if the ratio $L_{osc}/(L_0 \cos 2\theta) > 1$ is valid, i.e., if

$$1.5\times 10^{-7} \frac{E_\nu(\text{MeV})}{\Delta m^2(\text{eV}^2)} \rho_{Max}(\text{g cm}^{-3}) \frac{Z}{A} > 1 \, . \qquad (5.51)$$

Eq. (5.51) defines the adiabatic condition. Note that the vacuum mixing angle θ does not appear (a horizontal line in the usual Δm^2 vs. $\sin^2 2\theta$ plot). Also, neutrinos of higher energies will obey the condition more easily and will be converted more than the low energy neutrinos (see Figure 5.30).

To see when condition (b) above is obeyed one can evaluate P_x in the linear approximation (Landau-Zenner effect) to obtain

$$P_x = \exp\left\{ -\frac{\pi}{2} \frac{\sin^2 2\theta}{\cos 2\theta} \frac{\Delta m^2/2E_\nu}{\left| \frac{1}{\rho} \frac{d\rho}{dr} \right|} \right\} . \qquad (5.52)$$

Using the approximately exponential form of the density distribution in the Sun, Eq. (5.52) implies that for $P_x < 1$ one requires that

$$\frac{\sin^2 2\theta \Delta m^2}{E_\nu} \geqslant const \ , \tag{5.53}$$

where the *const* above depends on the logarithmic derivative of the solar density distribution. The condition (5.43) defines the "nonadiabatic solution" to the MSW effect. It represents a downsloping line in the Δm^2 vs. $\sin^2\theta$ plot. It also affects more the lower energy neutrinos than the higher energy ones. In Figure 5.30 we show the probability of reaching the Earth as an electron neutrino for electron neutrinos with different energies created near the solar center. Qualitatively, similar curves can be obtained for other values of the mass difference and vacuum mixing angle by numerical solution of the evolution equations (5.42').

In order to interpret the results of various solar neutrino experiments, one has to calculate the probability $P_{\nu_e \to \nu_e}$ in a grid of Δm^2 and $\sin^2 2\theta$ values. For each of these values the distribution of the source nuclei in the Sun as well as the appropriate folding over the neutrino energies must be included. In some cases the effect of neutrino propagation through the Earth (day-night effect, or summer-winter effect) should be also considered (Baltz & Weneser 87). In Section 5.10 we describe the consequences of the MSW effect for the experiments discussed in Section 5.8.

Figure 5.30. Probability that a ν_e originating near the solar center arrives at Earth as ν_e. The vacuum mixing angle is $\sin^2 2\theta = 0.03$. The curve A (adiabatic) is for $\Delta m^2 = 10^{-4}$ eV^2, the curve NA (nonadiabatic) is for $\Delta m^2 = 1.1 \times 10^{-6}$ eV^2. Both curves give the total suppression factor for the ^{37}Cl detector. (From Davis et al. 89.)

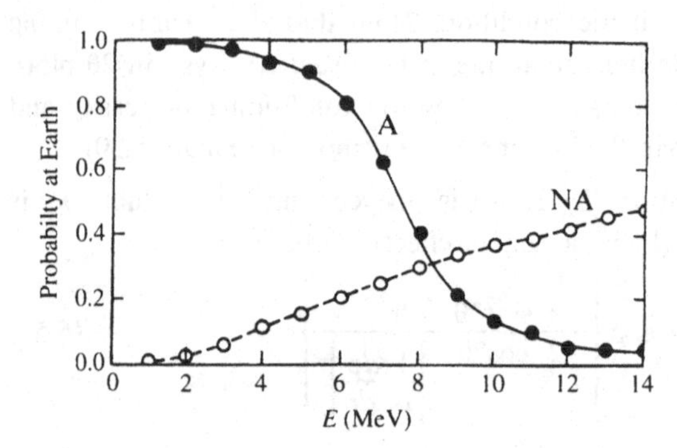

5.10 Solar Neutrino Data and the MSW Effect

In Section 5.8 we have shown that the measured solar neutrino flux in the ^{37}Cl and Kamiokande II detectors is reduced in comparison to the prediction of the standard solar model. The preliminary data of the SAGE detector, sensitive to the low energy, less model dependent *pp* neutrinos, seem to indicate a flux deficit as well. Then, in Section 5.9 we explained the phenomenon of the matter enhanced neutrino oscillations (MSW effect). Here, we shall see to what extent the MSW effect could explain the solar neutrino problem, and what values of the neutrino oscillation parameters Δm^2 and $\sin^2 2\theta$ would fit the data. Throughout, we shall assume that the treatment involving only two neutrino flavors is adequate. (Generally, for three flavors one has to consider two mass parameters, Δm_{12}^2 and Δm_{13}^2. However, if $\Delta m_{12}^2 << \Delta m_{13}^2$, as suggested by the see-saw mechanism, the consideration of three-flavor oscillations reduces to the two-flavor case.)

In Figure 5.31 we show the region in the Δm^2 and $\sin^2 2\theta$ plane that gives, due to the matter enhancement of the neutrino oscillations, the required reduction in the production of ^{37}Ar. All solar neutrino sources

Figure 5.31. Oscillation parameters compatible with the ^{37}Cl data. The shaded region gives the values of Δm^2 and $\sin^2 2\theta$ for which the ^{37}Cl experiment gives a rate of 1.5—2.7 SNU at 95% CL. (From Bahcall & Haxton 89.)

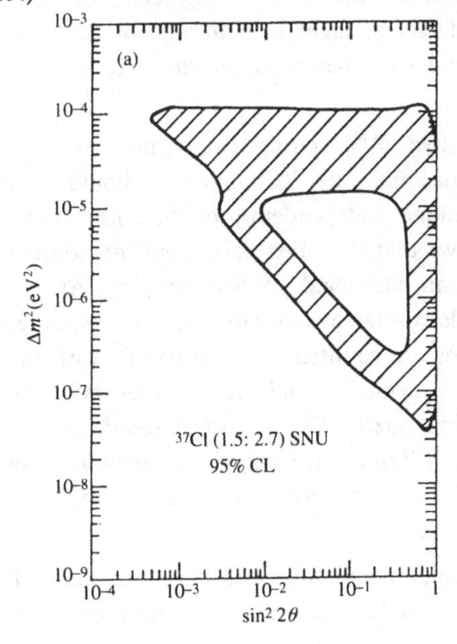

listed in Table 5.4 were included in the calculation. The "adiabatic" solution (horizontal band between Δm^2 of 10^{-5} and 10^{-4} eV^2), as well as the "nonadiabatic" solution (downsloping band), and the "large mixing angle" solution (vertical band near $\sin^2 2\theta = 1$) are clearly recognizable. Since the ^{37}Cl detector measures just one quantity, the ^{37}Ar production rate, it cannot distinguish between these three possibilities.

When considering the data from the Kamiokande II detector, one has to keep in mind several points. First, the threshold (9.3 MeV initially, later reduced to 7.5 MeV) is high enough so that the detector is sensitive exclusively to the neutrinos from the ^{8}B decay. Second, both the charged and neutral weak currents contribute to the neutrino scattering on electrons measured in the Kamiokande II experiment. *All* neutrinos contribute to the neutral current scattering, while only electron neutrinos contribute to the charged current scattering (and can change ^{37}Cl into ^{37}Ar). It turns out that it is very difficult to distinguish the charged and neutral current scattering by the recoil energy spectrum. Thus, as a practical rule, muon and tau neutrinos behave in the Kamiokande II detector as electron neutrinos, with the cross section reduced by a factor of $\approx$6. One has to include the irreducible neutral current effect in the analysis of the $\nu - e$ scattering in terms of the oscillation parameters. Finally, Kamiokande II is a "live" detector and thus one can look for the day-night effect caused by the passage of solar neutrinos through Earth. It follows from Eqs. (5.44)—(5.47) that interesting effects lie in the region $E_\nu/\Delta m^2 \simeq 10^6$—$10^7$ (Baltz & Weneser 87). When neutrinos pass through the Earth's interior a "regeneration" can occur, in which the probability of detecting ν_e is larger at night than during the day. No difference in signal between day and night has been detected (Hirata et al. 91).

Since Kamiokande II is sensitive to ^{8}B neutrinos only, one can infer, based on the Kamiokande II data, that the ^{37}Cl detector should have observed 2.8$\pm$0.5 SNU from ^{8}B alone, independently of the solar model. Moreover, from Table 5.4 it follows that the other solar neutrino sources, primarily ^{7}Be and *pep*, contribute an additional 1.8 SNU in the ^{37}Ar production rate according to the standard solar model. However, the measured rate, 2.20$\pm$0.24 SNU when all runs are counted, is incompatible with this prediction. This contradiction, taken literally, would exclude solar models as an explanation of the solar neutrino puzzle. The statistical significance of this conclusion is, unfortunately, still limited. If smaller experimental error bars confirm the incompatibility of the two experiments, neutrino oscillation would remain the only viable solution.

In Figure 5.32 we show the region compatible with the Kamiokande II data. Since the adiabatic solution would result in the almost complete suppression of the higher energy part of the spectrum (see Figure 5.30), contrary to the data (Figure 5.26), only the "nonadiabatic" and the "large mixing angle" solutions survive. (Even though, again, the statistical significance of this conclusion is not overwhelming.) In addition, the absence of observable day-night effect (Hirata et al. 91) removes part of the "large mixing angle" solution. The "nonadiabatic" solution represents, obviously, a possible common explanation of the data from both experiments.

Finally, in Figure 5.33 we show schematically the region in the Δm^2, $\sin^2 2\theta$ space that would lead to a significant suppression of the solar neutrino signal in the ^{71}Ga based detectors. Since the relevant neutrino energies are less than in the two previous experiments, the sensitive region involves correspondingly smaller values of Δm^2. Depending on the outcome of the *GALLEX* and *SAGE* experiments, it could be possible to pinpoint the relevant oscillation parameters with good accuracy.

Figure 5.32. Oscillation parameters compatible with the Kamiokande II data. The dotted region gives the values of Δm^2 and $\sin^2 2\theta$ compatible at 90% CL with the solar neutrino flux measured by the Kamiokande II experiment. The shaded region is excluded, also at 90% CL, by the absence of the day/night effect in that measurement. (From Hirata et al. 91.)

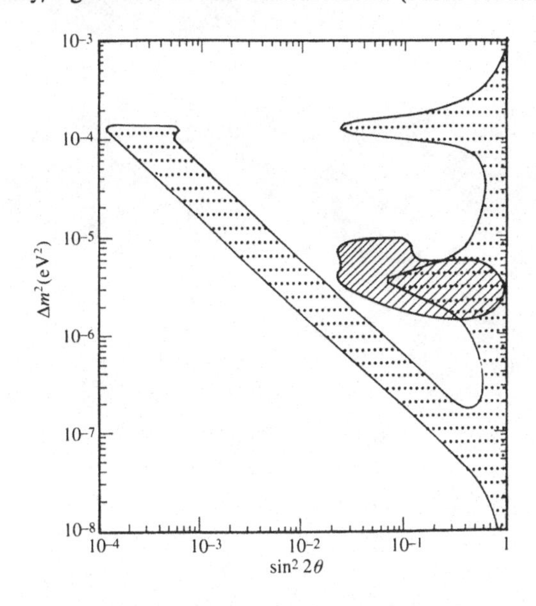

Figure 5.33. Regions of sensitivity of the ^{37}Cl and ^{71}Ga detectors. To the right and below the dashed curves are regions where the ν_e flux is reduced by more than 50% in the indicated detectors.

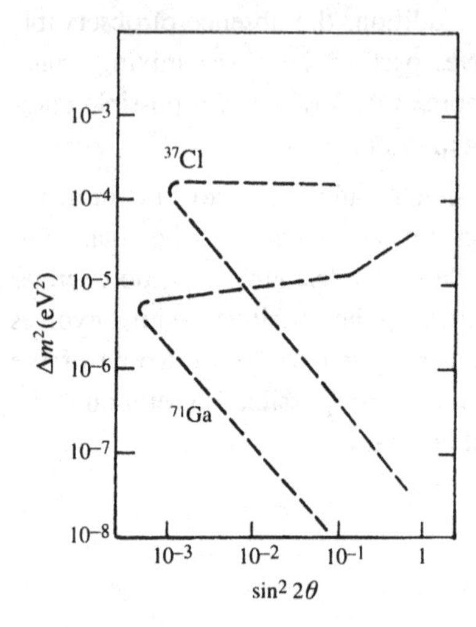

6

Double Beta Decay

This Chapter is devoted to the exploration of double beta decay, a remarkable rare transition between two nuclei with the same mass number, which involves a change of the nuclear charge number by two units. It has long been recognized that a particular mode of double beta decay, in which two electrons and no neutrinos are emitted, is a powerful tool for the study of lepton conservation in general, and neutrino properties in particular. The description of the neutrinoless double beta decay involves an intricate mixture of elementary particle physics and physics of the nucleus. Among the particle physics issues which we wish to address with the help of double beta decay are the questions: is the neutrino a Majorana particle and if so, what is its mass? and, is there evidence that the weak interactions contain leptonic currents with a small right-handed component? The principal nuclear physics issues have to do with the identification and evaluation of the nuclear matrix elements responsible for the decay rate. As it is our goal to arrive at a quantitative answer for the mentioned particle properties, we have no choice but to learn first how to understand the nuclear mechanisms.

Our principal concern, as everywhere in this book, is the problem of extracting information on neutrino mass from these nuclear experiments. Accordingly, we begin with Section 6.1 on the phenomenology of double beta decay and how it relates to neutrino mass. This is followed in Section 6.2 with a discussion on the nuclear physics aspects. We then proceed with a description of the experimental results in Section 6.3, both from laboratory work and from studies of geological samples. Finally, in Section 6.4 we discuss the implications as far as neutrino mass and right-handed current coupling are concerned.

The review articles by Haxton & Stephenson (84), Doi et al. (85), Vergados (86), Avignone & Brodzinski (88), Caldwell (89), Morales (89), and Tomoda (91) are helpful guides for additional studies.

6.1 Phenomenology of Double Beta Decay

6.1.1 *General introduction*

As double beta decay is closely related to ordinary nuclear beta decay, we begin this Section with a brief introduction to nuclear beta decay. In beta decay of the neutron, the most elementary semileptonic weak process, a free neutron transforms itself into a proton by emitting an electron and an antineutrino. The beta (or electron) decay of a nucleus is the analogue of free neutron decay, however, with the initial neutron and the final proton bound in a nucleus. While a free proton is stable against weak decays, nuclei are known to undergo positron decay (or electron capture) whereby the nuclear charge decreases by one unit (as if a proton were to decay into a neutron). Thus, nuclei have a richer variety of semileptonic weak decays than free particles.

The theoretical description of electron and positron decays (Commins & Bucksbaum 83; Konopinski 66) begins with the general weak semileptonic Lagrangian, and is greatly simplified by the fact that nucleons in nuclei are essentially nonrelativistic ($v^2/c^2 \sim 1/16$) and that the de Broglie wavelength of the outgoing leptons is much longer than the nuclear radius ($pR \sim 1/40$). As leptons are much lighter than nuclei, nuclear recoil can usually be neglected, and the available energy, therefore, is shared by the leptons only. For transitions in which the initial and final nuclear states have the same parity and differ at most by one unit of angular momentum, one can separate the nuclear and leptonic degrees of freedom. (This is the so called allowed approximation; later in this Chapter, when we discuss the effects of the leptonic right-handed currents on double beta decay, we will consider terms beyond this approximation.) The lepton spectra in allowed decays are determined only by the phase space of the leptons, and the Coulomb interaction of the charged leptons with the final nucleus must be properly included. Nuclear structure does not affect the shape of an allowed spectrum; it determines, however, the total decay probability through corresponding nuclear matrix elements.

The electron (positron) spectra are given by Eq. (2.3). The quantity $|M|^2$ there contains the Fermi $<1>$ and Gamow-Teller $<\sigma>$ nuclear matrix elements in the following combination,

$$|M|^2 = \delta_{I_i, I_f} |g_V|^2 <1>^2 + |g_A|^2 <\sigma>^2 , \tag{6.1}$$

where g_V and g_A are the vector and axial-vector weak interaction coupling constants, respectively. Similar expressions can be obtained for electron capture (EC). (The effect of neutrino mass on the spectrum shape (6.1) is of no concern to us here; it is, however, discussed in a different context in Chapter 2.)

Whether a nucleus is stable or undergoes weak decay has to do with the dependence of the atomic mass M_A of the isotope (Z,A) on the nuclear charge Z. This functional dependence near its minimum can be approximated by a parabola

$$M_A(Z,A) = const + 2b_{sym}\frac{(A/2-Z)^2}{A^2} + b_{Coul}\frac{Z^2}{A^{1/3}} + m_eZ + \delta \,, \tag{6.2}$$

where the symmetry energy coefficient is $b_{sym}{\sim}50$ MeV and the Coulomb energy coefficient is $b_{Coul}{\sim}0.7$ MeV. The m_eZ term represents the mass of the bound electrons; their binding energy, for our purposes, is small enough to be neglected. The last term δ describes nuclear pairing, the increase in binding as pairs of like nucleons couple to angular momentum zero. It is a small correction term and is given in a crude approximation by $\delta \sim \pm 12/A^{1/2}$ for odd N, odd Z, or even N, even Z, respectively, while δ=0 for odd A. (There are large fluctuations around these values; a somewhat better general formula is given by Vogel et al. (84).) Thus, for odd A nuclei, typically only one isotope is stable; nuclei with charge Z smaller than the stable nucleus decay by electron emission, while those with larger Z decay by electron capture or positron emission or by both these modes simultaneously.

For even A nuclei the situation is different. Due to the pairing term δ, the even-even nuclei form one parabola while the odd-odd nuclei form another one, at larger mass, as shown in Figure 6.1, using $A = 76$ as an example. Consequently, in a typical case there exist two even-even nuclei for a given A which are stable against both electron and positron (or EC) decays. As these two nuclei usually do not have the same mass, the heavier may decay into the lighter through a second order weak process in which the nuclear charge changes by two units. This process is called double beta decay.

Double beta decay, therefore, should proceed between two even-even nuclei. All ground states of even-even nuclei have spin and parity 0^+ and thus transitions $0^+ \rightarrow 0^+$ are expected in all cases. Occasionally, population of the low-lying excited states of the daughter nucleus is energetically possible, giving rise to $0^+ \rightarrow 2^+$ transitions or to transitions to the excited 0^+ state.

Double beta decay in the 2ν mode, in analogy to electron decay, is a transition with two electrons and two electron antineutrinos in the final state, i.e.,

$$(Z,A) \rightarrow (Z+2,A) + e_1^- + e_2^- + \bar{\nu}_{e1} + \bar{\nu}_{e2} \,, \tag{6.3}$$

and, of course, is subject to the condition

$$M_A(Z,A) > M_A(Z+2,A)\,, \tag{6.4}$$

with the supplementary practical requirement that single beta decay is absent, i.e., $M_A(Z,A) < M_A(Z+1,A)$, or that it is so much hindered (e.g., by the angular momentum selection rules) that it does not compete with double beta decay.

In analogy to positron decay and electron capture, there are three possible decay modes in which the nuclear charge decreases by two units and with two neutrinos in the final state:

$$(Z,A) \rightarrow (Z-2,A) + e_1^+ + e_2^+ + \nu_{e1} + \nu_{e2}\,, \tag{6.5a}$$

if $M_A(Z,A) > M_A(Z-2,A) + 4m_e$;

$$(Z,A) + e_b^- \rightarrow (Z-2,A) + e^+ + \nu_{e1} + \nu_{e2}\,, \tag{6.5b}$$

if $M_A(Z,A) > M_A(Z-2,A) + 2m_e + B_e$; and

$$(Z,A) + e_b^- + e_b^- \rightarrow (Z-2,A) + \nu_{e1} + \nu_{e2}\,, \tag{6.5c}$$

if $M_A(Z,A) > M_A(Z-2,A) + 2B_e$. (Here e_b^- is a bound atomic electron with binding energy B_e.)

While there are 11 known candidates (Table 6.1) for the decay (6.3) in which the Q value, i.e., the kinetic energy available to the leptons, is larger

Figure 6.1. Atomic masses of nuclei with $A = 76$. Parabolas connecting the even-even and odd-odd nuclear masses are indicated. ^{76}Ge and ^{76}Se are stable with respect to ordinary beta decay; ^{76}Ge, however, can decay by double beta decay.

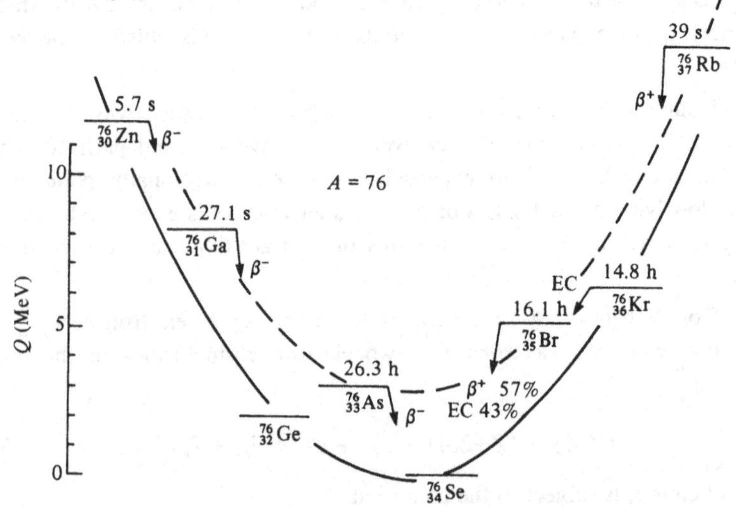

Table 6.1. *Characteristics of double beta decay candidates. The quantities $G^{2\nu}$ and $G^{0\nu}$ are defined in Eqs. (6.18) and (6.24), respectively.*

$\beta^-\beta^-$ candidates	T_0 (keV)	Abundance (%)	$(G^{2\nu})^{-1}$ (y)	$(G^{0\nu})^{-1}$ (y)
$^{46}Ca \rightarrow ^{46}Ti$	987 ±4	0.0035	8.71*E*21	7.16*E*26
$^{48}Ca \rightarrow ^{48}Ti$[a]	4271 ±4	0.187	2.52*E*16	4.10*E*24
$^{70}Zn \rightarrow ^{70}Ge$	1001 ±3	0.62	3.17*E*21	4.27*E*26
$^{76}Ge \rightarrow ^{76}Se$	2039.6±0.9	7.8	7.66*E*18	4.09*E*25
$^{80}Se \rightarrow ^{80}Kr$	130 ±9	49.8	8.20*E*27	2.34*E*28
$^{82}Se \rightarrow ^{82}Kr$	2995 ±6	9.2	2.30*E*17	9.27*E*24
$^{86}Kr \rightarrow ^{86}Sr$	1256 ±5	17.3	3.00*E*20	1.57*E*26
$^{94}Zr \rightarrow ^{94}Mo$	1145.3±2.5	17.4	4.34*E*20	1.57*E*26
$^{96}Zr \rightarrow ^{96}Mo$[a]	3350 ±3	2.8	5.19*E*16	4.46*E*24
$^{98}Mo \rightarrow ^{98}Ru$	112 ±7	24.1	1.03*E*28	1.49*E*28
$^{100}Mo \rightarrow ^{100}Ru$	3034 ±6	9.6	1.06*E*17	5.70*E*24
$^{104}Ru \rightarrow ^{104}Pd$	1299 ±2	18.7	1.09*E*20	8.32*E*25
$^{110}Pd \rightarrow ^{110}Cd$	2013 ±19	11.8	2.51*E*18	1.86*E*25
$^{114}Cd \rightarrow ^{114}Sn$	534 ±4	28.7	6.93*E*22	6.10*E*26
$^{116}Cd \rightarrow ^{116}Sn$	2802 ±4	7.5	1.25*E*17	5.28*E*24
$^{122}Sn \rightarrow ^{122}Te$	364 ±4	4.56	9.55*E*23	1.16*E*27
$^{124}Sn \rightarrow ^{124}Te$	2288.1±1.6	5.64	5.93*E*17	9.48*E*24
$^{128}Te \rightarrow ^{128}Xe$	868 ±4	31.7	1.18*E*21	1.43*E*26
$^{130}Te \rightarrow ^{130}Xe$	2533 ±4	34.5	2.08*E*17	5.89*E*24
$^{134}Xe \rightarrow ^{134}Ba$	847 ±10	10.4	1.16*E*21	1.30*E*26
$^{136}Xe \rightarrow ^{136}Ba$	2479 ±8	8.9	2.07*E*17	5.52*E*24
$^{142}Ce \rightarrow ^{142}Nd$	1417.6±2.5	11.1	1.38*E*19	2.31*E*25
$^{146}Nd \rightarrow ^{146}Sm$[b]	56 ±5	17.2	2.06*E*29	7.05*E*27
$^{148}Nd \rightarrow ^{148}Sm$[b]	1928.3±1.9	5.7	9.35*E*17	7.84*E*24
$^{150}Nd \rightarrow ^{150}Sm$	3367.1±2.2	5.6	8.41*E*15	1.25*E*24
$^{154}Sm \rightarrow ^{154}Gd$	1251.9±1.5	22.6	2.44*E*19	2.38*E*25
$^{160}Gd \rightarrow ^{160}Dy$	1729.5±1.4	21.8	1.51*E*18	7.99*E*24
$^{170}Er \rightarrow ^{170}Yb$	653.9±1.6	14.9	1.82*E*21	6.92*E*25

Table 6.1, *continued*

$\beta^-\beta^-$ candidates	T_0 (keV)	Abundance (%)	$(G^{2\nu})^{-1}$ (y)	$(G^{0\nu})^{-1}$ (y)
$^{176}Yb \to {}^{176}Hf$	1078.8 ± 2.7	12.6	3.26*E*19	1.75*E*25
$^{186}W \to {}^{186}Os^b$	490.3 ± 2.2	28.6	7.68*E*21	6.95*E*25
$^{192}Os \to {}^{192}Pt$	417 ± 4	41.0	1.98*E*22	7.70*E*25
$^{198}Pt \to {}^{198}Hg$	1048 ± 4	7.2	1.63*E*19	8.74*E*24
$^{204}Hg \to {}^{204}Pb$	416.5 ± 1.1	6.9	1.23*E*22	5.06*E*25
$^{232}Th \to {}^{232}U^b$	858.2 ± 6	100	1.68*E*19	3.97*E*24
$^{238}U \to {}^{238}Pu^b$	1145.8 ± 1.7	99.27	1.47*E*18	1.68*E*24
$\beta^+\beta^+$ candidates	T_0 (keV)	Abundance (%)	$(G^{2\nu})^{-1}$ (y)	$(G^{0\nu})^{-1}$ (y)
$^{78}Kr \to {}^{78}Se$	838	0.35	2.56*E*24	1.8*E*29
$^{96}Ru \to {}^{96}Mo$	676	5.5	3.34*E*25	8.8*E*29
$^{106}Cd \to {}^{106}Pd$	738	1.25	1.69*E*25	7.4*E*29
$^{124}Xe \to {}^{124}Te$	822	0.10	7.57*E*24	5.9*E*29
$^{130}Ba \to {}^{130}Xe$	534	0.11	6.92*E*26	6.4*E*30
$^{136}Ce \to {}^{136}Ba$	362	0.19	5.15*E*28	6.1*E*31

EX signifies 10^x

[a] The single beta decay is kinematically allowed.

[b] The daughter nucleus is unstable against alpha decay.

than 2 MeV, there is only one candidate for the decay (6.5a), $^{124}Xe \to {}^{124}Te$, in which the Q value is more than 1 MeV. As we shall see shortly, the decay rate increases rapidly with Q and chances of observing the decays (6.5) or their neutrinoless analogues are quite remote. Consequently, we concentrate our attention on double beta decays in which the nuclear charge *increases* by two units.

The double beta decays (6.3) and (6.5) were first considered more than half a century ago by Goeppert-Mayer (35). Today, several experiments described below in Section 6.3 result in confirmed observation of the decay (6.3) in a handful of cases. In the early days of nuclear physics, in the late thirties, Furry (39), following the classical papers by Majorana (37) and Racah (37), already realized that decays (6.3) and (6.5) are not the only modes possible. Let us consider the "Racah sequence"

$$n_1 \rightarrow p_1 + e^- + "\nu" ,$$

$$"\nu" + n_2 \rightarrow p_2 + e^- ,$$

which may occur provided there is no difference between ν_e and $\bar{\nu}_e$ (indicated by the notation "ν" above), i.e., if the electron neutrino is a Majorana particle. According to this sequence, the *neutrinoless* or 0ν mode of the double beta decay is possible, differing from (6.3) and (6.5) by the absence of neutrinos (or antineutrinos) in the final state. Thus, for the transition in which the nuclear charge increases by two units, we have

$$(Z,A) \rightarrow (Z+2,A) + e_1^- + e_2^- . \tag{6.6}$$

The decay proceeds by the "Racah sequence" whereby the neutrino emitted in the first step and absorbed in the second one is virtual. Observation of the 0ν $\beta\beta$ decay was recognized as a convincing proof that neutrinos are Majorana particles. Moreover, we shall show below that, under very general circumstances, observation of the 0ν $\beta\beta$ decay implies the existence of *massive* Majorana neutrinos.

6.1.2 *The* 2ν *decay rate formula*

We shall now derive the rate formula for the 2ν mode (6.3), a standard second order weak decay which is independent of neutrino properties. A typical nuclear energy level scheme, with ^{76}Ge chosen as an example, is shown in Figure 6.2. The decay $^{76}\text{Ge} \rightarrow {}^{76}\text{Se} + 2e^- + 2\bar{\nu}$ could proceed according to the diagram in Figure 6.3, where the states in the intermediate odd-odd nucleus $Z+1$ are virtual. This diagram represents the so called two-nucleon mechanism of double beta decay. We shall mention some more exotic possibilities later.

The mathematical derivation of the double beta decay rate formula is analogous to the treatment of ordinary beta decay outlined above. It begins with the Fermi golden rule for second order weak decay

$$d\lambda = 2\pi\delta(E_0 - \sum_f E_f) \left| \sum_{m,\beta} \frac{\langle f | H_\beta | m \rangle \langle m | H^\beta | i \rangle}{E_i - E_m - p_\nu - E_e} \right|^2 . \tag{6.7}$$

Taking into account that the weak Hamiltonian is the product of the nuclear and lepton currents, we obtain

$$d\lambda = 8\pi G_F^4 \cos^4\theta_C \delta(E_0 - \sum_f E_f) \left| \sum_{m,\alpha,\beta} M_{f,m}^{\alpha} M_{m,i}^{\beta} \sum_{n_e n_\nu} (-)^{n_e+n_\nu} \frac{J_\alpha^{n'_e n'_\nu} J_\beta^{n_e n_\nu}}{E_i - E_m - p_{n_\nu} - E_{n_e}} \right|^2 , \tag{6.8}$$

where M^α and J_α denote the four-vectors of the nuclear and lepton currents, respectively, and n_e=1,2 the first and second electron. The index n' denotes the complement of n (if n=1 then n'=2), and m labels the intermediate states. Next, we perform the summation over the lepton polarizations, taking into account the indistinguishability of the final lepton pairs. Because we are interested in the rate formula, we neglect terms linear in $\vec{p}_e$ and $\vec{p}_\nu$ which disappear after integration over angles. (The angular distribution of the electrons is of the form $(1 - \vec{\beta}_1 \cdot \vec{\beta}_2)$ for the $0^+ \rightarrow 0^+$ transitions and $(1 + \vec{\beta}_1 \cdot \vec{\beta}_2/3)$ for the $0^+ \rightarrow 2^+$ transitions where $\vec{\beta} = \vec{p}/E$.) For compactness we introduce the following notation:

$$K_m = \frac{1}{E_m + E_{e1} + p_{\nu 1} - E_i} + \frac{1}{E_m + E_{e2} + p_{\nu 2} - E_i} ,$$

$$M_m = \frac{1}{E_m + E_{e1} + p_{\nu 1} - E_i} - \frac{1}{E_m + E_{e2} + p_{\nu 2} - E_i} ,$$

$$L_m = \frac{1}{E_m + E_{e1} + p_{\nu 2} - E_i} + \frac{1}{E_m + E_{e2} + p_{\nu 1} - E_i} ,$$

Figure 6.2. States participating in the double beta decay of ^{76}Ge. Principal decay branches are indicated.

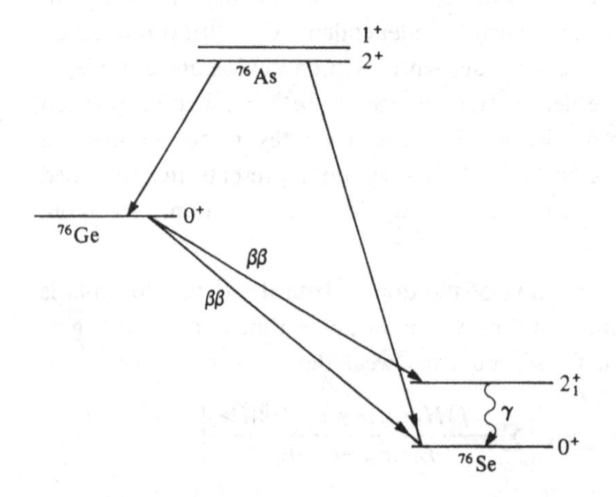

$$N_m = \frac{1}{E_m+E_{e1}+p_{\nu 2}-E_i} - \frac{1}{E_m+E_{e2}+p_{\nu 1}-E_i} .$$

After summing (6.8) over α and β we obtain an expression, denoted by X, which contains all quantities depending on the nuclear states. For the transition between $|i\rangle = 0^+$ and $|f\rangle = 0^+$ we find

$$X=\frac{1}{4}\sum_{m,m'}[\, g_V^4 \langle f|\tau^+|m\rangle\langle m|\tau^+|i\rangle\langle f|\tau^+|m'\rangle^*\langle m'|\tau^+|i\rangle^*$$
$$\times\, [K_mK_{m'}+L_mL_{m'}-\frac{1}{2}\,(K_mL_{m'}+L_mK_{m'})]$$
$$-\, g_V^2g_A^2\mathrm{Re}(\langle f|\tau^+|m\rangle\langle m|\tau^+|i\rangle\langle f|\sigma\tau^+|m'\rangle^*\langle m'|\sigma\tau^+|i\rangle^*$$
$$\times\,(K_mL_{m'}+L_mK_{m'})$$
$$+\, g_A^4\langle f|\sigma\tau^+|m\rangle\langle m|\sigma\tau^+|i\rangle\langle f|\sigma\tau^+|m'\rangle^*\langle m'|\sigma\tau^+|i\rangle^*$$
$$\times\,[K_mK_{m'}+L_mL_{m'}+\frac{1}{2}\,(K_mL_{m'}+L_mK_{m'})\,]\,] . \tag{6.9}$$

For the other situation of interest, namely when $|i\rangle = 0^+$ and $|f\rangle = 2^+$ we obtain

$$X=\frac{1}{4}g_A^4\sum_{m,m'}\langle f|\sigma\tau^+|m\rangle\langle m|\sigma\tau^+|f\rangle\langle f|\sigma\tau^+|m'\rangle^*\langle m'|\sigma\tau^+|i\rangle^*$$
$$\times\,(K_m-L_m)(K_{m'}-L_{m'}) . \tag{6.10}$$

To calculate the total rate we have to multiply X by the lepton phase space volume and perform the corresponding integration over the unobserved variables. Finally, we obtain the decay rate as

Figure 6.3. Schematic diagram for the 2ν mode of double beta decay in the two-nucleon mechanism.

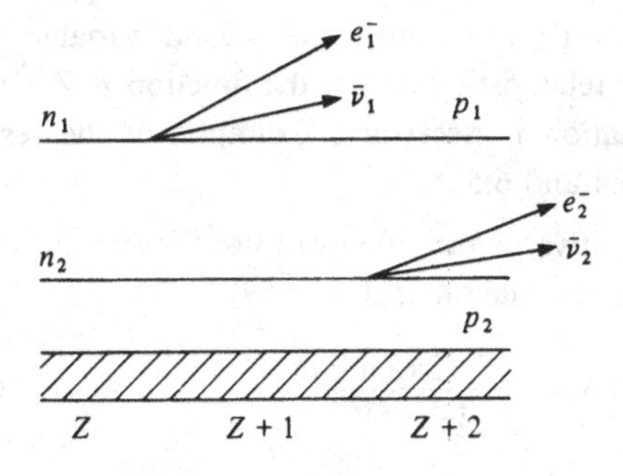

$$\omega_{2\nu} = \frac{G_F^4 \cos^4\theta_C}{8\pi^7} \int_{m_e}^{E_0-m_e} F(Z,E_{e1}) p_{e1} E_{e1} dE_{e1} \int_{m_e}^{E_0-E_{e1}} F(Z,E_{e2}) p_{e2} E_{e2} dE_{e2}$$

$$\times \int_0^{E_0-E_{e1}-E_{e2}} X p_{\nu 1}^2 \, (E_0-E_{e1}-E_{e2}-p_{\nu 1})^2 dp_{\nu 1} \, . \qquad (6.11)$$

The energy denominators in the factors K, M, L and N contain contributions of the nuclear energies $E_m - E_i$, as well as the lepton energies $E_e + p_\nu$. When calculating the $0^+ \rightarrow 0^+$ transitions, it is generally a very good approximation to replace these lepton energies by the corresponding average value, i.e., $E_e + p_\nu \sim E_0/2$. In that case

$$K_m \sim L_m \sim \frac{1}{E_m - E_i + E_0/2} = \frac{1}{E_m - (M_i + M_f)/2} \, , \qquad (6.12a)$$

$$M_m \sim N_m \sim 0 \, . \qquad (6.12b)$$

We note that this approximation is unacceptable, however, for the $0^+ \rightarrow 2^+$ transitions where we have to use

$$K_m - L_m = \frac{2(p_{\nu 1} - p_{\nu 2})(E_{e1} - E_{e2})}{(E_m - E_i + E_0/2)^3} \, . \qquad (6.13)$$

The antisymmetry in the electron and antineutrino energies suggests strong suppression of the $0^+ \rightarrow 2^+$ 2ν $\beta\beta$ decay (Haxton & Stephenson 84 and Doi et al. 85) and for that reason we shall not discuss this transition further.

With the approximation (6.12) the nuclear and leptonic degrees of freedom are separated from each other. The last integral in (6.11) is now proportional to

$$\frac{1}{30} (E_0 - E_{e1} - E_{e2})^5 \, .$$

The single electron spectrum is obtained by performing in (6.11) integration over $dp_{\nu 1}$ as well as integration over dE_{e2}, while the spectrum of summed electron energies is obtained by changing the variables to $E_{e1} + E_{e2}$ and $E_{e1} - E_{e2}$ and performing the integration over the second variable. If an accurate result is required, the relativistic form of the function $F(Z,E)$ must be used and numerical evaluation is necessary. Examples of the resulting spectra are shown in Figures 6.4 and 6.5.

For a qualitative and intuitive picture, one can use the simplified non-relativistic Coulomb expression (Primakoff & Rosen 59)

$$F(Z,E) = \frac{E}{p} \frac{2\pi Z\alpha}{1 - e^{-2\pi Z\alpha}} \, . \qquad (6.14)$$

This approximation allows us to perform the required integrals analytically. The single electron spectrum is now of the form

$$\frac{dN}{dT_e} \sim (T_e+1)^2 \, (T_0-T_e)^6 \, [(T_0-T_e)^2 + 8(T_0-T_e) + 28] \, , \qquad (6.15)$$

where T_e is the electron kinetic energy in units of electron mass and $T_0 = E_0 - 2$ is the maximal kinetic energy. The sum electron spectrum, which is of primary interest from the experimental point of view, is of the form

$$\frac{dN}{dK} \sim K(T_0-K)^5 \left[1 + 2K + \frac{4K^2}{3} + \frac{K^3}{3} + \frac{K^4}{30} \right] , \qquad (6.16)$$

where K is the sum of the kinetic energies of both electrons, again in units of electron mass. Finally, by integrating (6.15) over T_e, we find the dependence of the total rate on T_0, which in this approximation is independent of the nuclear charge Z,

Figure 6.4. Electron spectrum for 2ν ββ decay of ^{136}Xe. The exact spectrum is shown by the full curve, the Primakoff-Rosen approximation (6.16) by the dashed curve. Both spectra are normalized to the same unit area. The independent variable is the total kinetic energy K of the electrons divided by its maximum value T_0.

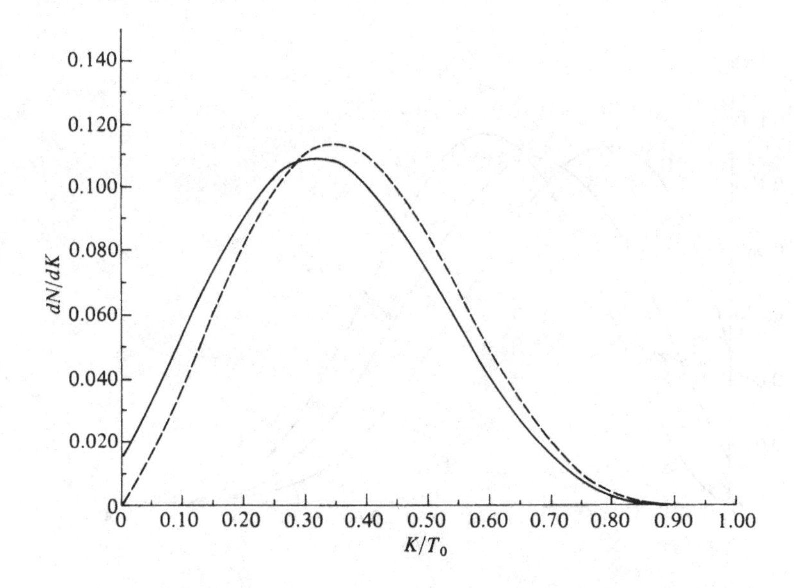

$$\omega_{2\nu} \sim T_0^7 \left(1 + \frac{T_0}{2} + \frac{T_0^2}{9} + \frac{T_0^3}{90} + \frac{T_0^4}{1980}\right) . \tag{6.17}$$

Examples of the sum spectra are shown in Figures 6.5 and 6.4 where the exact and approximate spectra are compared. In Figure 6.6 we show, for comparison, the single electron spectra. Note that the Primakoff-Rosen approximation (6.14) describes the shapes of the single electron and sum spectra reasonably well. It fails, however, in the absolute rate calculation.

Now we are in a position to use (6.12a) and express the (half-life)$^{-1}$ of the decay as

$$\left[T_{\frac{1}{2}}^{2\nu}\,(0^+ \rightarrow 0^+)\right]^{-1} = G^{2\nu}(E_0,Z) \left| M_{GT}^{2\nu} - \frac{g_V^2}{g_A^2} M_F^{2\nu} \right|^2 , \tag{6.18}$$

where the function $G^{2\nu}(E_0,Z)$ results from lepton phase space integration and contains all relevant constants. The quantity tabulated in Table 6.1 is $(G^{2\nu})^{-1}$. (The units are such that the half-life is in years provided the energy denominators in (6.19a) and (6.19b) are in units of electron mass.) The

Figure 6.5. Electron spectra for 2ν $\beta\beta$ decay of ^{48}Ca ($T_0 = 4.3$ MeV), ^{136}Xe ($T_0 = 2.5$ MeV), and ^{238}U ($T_0 = 1.1$ MeV) in terms of the total kinetic energy K of the two electrons divided by its maximum value T_0. All spectra are normalized to the same area. Exact spectra are shown; the differences in shape are partially due to different T_0 values and to differences in Coulomb effects.

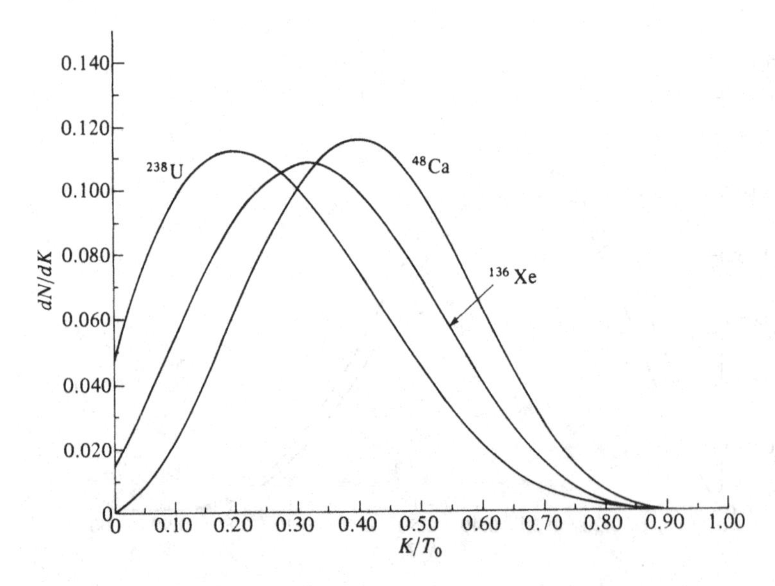

corresponding quantity for the double positron decay candidates is shown at the bottom of Table 6.1.

Besides the double positron decay, Eq. (6.5a), the decays (6.5b) and (6.5c) also result in the transition $Z \rightarrow Z-2$. The phase space factors for these two processes, which are accompanied by the K-electron capture, are given in Table 6.2. The entries were calculated for $g_A = 1.25$ using the relativistic treatment, to the order $1-(1-(Z\alpha)^2)^{\frac{1}{2}}$, of the bound K-electron and the outgoing positron. (We are not aware of similar complete calculation elsewhere.) Comparing the entries in Table 6.1 to the corresponding ones in Table 6.2 one can see that the 2ν decays involving the K-electron capture should be much faster than the ones with the double positron emission. The experimental difficulties associated with the observation of the processes (6.5b) and (6.5c) are, however, obvious.

The nuclear structure information is contained in the matrix elements $M_{GT}^{2\nu}$ and $M_F^{2\nu}$,

$$M_{GT}^{2\nu} = \sum_m \frac{\langle 0_f^+ | \sum_l \vec{\sigma}_l \tau_l^+ | m \rangle \cdot \langle m | \sum_k \vec{\sigma}_k \tau_k^+ | 0_i^+ \rangle}{E_m - (M_i + M_f)/2} , \tag{6.19a}$$

Figure 6.6. Single electron spectra for the indicated 2ν $\beta\beta$ decays. Exact spectra, all normalized to the same area are shown. The independent variable is the kinetic energy divided by its maximum value.

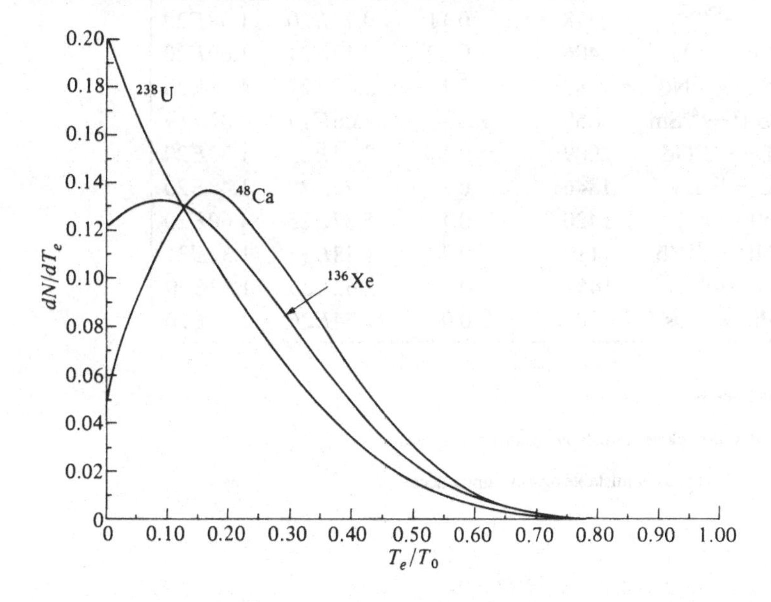

Table 6.2. *Phase space factors for the* 2ν *decay involving positron emission and K electron capture. (Only candidate nuclei with* $\Delta M^{atomic} \geqslant 2m_e$ *are listed.) The half-life for the corresponding decay branch is given by Eq. (6.18).*

$Z \rightarrow Z-2$ candidates	ΔM^{atomic} (keV)	Abund. (%)	$(G^{2\nu}_{\beta^+ K})^{-1}$ (y)	$(G^{2\nu}_{KK})^{-1}$ (y)
$^{50}Cr \rightarrow {}^{50}Ti$	1174	4.4	3.89*E*29	4.62*E*24
$^{58}Ni \rightarrow {}^{58}Fe$	1928	68.1	4.08*E*23	1.36*E*23
$^{64}Zn \rightarrow {}^{64}Ni$	1097	48.6	1.07*E*32	1.46*E*24
$^{74}Se \rightarrow {}^{74}Ge$	1209	0.9	4.88*E*28	3.81*E*23
$^{78}Kr \rightarrow {}^{78}Se$	2882	0.354	1.05*E*21	3.00*E*21
$^{84}Sr \rightarrow {}^{84}Kr$	1790	0.56	6.65*E*23	2.29*E*22
$^{92}Mo \rightarrow {}^{92}Zr$	1649	14.8	2.50*E*24	1.71*E*22
$^{96}Ru \rightarrow {}^{96}Mo$	2720	5.5	1.17*E*21	9.54*E*20
$^{102}Pd \rightarrow {}^{102}Ru$	1175	1.0	5.72*E*29	4.48*E*22
$^{106}Cd \rightarrow {}^{106}Pd$	2782	1.25	6.28*E*20	3.65*E*20
$^{112}Sn \rightarrow {}^{112}Cd$	1920	1.0	9.45*E*22	1.74*E*21
$^{120}Te \rightarrow {}^{120}Sn$	1698	0.1	7.81*E*23	2.37*E*21
$^{124}Xe \rightarrow {}^{124}Te$	2866	0.10	2.91*E*20	1.17*E*20
$^{130}Ba \rightarrow {}^{130}Xe$	2578	0.11	9.76*E*20	1.58*E*20
$^{136}Ce \rightarrow {}^{136}Ba$	2406	0.19	2.13*E*21	1.69*E*20
$^{144}Sm \rightarrow {}^{144}Nd^{b}$	1782	3.1	2.07*E*23	4.17*E*20
$^{148}Gd^{a} \rightarrow {}^{148}Sm$	3068	—	7.06*E*19	1.82*E*19
$^{156}Dy \rightarrow {}^{156}Gd$	2009	0.1	2.03*E*22	1.32*E*20
$^{162}Er \rightarrow {}^{162}Dy$	1846	0.1	7.75*E*22	1.48*E*20
$^{168}Yb \rightarrow {}^{168}Er$	1420	0.1	5.37*E*25	4.60*E*20
$^{174}Hf^{a} \rightarrow {}^{174}Yb$	1110	0.2	1.48*E*37	1.32*E*21
$^{184}Os^{a} \rightarrow {}^{184}W^{b}$	1454	0.02	2.62*E*25	1.83*E*20
$^{190}Pt^{a} \rightarrow {}^{190}Os$	1380	0.01	1.74*E*26	1.79*E*20

EX signifies 10^x

[a] The initial nucleus is unstable against alpha decay.

[b] The final nucleus is unstable against alpha decay.

$$M_F^{2\nu} = \sum_m \frac{<0_f^+|\sum_l \tau_l^+|m><m|\sum_k \tau_k^+|0_i^+>}{E_m-(M_i+M_f)/2} . \quad (6.19b)$$

We shall discuss the problems associated with the evaluation of the nuclear matrix elements in Section 6.2.

6.1.3 *Rate formula for* 0ν *decay*

In analogy to the diagram in Figure 6.3 for the 2ν mode, we show in Figure 6.7 how neutrinoless double beta decay proceeds in the two-nucleon mechanism. Within the minimum standard model this neutrinoless "Racah sequence" is strictly forbidden, even for Majorana neutrinos. As evident from Figure 6.7(a), the emitted "neutrino" is dominantly right-handed while the absorbed "neutrino" is dominantly left-handed. Hence we are dealing with a helicity mismatch and the 0ν decay cannot take place. Similarly, in Figure 6.7(b) the second vertex, representing coupling to the right-handed lepton current, is absent in the minimal standard model.

Thus, the 0ν ββ decay exists only if *the lepton number is not conserved*, i.e., if $\nu_e = \bar{\nu}_e$ and, at the same time, if *both neutrinos have the same helicity component*. The lepton sector of the minimal standard model must then be modified. The helicity matching condition could be satisfied in two ways:

First, if neutrinos are massive, there is a "wrong" helicity component with an amplitude proportional to m_ν / E_ν . With $E_\nu \sim m_e$ the rate for 0ν

Figure 6.7. Schematic diagram of the 0ν mode of double beta decay in the two-nucleon mechanism. Arrows represent the main neutrino helicity in each vertex. Part (a) is for left-handed lepton current in both vertices, part (b) is for left-right current interference.

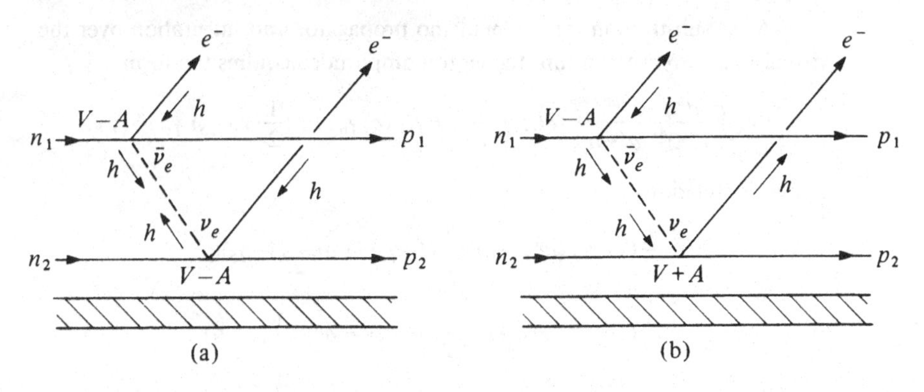

decay would be proportional to m_ν^2/m_e^2. (The precise meaning of the mass parameter m_ν depends on the neutrino mixing and will be defined later.)

Second, the helicity matching condition could also be satisfied if there is a right-handed-current weak interaction as indicated in Figure 6.7(b). The rate for 0ν $\beta\beta$ decay is then proportional to the square of some characteristic coupling strength of the right-handed current. However, existence of the right-handed current is not sufficient because, generally, the emitted and absorbed neutrinos are different. Hence the decay rate, besides being dependent on parameters characteristic of the right-handed current (such as those related to the properties of the hypothetical intermediate vector boson W_R which couples to the right-handed current) will also depend on the corresponding neutrino mixing parameters, and will vanish in the case of unmixed neutrinos.

Case of $m_\nu \neq 0$. We shall now derive the electron spectra and decay rates associated with the nonvanishing value of m_ν. The decay rate is of the general form

$$\omega_{0\nu} = 2\pi \sum_{spin} |R_{0\nu}|^2 \delta(E_{e1} + E_{e2} + E_f - M_i)\, d^3p_{e1} d^3p_{e2} \,, \qquad (6.20)$$

where E_f is the energy of the final nucleus and $R_{0\nu}$ is the transition amplitude including both the lepton and nuclear parts.

The lepton part of the amplitude is written as a product of two left- or right-handed currents

$$\bar{e}(x)\gamma_\rho \frac{1}{2}(1 \pm \gamma_5)\, \nu_j(x) \bar{e}(y)\gamma_\sigma \frac{1}{2}(1 \pm \gamma_5)\, \nu_k(y) \,,$$

where, ν_j, ν_k represent neutrinos of flavor j and k, and there is a contraction over the two neutrino operators. The $\pm$ signs correspond to the right- and left-handed lepton currents, respectively. The contraction above is allowed only if the neutrinos are Majorana particles.

After substitution for the neutrino propagator and integration over the virtual neutrino momentum, the lepton amplitude acquires the form

$$-i\delta_{jk} \int \frac{d^4q}{(2\pi)^4} \frac{e^{-iq(x-y)}}{q^2 - m_j^2} \bar{e}(x)\gamma_\rho \frac{1}{2}(1 \pm \gamma_5)\, (q^\mu \gamma_\mu + m_j) \frac{1}{2}(1 \pm \gamma_5)\, \gamma_\sigma\, e^C(y) \,.$$

Using the relations

$$\frac{1}{2}(1-\gamma_5)\, (q^\mu\gamma_\mu + m_j) \frac{1}{2}(1-\gamma_5) = m_j \frac{1}{2}(1-\gamma_5) \,,$$

$$\frac{1}{2}(1-\gamma_5)\, (q^\mu\gamma_\mu + m_j) \frac{1}{2}(1+\gamma_5) = q^\mu \gamma_\mu \frac{1}{2}(1+\gamma_5) \,,$$

we see that for purely left-handed currents (the upper expression) only the m_ν part of the neutrino propagator contributes, while for the left-right current interference (the lower expression) the lepton amplitude contains the part of the neutrino propagator proportional to the virtual four-momentum q.

Integration over the virtual neutrino energy leads to replacement of the propagator $(q^2-m_j^2)^{-1}$ by the residue π/ω_j $(\omega_j = (\vec{q}^2+m_j^2)^{1/2})$. For the remaining integration over the space part $d\vec{q}$ we have to consider, besides this denominator ω_j, the energy denominators of the second order perturbation expression, analogous to (6.8). Denoting

$$A_{n_e} = E_m - E_i + E_{n_e} ,$$

we find that integration over $d\vec{q}$ leads to an expression representing the effect of the neutrino propagation between the two nucleons. This expression has the form of a "neutrino potential" and appears in the corresponding nuclear matrix elements, introducing dependence of the transition operator on the coordinates of the two nucleons, as well as a weak dependence on the excitation energy $E_m - E_i$ of the virtual state. The neutrino potential is of the form

$$H_n(r,E_m) = \frac{R}{2\pi^2} \int \frac{d\vec{q}}{\omega} \frac{1}{\omega+A_n} e^{i\vec{q}\cdot\vec{r}} = \frac{2R}{\pi r} \int_0^\infty dq \frac{q\sin(qr)}{\omega\,(\omega+A_n)} . \tag{6.21}$$

(We have added the nuclear radius $R=1.2A^{1/3}$ as an auxiliary factor so that H becomes dimensionless.) The first factor ω in the denominator of (6.21) is the residue, while the factor $\omega+A_n$ is the energy denominator of perturbation theory. To obtain the final result one has to treat properly the antisymmetry between the identical outgoing electrons (see Doi et al. 85).

We expect that the momentum of the virtual neutrino is determined by the uncertainty relation $q \sim 1/r \sim 100$ MeV, where r is a typical spacing between two nucleons. Thus, to a reasonable approximation we can neglect A_n with respect to ω in (6.21) and obtain

$$H_n(r,E_m) \equiv H(r) = \frac{R}{r} e^{-rm_j} , \tag{6.21a}$$

independent of the energy of the intermediate nucleus E_m and the electron energy E_{n_e}. Note that in the present case, $m_j \neq 0$, the contributions of both electrons $n=1$ and 2 add coherently. The potential H depends weakly on the neutrino mass m_j for values below ~ 10 MeV, and is considerably damped for larger neutrino masses. (For very heavy neutrinos, $\gtrsim 1$ GeV, a different procedure is necessary.)

From the discussion above it is seen that for the case of left-right current interference the "neutrino potentials" contain extra factors ω or $\vec{q}$ in the integrals analogous to (6.21). The part proportional to ω does not represent difficulties in principle. The $\vec{q}$-part is more complicated as it represents an operator which is odd under parity transformation. Consequently, to obtain a nonvanishing value for the corresponding matrix elements we have to consider either the p-wave of the outgoing leptons or nuclear recoil as discussed in the following Subsection.

We now have all the ingredients needed to calculate the 0ν $\beta\beta$ decay rate associated with nonvanishing neutrino Majorana mass. The nuclear matrix elements have similar structures to those in (6.19) for the 2ν mode. Now, however, it is a better approximation to neglect variations of the excitation energies in the intermediate nucleus compared to the rather high energy of the virtual neutrino. Consequently, the sum over intermediate states can be performed by closure and we obtain

$$M_{GT}^{0\nu} = \langle f|\sum_{lk} \vec{\sigma}_l \cdot \vec{\sigma}_k \tau_l^+ \tau_k^+ H(r_{lk},\bar{E}_m)|i\rangle \; , \tag{6.22a}$$

$$\sim \langle f|R \sum_{lk} \vec{\sigma}_l \cdot \vec{\sigma}_k \tau_l^+ \tau_k^+ / r_{lk}|i\rangle \; , \tag{6.22b}$$

as well as

$$M_{F}^{0\nu} = \langle f|\sum_{lk} \tau_l^+ \tau_k^+ H(r_{lk},\bar{E}_m)|i\rangle \; , \tag{6.23a}$$

$$\sim \langle f|R \sum_{lk} \tau_l^+ \tau_k^+ / r_{lk}|i\rangle \; . \tag{6.23b}$$

In Eqs. (6.22) and (6.23) the l,k summation is over all nucleons of the nucleus, r_{lk} is the distance between the nucleons l and k, and $\bar{E}_m$ is an average excitation energy. The expressions (6.22b) and (6.23b) are valid in the approximations discussed above ($\omega >> A_{n_e}$ and for neutrino masses $m_\nu <$ 10 MeV). With these definitions, the (half-life)$^{-1}$ is of the form

$$\left[T_{\frac{1}{2}}^{0\nu}(0^+ \rightarrow 0^+)\right]^{-1} = G^{0\nu}(E_0,Z)\left| M_{GT}^{0\nu} - \frac{g_V^2}{g_A^2} M_F^{0\nu} \right|^2 \langle m_\nu\rangle^2 , \tag{6.24}$$

where the function $G^{0\nu}(E_0,Z)$ is the result of the two-electron phase space integration and is proportional to

$$G^{0\nu} \sim \int F(Z,E_{e1})F(Z,E_{e2})p_{e1}p_{e2}E_{e1}E_{e2}\, \delta(E_0-E_{e1}-E_{e2})dE_{e1}dE_{e2} \; . \tag{6.25}$$

Again, in the Primakoff-Rosen approximation (6.14), $G^{0\nu}$ is independent of Z and

$$G_{PR}^{0\nu} \sim \left(\frac{E_0^5}{30} - \frac{2E_0^2}{3} + E_0 - \frac{2}{5} \right) . \tag{6.25a}$$

Exact values of the phase space function $G^{0\nu}$ which contain all the relevant coupling constants are given in Table 6.1. (The tabulated quantity is $(G^{0\nu})^{-1}$, and the units are such that if the neutrino mass parameter $<m_\nu>$ is in eV, the half-life is in years.) The entries in Table 6.1, for both double beta decay modes, can be used as a convenient yardstick to judge the sensitivity of the different double beta decay candidates to the neutrino mass $<m_\nu>$, as well as the expected half-life for the 2ν mode. We use the notation $<m_\nu>$ to stress that this quantity is an effective neutrino Majorana mass.

The sum electron spectrum of the 0ν mode is very simple; it is just a δ-function peak at the endpoint energy E_0. This feature makes distinction between the 0ν and 2ν experimentally feasible. Each of the two electrons observed separately will have an energy spectrum determined by the phase space

$$\frac{dN}{dT_{e1}} \sim F(Z,E_{e1})F(Z,E_{e2})p_{e1}p_{e2}E_{e1}E_{e2} , \tag{6.26}$$

where $E_{e2} = E_0 - E_{e1}$. In the Primakoff-Rosen approximation (6.14) we obtain

$$\frac{dN}{dT_e} \sim (T_e + 1)^2(T_0 + 1 - T_e)^2 . \tag{6.26a}$$

In Figure 6.8 we show examples of the exact single electron spectra. Again, the Primakoff-Rosen approximation describes the spectrum shape reasonably well, but is less reliable for the calculation of the absolute decay rate.

It is of interest to contrast the 0ν and 2ν decay modes from the point of view of the phase space integrals. The 0ν mode has the advantage of the two-lepton final state, with the characteristic E_0^5 dependence in (6.25a) compared to the four-lepton final state with E_0^{11} dependence in (6.17) for the 2ν mode. In addition, the large average momentum of the virtual neutrino, compared with the typical nuclear excitation energy also makes the 0ν decay faster. Thus, if $<m_\nu>$ were to be of the order of m_e, the 0ν decay would be $\sim 10^5$ times faster than the 2ν decay. It is this phase space advantage which makes the 0ν ββ decay a sensitive probe for Majorana neutrino mass.

To conclude this discussion, let us point out that the two-nucleon mechanism of double beta decay used here is not necessarily the only one possible. Straightforward generalization involves internal excitation of nucleons, the Δ isobars, which are known to play an important role in nuclei. We discuss the possible role of Δ isobars in Section 6.2. Other mechanisms have also been considered, for example, transitions involving virtual Higgs scalars. (Such neutrinoless double beta decay would not involve exchange of a virtual neutrino.) It is also possible that, in addition to the two electrons, a light boson, the so called majoron, is emitted. This transition would have three-body phase space, giving rise to a continuous spectrum peaked at approximately three quarters of the decay energy T_0. We refer to Doi et al. (85) for a discussion of these topics.

Case of right-handed currents. We shall consider now the 0ν decay caused by weak interactions containing coupling to the right-handed lepton currents. We begin with a proper parametrization of the weak interaction Hamiltonian. The general Hamiltonian describing semileptonic weak interactions at low (below the W boson mass) energies can be written as

$$H_W = \frac{G}{\sqrt{2}}\left[J_L^\alpha (M_{L\alpha}^+ + \kappa M_{R\alpha}^+) + J_R^\alpha (\eta M_{L\alpha}^+ + \lambda M_{R\alpha}^+)\right] + h.c. \,, \quad (6.27)$$

where $J_{L(R)}^\alpha$ and $M_{L(R)}^\alpha$ are lepton and quark left (right)-handed current four-vectors, respectively. The deviation from the minimal standard model is characterized by the dimensionless parameters η, λ, and κ. The parameter η describes the coupling between the right-handed lepton current and left-handed quark current, λ describes the coupling between right-handed lepton current and right-handed quark current, and κ describes the coupling between the right-handed quark current and left-handed lepton current. (Note that the standard coupling of the left-handed quark and lepton currents has by definition a coupling constant equal to unity.) In more general models the parameters λ, η, and κ have nonvanishing and specific values. For example, in the left-right symmetric model $SU(2)_L \times SU(2)_R \times U(1)$ they are

$$\kappa = \eta = -\tan\zeta \quad ; \quad \lambda \sim (M_{WL}/M_{WR})^2 \,,$$

where ζ is the mixing angle of the intermediate vector bosons W.

When the quark current is transformed (using the impulse approximation) to the nucleon current, it turns out that the parameter κ gives a negligible contribution to the double beta decay and only the parameters λ and η are relevant, i.e., as expected, one can obtain information on the parameters involving right-handed lepton currents only.

To proceed further we have to consider the $0^+ \to 0^+$ and $0^+ \to 2^+$ transitions separately. As mentioned earlier, in both cases there is an additional factor in the integration over $d\vec{q}$ in the integrals analogous to (6.21), either a factor ω (the so called ω-part) or $\vec{q}$ (the so called q-part).

The ω-part has the same selection rules as in the m_ν case, i.e., both emitted electrons are in the s-state and only $0^+ \to 0^+$ transitions are possible. Unlike the m_ν case, however, the resulting expression is odd under the exchange of the two electrons. Consequently, the corresponding phase space integral contains the factor $(E_{e1}-E_{e2})^2$. In Figure 6.9 we show an example of the single electron spectrum associated with the ω-part. (Note the node in the middle when the two electrons share the available energy equally.) Analysis (Doi et al. 85) suggests that the ω-part would give major contributions only if the parameter λ in the Hamiltonian (6.27) were responsible for the double beta decay. Besides the nuclear matrix elements similar to (6.22a) and (6.23a), with slightly modified radial dependence, the ω-part depends also on the "tensor" matrix element

$$M_T^{0\nu} = \langle f | \sum_{lk} \frac{R}{r_{lk}^3} (\vec{\sigma}_l \cdot \vec{r}_{lk}) (\vec{\sigma}_k \cdot \vec{r}_{lk}) \tau_l^+ \tau_k^+ | i \rangle \ . \qquad (6.28)$$

The q-part contains a vectorlike neutrino potential which changes parity. Therefore, it is no longer possible to stay within the allowed approximation. Two principal terms have to be considered for the $0^+ \rightarrow 0^+$ transitions. In the first of these terms one of the electrons is in the p-state. Nominally, the p-state is characterized by a small parameter $p_e R \sim 1/40$. However, for the $p_{1/2}$ electron states, this term is strongly enhanced by the Coulomb interaction, leading to the replacement of the parameter $p_e R$ by the much

Figure 6.8. Single electron spectra of the indicated 0ν $0^+ \rightarrow 0^+$ ββ decays for the case of $\langle m_\nu \rangle \neq 0$. The spectra are normalized to the same value. The independent variable is the electron kinetic energy T_e divided by its maximum value T_0.

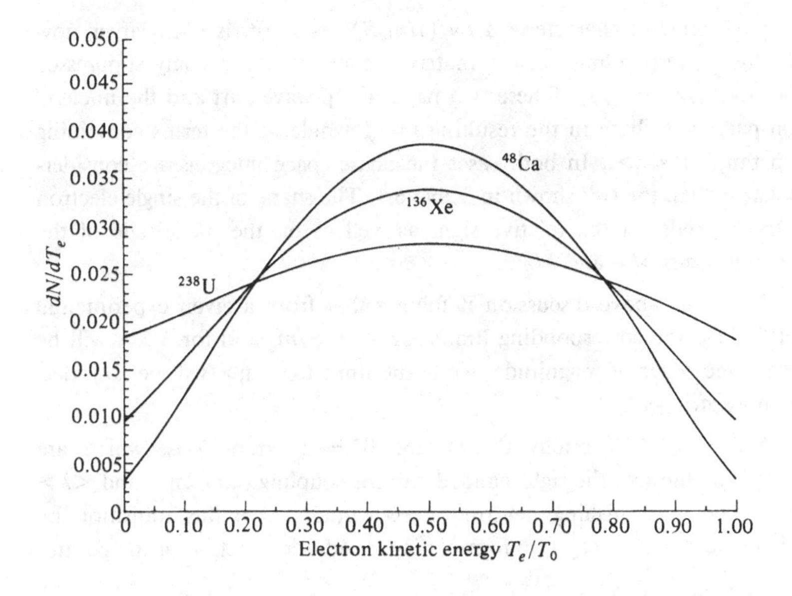

larger factor $Z\alpha$. The corresponding "neutrino potential" has almost the same form as in Eq. (6.21), i.e.,

$$H' \sim -r\frac{d}{dr}H \sim \frac{R}{r} ,$$

and the nuclear matrix element is of the form

$$M_P^{0\nu} = g_V < f | \sum_{lk} iH'(r_{lk}) \frac{|\vec{r}_l + \vec{r}_k|}{2r_{lk}} \left[(\vec{\sigma}_l - \vec{\sigma}_k) \cdot \frac{\vec{r}_{lk}}{r_{lk}} \times \frac{\vec{r}_l + \vec{r}_k}{|\vec{r}_l + \vec{r}_k|} \right] \tau_l^+ \tau_k^+ | i > . \quad (6.29)$$

The enhancement, in comparison to the ω-part, is characterized by the factor $(Z\alpha/m_e R)^2 \sim 18^2$ (Z=32) in the corresponding phase space integrals. The other principal contribution to the q-term is associated with the nucleon recoil, enhanced by the relatively large value of the virtual neutrino momentum. The corresponding "neutrino potential" is now

$$H_R \sim -\frac{R}{M}\frac{d^2}{dr^2}H(r) , \quad (6.30)$$

and the nuclear matrix element is

$$M_R^{0\nu} = g_V < f | \sum_{lk} H'(r_{lk}) \frac{R}{2r_{lk}} \left[\frac{\vec{r}_{lk}}{r_{lk}} \cdot (\vec{\sigma}_l \times \vec{D}_k + \vec{D}_l \times \vec{\sigma}_k) \right] \tau_l^+ \tau_k^+ | i > . \quad (6.31)$$

The vector $\vec{D}_k$ depends on the initial and final momenta $\vec{P}_k$, $\vec{P}_k'$ of the nucleon k through

$$\vec{D}_k = (\vec{P}_k + \vec{P}_k') - [1 - i(\mu_p - \mu_n) \vec{\sigma}_k \times (\vec{P}_k - \vec{P}_k')] / (2M) .$$

The enhancement, characterized by $(1/m_e R)^2 \sim (75)^2$, is even larger now while the corresponding nuclear matrix element is not strongly suppressed ($|D| \sim 4.7q/M \sim 1/5$). These two parts, the p-wave part and the nucleon recoil part, contribute in the resulting rate formula to the terms containing the parameter $<\eta>$. In both cases the phase space integrals are considerably larger than the $G^{0\nu}$ shown in Table 6.1. The shape of the single electron spectra depends on the relative sign, as well as on the magnitude of the matrix elements $M_P^{0\nu}$ and $M_R^{0\nu}$.

From the above discussion it follows that from a given experimental limit for $T_{0\nu}$ the corresponding limits for $<m_\nu>/m_e$ and for $<\lambda>$ will be of the same order of magnitude, while the limit for $<\eta>$ will be considerably more stringent.

Now, we shall briefly discuss the $0^+ \rightarrow 2^+$ transitions, which are present only through the right-handed current coupling (the $<\eta>$ and $<\lambda>$ terms). The nonvanishing neutrino mass cannot cause this transition by itself. The $0^+ \rightarrow 2^+$ decay rate formula is a quadratic function of the parameters $<\eta>$ and $<\lambda>$ with coefficients depending on new nuclear matrix elements and on new phase space integrals. Dominant contributions come from the case in which one of the emitted electrons is in the $p_{3/2}$-state and the other one in the s-state. The phase space integrals have a characteristic E_0^7 dependence in the Primakoff-Rosen approximation. There are four nuclear matrix elements (see Haxton & Stephenson (84); Doi et al. (85)). The corresponding operators are now tensors of the second rank. For example, one of these matrix elements is

$$M_2^{0\nu} = <2_f^+|\sum_{lk} H'(r_{lk})\, \vec{\sigma}_l \cdot \vec{\sigma}_k\, [\vec{r}_{kl}\times\vec{r}_{kl}]^{(2)}/r_{kl}^2\, \tau_k^+\tau_k^+|0_i^+> \,. \qquad (6.32)$$

Observation of the $0^+ \rightarrow 2^+$ transition would be a clear demonstration of the presence of right-handed lepton currents. In Section 6.4 we shall show that, in addition, neutrino mixing is required in order that the parameters $<\eta>$ and $<\lambda>$ have nonvanishing values.

Summary of results. Combining contributions from the mass and right-handed current term, we arrive at a general formula of the form

Figure 6.9. Single electron spectra of the indicated 0ν $0^+ \rightarrow 0^+$ $\beta\beta$ decays for the "ω-part" of the right-handed current mechanism. Exact spectra, all normalized to the same value, are shown. The independent variable is the electron kinetic energy T_e divided by its maximum value T_0.

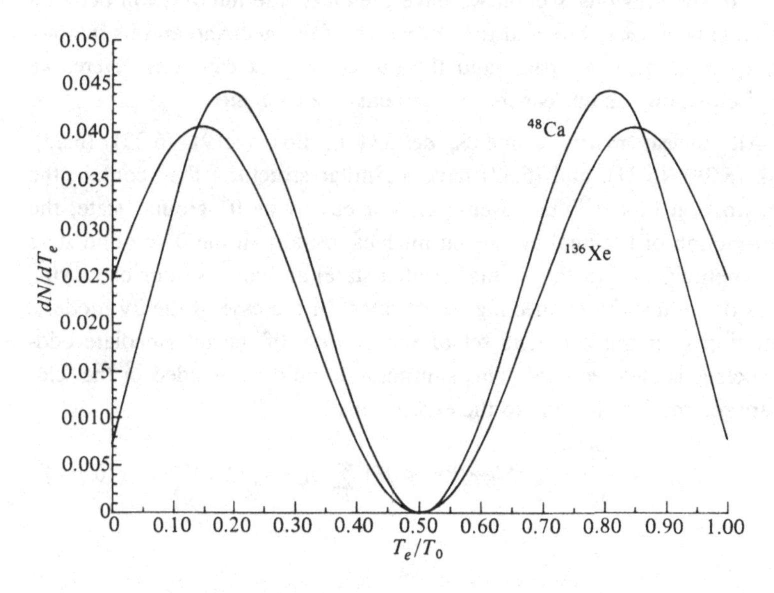

$$\left[T_{\frac{1}{2}}^{0\nu}(0^+\rightarrow 0^+)\right]^{-1}=C_1\frac{<m_\nu>^2}{m_e^2}+C_2<\lambda>\frac{<m_\nu>}{m_e}\cos\psi_1+C_3<\eta>\frac{<m_\nu>}{m_e}\cos\psi_2$$

$$+C_4<\lambda>^2+C_5<\eta>^2+C_6<\lambda><\eta>\cos(\psi_1-\psi_2)\,. \qquad (6.33)$$

Here ψ_1 and ψ_2 are phase angles between the generally complex numbers m_ν and λ, and m_ν and η, respectively (if *CP* invariance is assumed, ψ_1 and ψ_2 are 0 or π), while $< >$ denotes the absolute value of these parameters. We wish to emphasize that $<m_\nu>$, $<\eta>$, and $<\lambda>$ are the quantities which could be extracted from the analysis of experiments on double beta decay. They depend not only on the neutrino mass value (in the case of $<m_\nu>$) or on the coupling constants of the Hamiltonian (6.27), but also on parameters describing neutrino mixing. The functions C_n contain the nuclear matrix elements and the phase space integrals, e.g.,

$$C_1=\left|M_{GT}^{0\nu}-\frac{g_V^2}{g_A^2}M_F^{0\nu}\right|^2 G^{0\nu}(E_0,Z)m_e^2\,. \qquad (6.34)$$

The definitions of the other C_n can be found in Doi et al. (85). Let us stress that the phase space integrals do not represent any difficulty and can be evaluated exactly. On the other hand, the nuclear matrix elements, to be discussed in the next Section 6.2, are difficult to calculate.

6.2 Double Beta Decay and Nuclear Structure

In the previous Section we have presented the interrelation between the double beta decay rate and the properties of the neutrino and its interactions, the lepton phase space, and the nuclear matrix elements. Here, we shall discuss how the nuclear matrix elements are evaluated.

All nuclear matrix elements, defined in Eqs. (6.19), (6.22), (6.23), (6.28), (6.29), (6.31), and (6.32) have a similar structure; they contain the wave function of the initial even-even nucleus in its 0^+ ground state, the wave function of the final even-even nucleus, usually in the 0^+ ground state (but sometimes also in the 2^+ first excited state) and an operator of varying degrees of complexity connecting these states. In the case of the 2ν mode, a summation over the complete set of states $|m>$ of the intermediate odd-odd nucleus is also required. This summation could be avoided in the "closure approximation" leading to the expressions

$$M_{GT}^{2\nu}=\frac{M_{GT}^{(clos)}}{\Delta\bar{E}_{GT}}\ ;\ M_{GT}^{(clos)}=<0_f^+|\sum_{kl}\vec{\sigma}_k\cdot\vec{\sigma}_l\tau_k^+\tau_l^+|0_i^+>\,, \qquad (6.19c)$$

and

$$M_F^{2\nu} = \frac{M_F^{(clos)}}{\Delta \bar{E}_F} \; ; \; M_F^{(clos)} = <0_f^+| \sum_{kl} \tau_k^+ \tau_l^+ |0_i^+> , \qquad (6.19d)$$

where $\Delta\bar{E}$ is the average energy denominator. In the closure approximation the structures of all matrix elements for both 2ν and 0ν modes are the same and only wave functions of the initial and final states are required. As stated earlier, closure can be well justified for 0ν decay where the virtual neutrino has relatively high "energy". (The adequacy of the closure approximation for the 0ν mode has been tested by Suhonen et al. 90.) For the 2ν decay, on the other hand, closure appears to be a poorer approximation. Therefore, it will be necessary to evaluate explicitly the sum over all intermediate states.

Ideally, one should solve the nuclear many-body problem with a realistic nucleon-nucleon interaction and with a minimum number of approximations and find the required wave functions. Unfortunately, such a procedure is not technically feasible, except in the case of the lightest double beta decay candidates, primarily ^{48}Ca. In all other cases, severe approximations as to the number of configurations accepted, and the complexity of the nuclear Hamiltonian, are required. It is, therefore, important to test the calculation by using an identical procedure for evaluating not only the double beta matrix elements, but also other quantities which depend on operators of a similar structure and which may be accessible to experimental verification. Alternatively, one may consider 2ν ββ decay as a test case for some of the matrix elements in 0ν decay.

Shell model evaluation of the decay rate of ^{48}Ca. The double beta decay ^{48}Ca → ^{48}Ti is the simplest case from the point of view of nuclear structure. (The single beta decay, ^{48}Ca → ^{48}Sc, is energetically possible, with $Q = 278$ keV. A large angular momentum change, ΔI=5 or 6, in the decay of ^{48}Ca to the lowest states of ^{48}Sc causes this decay to be very slow. Single beta decay has not been observed so far, and the experimental lifetime limit is 6×10^{18} y (Alburger & Cumming 85).) The nuclear structure aspects for this decay have been studied extensively, and our discussion will illustrate the problems one encounters. In all calculations performed so far the doubly magic core of ^{40}Ca (Z=20, N=20) was treated as inert, and nucleons were allowed to occupy states in the subshells $f_{7/2}$, $f_{5/2}$, $p_{3/2}$, and $p_{1/2}$. These subshells are filled as protons and neutrons are added to the ^{40}Ca core.

We begin with the analysis of the 2ν matrix elements. The Fermi matrix element $M_F^{(clos)}$ is negligibly small because it does not connect states with different isospin. Although the isospin in nuclei is not conserved exactly, the admixtures of states with "wrong" isospin to the ground state are known to be very small. In the following, we shall therefore neglect $M_F^{(clos)}$ as well as $M_F^{2\nu}$ when considering the 2ν $\beta\beta$ decay.

As an illustration, we consider first the simplest case where neutrons and protons are allowed to occupy only the $f_{7/2}$ subshell. The wave functions are then:

$$\Psi(^{48}Ca)_{0^+} = (n f_{7/2})^8_{0^+} ,$$

$$\Psi(^{48}Ti)_{0^+} = \sum_{L-even} C(L)\Big[(nf_{7/2})^6_L(pf_{7/2})^2_L\Big]_{0^+} .$$

Here, the coefficients $C(L)$ have to be calculated and depend on the internucleon interaction. The resulting $M_{GT}^{(clos)}$ is obtained as a sum over the angular momentum L of the two protons (or six neutrons) in the final nucleus ^{48}Ti. Brown (85) obtained $M_{GT}^{(clos)} = 0.38$ with contributions from $L=0$ of 2.33, from $L=2$ of -1.98, from $L=4$ of 0.44, and a negligible contribution from $L=6$. Thus we see that the $L=0$ and $L=2$ parts almost exactly cancel, leading to a considerable suppression of the double beta decay rate. More complete calculations of $M_{GT}^{(clos)}$ and $M_{GT}^{2\nu}$ (including the full f,p shell, but often not all possible configurations) have been performed by a number of authors, and the results are summarized in Table 6.3; the cancellation persists in the more complete calculations.

It should be noted that most of the evaluations of $M_{GT}^{2\nu}$ in Table 6.3 lead to a lifetime that is shorter than the experimental limit of 3.6×10^{19} y (Bardin et al. 70). Thus, it appears that some important ingredient might still be missing in the calculations listed in Table 6.3.

For the estimate of the lifetime of the 0ν mode, we need the matrix elements $M_{GT}^{0\nu}$ and $M_F^{0\nu}$ (if we assume, for simplicity, that $\langle m_\nu \rangle \neq 0$ and that there are no right-handed currents). Our previous remark about the smallness of the Fermi matrix element is no longer valid because the factor $H(r,E)$ in Eq. (6.23a) or $1/r_{lk}$ in Eq. (6.23b) now allows change of isospin. In fact, the Fermi matrix element is only about four times smaller than, and of opposite sign to, the Gamow-Teller matrix element, enhancing the decay rate.

Calculation of $M_{GT}^{2\nu}$ in heavier nuclei. The approach closest to the canonical shell model was adopted by Haxton et al. (81, 82) and summarized by Haxton & Stephenson (84), who calculated $M_{GT}^{(clos)}$ as well as the full set of matrix elements for the 0ν decay. Their calculation is based on a version of the weak coupling approximation. The method allows calculation of all nuclear matrix elements of interest, including those in Table 6.4 as well as those utilized in Table 6.6. However, it is difficult to estimate the uncer-

Table 6.3 *Nuclear matrix elements and calculated half-lives for* 2ν and 0ν ββ *decay for* $^{48}Ca \rightarrow {}^{48}Ti$.

Reference	$M_{GT}^{(clos)}$	$M_{GT}^{2\nu}$ (m_e^{-1})	$T_{\frac{1}{2}}^{2\nu}(y)$	$\lvert M_{GT}^{0\nu} - \frac{g_V^2}{g_A^2} M_F^{0\nu} \rvert$	$T_{\frac{1}{2}}^{0\nu}(y)^a$
Haxton et al. (82)	0.44		2.9×10^{19}	1.15	3.2×10^{24}
Zamick & Auerbach (82)	0.36		4.3×10^{19}		
Skouras & Vergados (83)	0.25	0.028	3.2×10^{19}	1.05	3.7×10^{24}
Tsuboi et al. (84)	0.46	0.065	6.1×10^{18}		
Brown (85)	0.47^b	0.059	7.2×10^{18}	1.42^c	2.0×10^{24}
Ogawa & Horie (89)		0.045^d	1.3×10^{19}		
Zhao et al. (90)	0.20^e	0.036^e	1.9×10^{19}		
Caurier et al. (90)	$0.11^{d,e}$	$0.021^{d,e}$	5.5×10^{19}		

[a] Assuming $\langle m_\nu \rangle = 1$ eV and no right-handed currents.

[b] Configurations up to 2p2h included.

[c] $M_F^{(0\nu)}$ not included

[d] No restrictions, full f,p shell included

[e] Reduced by the "effective axial charge" $(1/1.30)^2$

tainty associated with the closure approximation and the omission of spin-orbit partners and higher lying configurations.

Most other calculations include only the simplest configurations possible, i.e., primarily the configurations directly connected to the unperturbed ground state by the Gamow-Teller operator. However, at the same time these calculations are capable of describing explicitly the intermediate odd-odd nucleus and thus allow a number of additional tests. Owing to the vector-isovector nature of the Gamow-Teller operator $\sigma\tau$, the summation over the intermediate states $|m\rangle$ in (6.19a) involves only states with the spin and parity $I^{\pi}=1^{+}$ for the $0^{+} \rightarrow 0^{+}$ and $0^{+} \rightarrow 2^{+}$ transitions. Thus, the task is to determine the spectrum of 1^{+} states in the odd-odd nucleus and the matrix elements of the Gamow-Teller operator $\sigma\tau$ connecting them with the 0^{+} ground states of the initial and final nucleus, respectively.

Additional tests of the calculation are possible because the factor $\langle 1_{m}^{+}|\sum \vec{\sigma}_{k}\tau_{k}^{+}|0_{i}^{+}\rangle$ in (6.19a) determines, at the same time, the strength of the β^{-} transition between the initial and the intermediate nucleus (if such a transition were to be energetically possible), while the factor $\langle 0_{f}^{+}|\sum \vec{\sigma}_{l}\tau_{l}^{+}|1_{m}^{+}\rangle$ similarly determines the strength of the β^{+} transition between the final and the intermediate nucleus. It has been realized (see, e.g., Gaarde 83) that these same factors also determine the cross section of the charge exchange reactions (p, n) and (n, p), respectively, in selected kinematic regions. In Figure 6.10 we show examples of the forward cross section of the (p, n) reaction at 200 MeV, where the proportionality of the cross section to the Gamow-Teller strength holds. The spectra are dominated by single peaks, the "giant Gamow-Teller resonance". Calculations obviously must reproduce the position and strength of this resonance.

A further constraint is obtained from the model independent sum rule which relates the total β^{-} ($\sigma\tau^{+}$ operator) and β^{+} ($\sigma\tau^{-}$ operator) strengths. This sum rule states that

$$\sum_{m} |\langle 1_{m}^{+}|\sum \vec{\sigma}_{k}\tau_{k}^{+}|0_{i}^{+}\rangle|^{2} - \sum_{n} |\langle 1_{n}^{+}|\sum \vec{\sigma}_{k}\tau_{k}^{-}|0_{i}^{+}\rangle|^{2} = 3(N-Z)\,,$$

and is valid as long as we consider only nucleonic degrees of freedom in the nucleus. In nuclei with a large neutron excess (most double beta decay candidates), the second term in the sum rule is much smaller than the first one, and thus the sum rule represents an important constraint for the total β^{-} strength.

Role of the Δ isobar. From experiment it is found that the giant Gamow-Teller resonance depicted in Figure 6.10 does not exhaust the sum rule; it contains only 50—60% of the expected strength. The coupling to the Δ isobars offers at least partial resolution of this "missing strength" problem (Gaarde 83). In the simplest quark model, Δ isobars are excited three-quark systems having all quarks in the s-state, coupled to spin S=3/2 and isospin T=3/2; they differ from a nucleon by a spin and isospin flip of one of the quarks. Thus the Gamow-Teller operator $\sigma\tau$, properly generalized to quarks, connects nucleons with the different charge states of the Δ. Some of the strength of the low-lying nuclear states is then shifted to the states containing Δ, at much higher energy (~300 MeV), and is unobservable in (p,n) reactions. Within the simplest quark model, inclusion of these states indeed explains the "missing strength" problem (Bohr & Mottelson 81). The matrix element $M_{GT}^{2\nu}$ is not, however, directly affected by these high-lying states (Vogel & Fisher 85); although, through mixing with the low-lying 1^+ states, the decay rate is reduced by a factor of ~2. To correct in a crude way for the "missing strength" one often uses g_A = 1 instead of the free nucleon value g_A = 1.26. We use this approach in Table 6.4 to relate the experimental lifetime to the Gamow-Teller matrix elements, Eq. (6.19a). (This is equivalent to assigning an "axial charge" 1/1.26 to the Gamow-Teller operator $\sigma\tau$, see also the last two lines in Table 6.3.)

There have been suggestions that the double beta decay could be mediated by Δ isobars (Primakoff & Rosen 59). If the initial or final nucleus con-

Figure 6.10. Neutron spectra of the (p,n) reaction for forward going neutrons. The spectra are normalized to show relative cross sections.

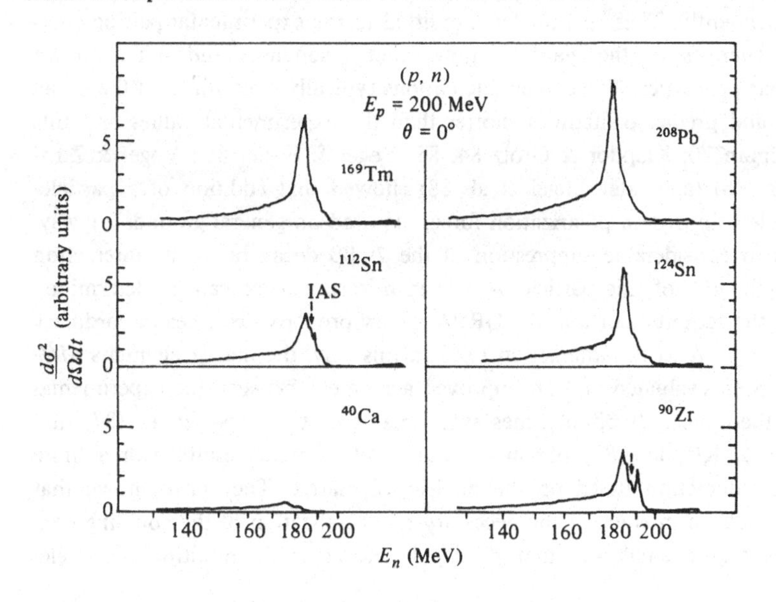

tains admixtures of the Δ-nucleon hole type (estimated probability of ~ 1%), the double beta decay could proceed by either

$$\Delta^- \to p + 2e^- \, (+2\bar{\nu}) \, , \text{ or } \; n \to \Delta^{++} + 2e^- \, (+2\bar{\nu}) \, .$$

However, angular momentum selection rules forbid this process in the most important $0^+ \to 0^+$ transitions (Doi et al. 85). Alternatively, the Δ could be already present in the initial nucleus as well as left in the final one. This possibility does not violate selection rules, but is expected to lead to low transition rates due to the small probability of finding a Δ in both nuclei simultaneously.

In a different context, it has been suggested that, owing to the effect of the intermediate Δ-nucleon hole states, there is a deep analogy between double beta decay and the pion double charge exchange reaction. Again, there is no consensus on the significance of this relation.

QRPA calculations of $M_{GT}^{2\nu}$. Quasi-particle random phase approximation (QRPA) offers a systematic procedure for calculating the matrix elements $M_{GT}^{2\nu}$. QRPA takes into account the two essential ingredients of nuclear structure: pairing, responsible for the extra binding of even-even nuclei, and spin-isospin polarization, responsible for the emergence of the giant Gamow-Teller resonance. The method often employs few adjustable parameters which allow one to change the strength of the effective nucleon-nucleon interaction in the particle-hole and particle-particle channels independently. These parameters are fitted to the experimental pairing energies, energies of the giant Gamow-Teller resonance, and other known nuclear properties. The earlier calculations typically overestimated the decay rates and predicted lifetimes shorter than the experimental values or limits (Huffman 70; Klapdor & Grotz 84, 85; Vogel & Fisher 85). Vogel & Zirnbauer (86) (and later Engel et al. 88) showed that addition of a particle-particle spin-isospin polarization force, required on general grounds anyway, leads to considerable suppression of the 2ν ββ decay rate. The interaction strength, g^{pp}, of this particle-particle spin-isospin force can be determined from the requirement that the QRPA theory properly describes the ordinary β^+ decay. With the interaction fixed in this way, the matrix elements $M_{GT}^{2\nu}$ have been evaluated and an improved agreement between the experimental and theoretical 2ν ββ lifetimes was obtained. Civitarese et al. (87) and Muto & Klapdor (88) obtained qualitatively similar results with a finite range interaction based on the nuclear G matrix. They have shown that with such an interaction the necessary renormalization of the coupling constants is quite small, i.e., that g^{pp} *app* 1. However, the resulting matrix ele-

Table 6.4 *Experimental and calculated matrix elements $M_{GT}^{2\nu}$ (energy denominators in units of $m_e c^2$).*

Initial nucleus	Exp. value[a]	Haxton et al. 81,82	Klapdor & Grotz 85[b]	Vogel & Fisher 85	Engel et al. 88[c]
^{76}Ge	0.13[d]	0.14	0.27 (0.19)	0.07	0.12
^{82}Se	0.072[d]	0.092	0.23 (0.13)	0.08	0.069
^{100}Mo	0.15[d]		0.26	0.23	0.21
^{128}Te	0.022[e]	0.11	0.11 (0.045)	0.10	0.073
^{130}Te	0.013[f]	0.11	0.11 (0.042)	0.10	0.049
^{136}Xe	<0.1[d]		0.07 (0.081)	0.10	0.025
^{150}Nd	<0.025[d]		0.15	0.06	

[a] From Eq. (6.18) with $G^{2\nu}$ from Table 6.1 reduced by $g_A^4 = 1.26^4$ to account for the "missing strength"

[b] In parentheses results of Klapdor & Grotz (84).

[c] For $\alpha'_1 = -390$ MeV·fm^3

[d] Based on laboratory experiments described in Section 6.3.

[e] Based on the ^{128}Te /^{130}Te lifetime ratio of Lee et al (91).

[f] Based on geochemical measurements of Kirsten et al (86).

ments depend so sensitively on the parameter g^{pp} that it is difficult to predict them reliably.

Some of the calculated matrix elements $M_{GT}^{2\nu}$ are collected in Table 6.4 and compared to the values deduced from the experiments. The agreement between columns 2 and 6 is quite good, although the overestimate of $M_{GT}^{2\nu}$ in ^{130}Te is still present. It is important to realize that a "natural unit" of $M_{GT}^{2\nu}$, based on the sum rule, is $\approx$3 for ^{130}Te, while experiment requires a value of $\approx$0.01, for a hindrance of a factor of $\sim$300. As often in nuclear physics, an accurate description of strongly hindered phenomena is difficult. Indeed, we see that QRPA is capable of reproducing most, if not all, of the considerable reduction of this matrix element.

Among the many later calculations of the rate of the 2ν double beta decay let us mention the QRPA based works of Civitarese et al. (91) and Staudt et al. (90), and of Hirsh & Krmpotic (90), the application of the $SU(4)$ symmetry by Bernabeu et al. (90), as well as the application of the QRPA to nuclear matter by Alberico et al. (88). Ching et al. (89) proposed the "Operator Expansion Method" that avoids the summation over the intermediate states in Eq. (6.19), but uses other approximations that are difficult to access. Discussion of these and similar works goes beyond the scope of the present book.

Matrix elements for the 0ν *mode.* There are several significant differences between the matrix elements $M_{GT}^{2\nu}$ and the matrix elements (6.22), (6.23), (6.28), (6.29), and (6.31) determining the 0ν decay rate. First, the radial dependence of the operators ($1/r$ for the simplest case of (6.22b) and (6.23b)) means that all virtual intermediate states contribute, rather than just the 1^+ states as in the 2ν mode. Thus, one expects that a larger set of single particle states is required. Second, the isospin selection rule, which caused $M_F^{2\nu}$ to vanish is no longer operative, and the corresponding Fermi matrix elements must be included. Third, the closure approximation is expected to be valid and thus the summation over intermediate states could be avoided. Nevertheless, the ground states of both the initial and the final nucleus must be described correctly (including the effect of zero point motion corresponding to all nuclear collective states).

A radical simplification has been suggested by Primakoff & Rosen (59), who argued that the $1/r$ factor in matrix elements (6.22b) and (6.23b) can be replaced by a constant. Thus, the matrix elements of the 0ν mode can be related to the closure matrix elements of the 2ν mode by scaling. However, in view of the substantial differences in the nuclear aspects of the 0ν and 2ν modes, one does not expect such a scaling to hold (see also Grotz & Klapdor 85).

QRPA represents a general method that can be applied to evaluation of all matrix elements responsible for the $\beta\beta$ decay, including all those relevant for the 0ν decay, whether by the mass mechanism or by the right-handed current mechanism. The calculations were performed by many authors who, as in the 2ν case above, use different forms and parametrization of the nucleon-nucleon interaction, treatment of pairing, treatment of the short range nucleon repulsion, finite nucleon size, etc. Here we mention just a few representative examples: Tomoda & Faessler (87), Engel et al. (88), Muto et al. (89), Staudt et al. (90), and Suhonen et al. (90, 91). As stated earlier, due to the neutrino propagator, all states in the intermediate nucleus contribute to the 0ν decay, not just the 1^+ states of the 2ν case. There is a general consensus that the ground state correlations, caused by the particle-particle force and characterized by its strength g^{pp}, affect the 1^+ states more than the states of other multipolarities. Therefore, when the particle-particle force is switched on, the parts of the resulting nuclear matrix elements associated with the intermediate 1^+ states are reduced considerably (or vanish) while the parts associated with other intermediate states are reduced less. Consequently, the final matrix elements are less sensitive functions of the parameter g^{pp} and, in particular, do not go through zero for the range of g^{pp} values of interest. This makes the evaluation of the matrix elements of 0ν decay less uncertain than the matrix elements of the 2ν decay. If the spread between the calculated matrix elements of various authors (or the same authors using various realistic approximations and parameters) could be used as a measure of the "theoretical uncertainty" one would conclude that the matrix elements that determine the value of the constant C_1 in Eq. (6.33), and consequently also of the mass parameter $<m_\nu>$, are known to within a factor of 2—3. The spread in matrix elements related to the microscopic parameter $<\eta>$ appears to be somewhat larger (see Table 6.6), due to the different ways in which the "recoil" and "p-wave" matrix elements are evaluated.

The generalized seniority method (Engel et al. 89) is a bridge between the shell model and QRPA approaches. It uses the insight of QRPA in order to select the important configurations, but uses the exact shell model diagonalization to obtain the corresponding wave functions. The results qualitatively confirm the conclusions of QRPA regarding the suppression of the 2ν matrix elements caused by the ground state correlations. In Table 6.5 we compare the results of all discussed methods for the matrix elements responsible for the mass mechanism of the 0ν ββ decay. Our previous conclusion about the magnitude of "theoretical uncertainty" seems to be confirmed.

Keeping in mind all the enumerated difficulties, we can nevertheless use some consistent set of matrix elements and find upper limits for the microscopic parameters $<m_\nu>$, $<\eta>$, and $<\lambda>$. To make such a procedure easier, we present in Table 6.6 the coefficients C_i of (6.33), which contain products of the nuclear matrix elements and the corresponding phase space integrals. To illustrate the spread between the theoretical values we list the results of Doi et al (85) (based on the calculations of Haxton & Stephenson 84), Muto et al. (89), and Suhonen et al. (91). In Section 6.4 we shall use these coefficients in an analysis of the experimental half-life limits.

6.3 Experimental Tests

Clearly, the most direct test of double beta decay is a search for the two outgoing electrons. Such a search can be accomplished with the help of

Table 6.5 *Matrix elements* $|M_{GT}^{0\nu} - M_F^{0\nu}|$, *Eqs. (6.22a) and (6.23a), calculated using different methods.*

	^{76}Ge	^{82}Se	^{130}Te
Gen. Seniority, Engel et al. 89	3.3 – 5.0	1.8 – 3.7	3.7 – 5.7
Shell Model, Haxton & Stephenson 84	5.5	4.4	6.8
QRPA, Engel et al. 88 $\alpha_1'=-390$ (MeV fm^3)	2.0	1.5	3.1
QRPA, Tomoda & Faessler 87	4.8	4.4	3.8

a beta spectrometer capable of the energy resolution required to distinguish between 0ν and 2ν ββ decays. A number of spectrometric counting experiments have been performed and will be discussed in Subsection 6.3.1.

Instead of exploring the electron spectrum, double beta decay can also be studied by identifying the daughter nucleus with the help of a chemical extraction or laser excitation followed by mass analysis. Experiments based on chemical and isotopic identification will be discussed in Subsection 6.3.2.

For a compilation of all the recent experiments and their results we refer the reader to the reviews by Moe (91) and Ejiri (91).

6.3.1 *Counting experiments*

In discussing direct counting experiments and their advantages as well as difficulties, it is useful to be reminded of the extremely low count rates characteristic of double beta decay. At the present time, the lower limit of the half-life of 0ν double beta decay in ^{76}Ge is 1.2×10^{24} y. The rate of decay, for example, of a 1 mole sample of an isotope, having a half-life of 10^{24} y, is 0.4 disintegrations per year. Clearly, the background rates in a

Table 6.6 *Coefficients* C_1 (in y^{-1}) and the ratios C_i/C_1 of Eq. (6.33). First entry from Doi et al. (85), second from Muto et al. (89), third from Suhonen et al. (91).

Microscopic parameter	Coefficient	^{48}Ca	^{76}Ge	^{82}Se	^{130}Te
$<m_\mu>^2$	C_1	8.4E-14	1.6E-13	4.6E-13	1.7E-12
			1.1E-13	4.3E-13	5.3E-13
			6.4E-14	1.8E-13	3.3E-13
$<m_\nu><\lambda>$	C_2/C_1	−0.35	−0.35	−0.42	0.38
			−0.37	−0.37	−0.42
			−0.62	−0.83	−1.32
$<m_\nu><\eta>$	C_3/C_1	−0.57	−10.6	−12.6	−10.4
			196.	147.	170.
			65.3	57.1	115.
$<\lambda>^2$	C_4/C_1	3.8	1.0	2.1	1.5
			1.2	2.3	2.0
			2.0	4.8	6.2
$<\eta>^2$	C_5/C_1	1.5	118.	272.	157.
			4.0E4	3.5E4	4.2E4
			4.2E3	4.9E3	9.3E3
$<\lambda><\eta>$	C_6/C_1	−4.2	−0.8	−2.0	−0.9
			−0.4	−0.9	−0.8
			−1.2	−3.4	−3.8

EX signifies 10^x

counting system are of crucial importance and warrant proper discussion in this Section.

Early counting experiments have attempted to observe coincidence events between the two electrons, but, owing to background counts, were limited in sensitivity to a decay lifetime of about 10^{15} y. Subsequent work with the cloud chamber allowed the recording of the tracks of the electrons and their curvatures in a magnetic field, and thus represented a much more specific search for the electron pair. Lower limits of lifetimes established in these early cloud chamber experiments in ^{48}Ca, ^{96}Zr, ^{116}Cd , and ^{124}Sn were 10^{16}—10^{17} y. For a description of these early experiments we refer the reader to the reviews of Zdesenko (80) and Haxton & Stephenson (84).

Background considerations. As we have seen, the background count rate in the detector remains one of the principal limitations of the sensitivity to be achieved in a double beta decay experiment. We now discuss the sources of various backgrounds in the detector system.

One of the most important background components comes from cosmic ray muons. It is, therefore, mandatory to install an experiment underground in a deep mine or a tunnel. We show in Figure 6.11 the cosmic ray muon flux as a function of depth in units of meter water equivalent (mwe). At sea level this flux is 5×10^9 m^{-2} y^{-1}. The reduction of the muon flux in an underground laboratory such as that in the Gran Sasso tunnel in Italy with an overburden of 1500 m of rock, or 3600 mwe, is about 2×10^{-7}, and that in the Gotthard tunnel laboratory in Switzerland (1200 m of rock, or 3000 mwe), is about 10^{-6}. At these attenuations, the background count rate from cosmic ray muons in a small detector is sufficiently suppressed to be negligible. For a larger detector, the muons must be recognized and dealt with. Below, we describe an experiment with a time projection chamber experiment, a detector which has a surface area of ≈ 1 m^2. There, the incident muons which strike at a rate of a few per day are recognized and separated from double beta decay events with the help of track analysis.

The next important background contribution comes from the decay of radioisotopes in the detector system and its shielding. Besides the isotopes associated with the natural ^{232}Th and ^{238}U radioactive decay chains, one finds radioisotopes induced by cosmic ray neutrons while the detector was manufactured and tested above ground. In addition, there are neutrons from spontaneous fission or (α,n) reactions in the decay of ^{238}U in the surrounding rock and shielding, and their reaction products.

As an example of such a background, we show in Figure 6.12 a gamma ray spectrum from a germanium detector located in a shielded cave in the Gotthard tunnel laboratory (Fisher 86). In this and other low background studies (Brodzinski et al. 85; Bellotti et al. 86), a number of gamma rays have been identified whose characteristics and origins are briefly the following.

The gamma rays emitted in the decay chain of ^{238}U (uranium-radium series) include ^{222}Rn, decaying with a half-life of 3.8 d into its decay products ^{214}Bi and ^{214}Pb (prominent high energy lines at 2204 and 2447 keV). Then there are the gamma rays from the decay of ^{232}Th (thorium series) including ^{228}Ac (prominent line at 911 keV) and ^{208}Tl (prominent lines at 510.8, 609.3, and, most conspicuously, 2614 keV).

Besides these uranium and thorium backgrounds, the spectrum of Figure 6.12 exhibits the 1461 keV line from ^{40}K, originating from components containing aluminum and glass, residing near the detector, such as circuit elements and circuit boards.

Figure 6.11. Cosmic ray muon intensity as a function of the overburden of rock in mwe. Laboratories that accommodate double beta decay and neutrino experiments are indicated.

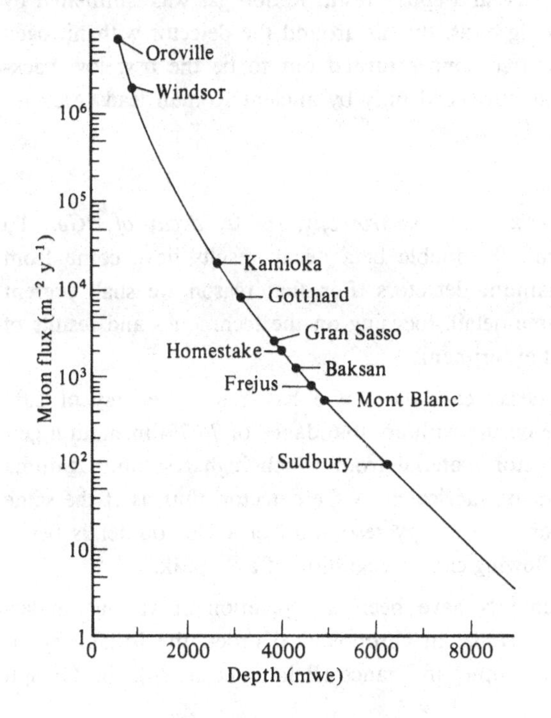

Additionally, a large number of gamma lines (see Figure 6.12) can be attributed to the decay of the isotopes $^{56,57,58,60}Co$, ^{54}Mn, and ^{68}Ge which were produced by cosmic ray neutron-activation of ^{63}Cu, a material very abundant in the massive copper shielding surrounding many detectors, and by neutron activation (via $(n,3n)$) of ^{70}Ge in the Ge detector. Owing to its long half-life of 288 d and large decay energy of 2.9 MeV (1.9 MeV positrons) of its daughter, ^{68}Ga, ^{68}Ge is one of the most serious contaminants. As this activation process took place before the apparatus went underground, the lines continue to decay in intensity leading to an improvement of backgrounds over time.

It should be remembered that in a spectrum such as that depicted in Figure 6.12, the full energy peak associated with each gamma ray is but a small part of the total detector response. The gamma ray interacting with the detector also gives rise to a broad Compton continuum below the full energy peak. Thus, the continuous background around 2 MeV, for example, is mainly caused by gamma rays of higher energy, such as the 2.6 MeV line from ^{208}Tl, or the gamma rays (with energy up to 6 MeV) emitted following capture of residual neutrons in the detector and shielding material.

In the experiments quoted below the contribution of these background lines was minimized by eliminating as far as possible materials such as steel, lead (solder), glass products, and epoxy resin. Radon gas was eliminated by displacing, on a continuing basis, the air around the detector with nitrogen gas. High purity oxygen-free copper turned out to be the best low background shielding material surpassed only by ancient Roman lead (Alessandrello et al. 91).

High resolution germanium spectroscopy: the 0ν *decay of* ^{76}Ge. To date, the most significant 0ν double beta decay results have come from experiments with germanium detectors. For that reason we shall present here the Ge work in some detail, focusing on the techniques and results of some of the more recent experiments.

The double beta decay candidate ^{76}Ge has a decay energy of 2.04 MeV. The isotope ^{76}Ge occurs with an abundance of 7.67% in natural germanium, the semiconductor material from which high resolution gamma ray and particle counters are fabricated. A Ge detector, thus, is at the same time source and detector. The energy resolution of a Ge counter is better than 3 keV at 2 MeV, allowing clear recognition of a 0ν peak.

Several Ge experiments have been in operation at various underground locations. They include, in chronological order, the Milan experiment in the Mont Blanc tunnel in France (Bellotti et al. 86), the Guelph experiment in the Windsor mine in Canada (Simpson et al. 84), the PNL-USC experiment in the Homestake mine in the South Dakota, USA (Avignone et al. 86), the ITEP-YEPI (Yerevan Physical Institute) experiment in the Avan salt mine in the USSR (Vasenko et al. 90, 91), and the Osaka experiment in the Kamioka mine in Japan (Ejiri et al. 87). We single out, for further elaboration, the UCSB-LBL experiment in the Oroville dam in California (Caldwell et al. 87, 91) and the Caltech-Neuchatel-PSI experiment in the Gotthard tunnel through the Swiss Alps (Fisher et al. 89; Reusser et al. 91a, 91b).

Among the largest Ge detectors used so far are the 1300 cm^3 (6.9 kg) detector of the UCSB-LBL collaboration and the 1100 cm^3 (5.8 kg) detector of the Caltech-Neuchatel-PSI collaboration. Both detectors are constituted of eight individual Ge crystals. The UCSB-LBL detector has 7 moles of ^{76}Ge, or 4×10^{24} atoms. In the absence of background counts, such a detector, after operating for one year and seeing zero counts, would allow us to place a half-life limit (68% CL) of 3×10^{24} y. This, clearly, is the sensitivity limit for the most ideal case. In reality, there is always a background component which prevents us from reaching this limit. In the presence of background, the actual half-life limit in years is given by

$$T_{1/2}(0\nu) > \ln 2\times10^{23}\left(\frac{Nt}{B\Delta E}\right)^{1/2} , \tag{6.35}$$

where N is the number of sample atoms of the desired isotope in the

Figure 6.12. Background gamma ray spectrum at the Gotthard laboratory (3000 mwe) measured with a shielded 90 cm^3 Ge detector. The gamma ray peaks are identified and labelled.

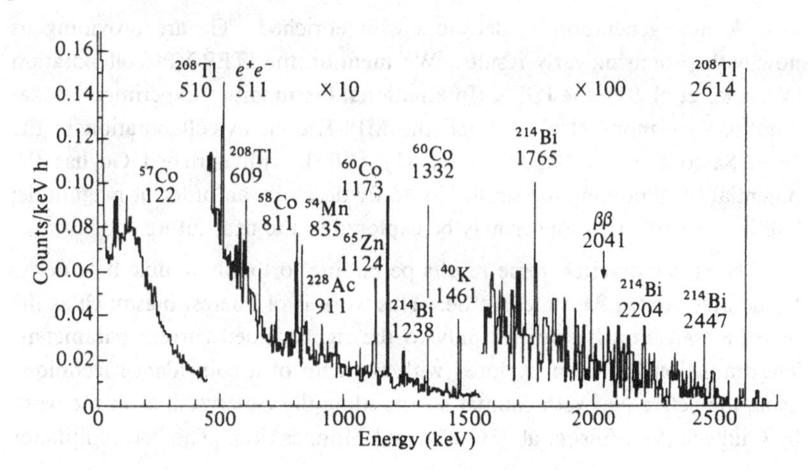

fiducial volume of the detector in units of 10^{23}, B is the number of background counts per 10^{23} source atoms in a 1 keV interval, ΔE is the full width in keV of the line at the decay energy E_0 (such as twice the detector resolution), and t is the measuring time in years.

The better the energy resolution of a Ge detector, the lower is the number of background counts in ΔE. The background count rate around 2 MeV (the 0ν-decay energy) in the Gotthard experiment using the 1100 cm^3 detector was found to be 3.7 counts/(keV kg y) at the beginning of the experiment, just after installation of the detector underground. At the end of an aggregate counting time of 10.0 kg y, this rate was found to be 1.6 counts/(keV kg y). The observed rate in an energy bin at 2.04 MeV averaged over that counting period was 2.4 counts/(keV kg y), entirely consistent with background alone. A fit to the data gives an upper limit for the half-life of $T_{1/2} > 6.6\times10^{23}$ y (68% CL) or 3.7×10^{23}y (90% CL). The UCSB-LBL experiment with an aggregate of 21 kg y has reached a background level of 1.2 counts/(keV kg y) and a half life limit of $T_{1/2} > 2.4\times10^{24}$ y (68% CL), or 1.2×10^{24} y (90% CL). To date, this is the most restrictive limit for the half-life of 0ν double beta decay. This ^{76}Ge result is also listed in Table 6.7. Neutrino mass and the parameters describing right handed current weak interaction derived from these results are discussed in Section 6.4.

In Figures 6.13 and 6.14 we show the detector and shielding arrangement for the Gotthard and the Oroville experiment, respectively. In the Oroville experiment, the Ge detector was surrounded by a NaI anticoincidence shielding. Figure 6.15 illustrates the data of the Oroville experiment (Caldwell 91) in the region of the 0ν energy. Both experiments have now been discontinued as further data taking would bring only little gain in sensitivity.

A new generation of detectors with enriched ^{76}Ge are providing us now with promising early results. We mention the ITEP-YPI collaboration (Vasenko et al. 91), the IGEX (International Germanium Experiment) collaboration (Avignone et al. 91) and the MPI-Kurchatov-collaboration in the Gran Sasso tunnel (Klapdor et al. 91). Work with enriched Ge has the potential of improving the sensitivity to 0ν decay by an order of magnitude; half-lives of 10^{25} y or longer may be explored in the near future.

Next, we describe some results pertaining to the 0ν double beta decay to the first excited 2+ state in ^{76}Se. This work is of interest inasmuch as the $0+\rightarrow2+$ transition is sensitive only to the right-handed current parameters. The transition has been explored with the help of a coincidence technique using the active NaI veto counter surrounding the Ge crystal, as in the work by Caldwell (91); Ejiri et al. (86, 87), and Morales (91). The NaI scintillator

Table 6.7 *Some experimental half-lives (in y) for double beta decay (limits are 90% CL). For references see text. Corresponding upper limits of the neutrino mass parameter* $<m_\nu>$ *(in eV) are also shown.*

	Spectroscopic experiments				
	^{48}Ca	^{76}Ge	^{82}Se	^{100}Mo	^{136}Xe
$T_{1/2}(0\nu)$	$>2\times10^{21}$	$>1.2\times10^{24}$	$>7\times10^{21}$	$>2.6\times10^{21}$	$>2.5\times10^{23}$
$<m_\nu>^a$		<1.4	<9.3	<22.1	<3.0
$<m_\nu>^b$		<1.8	<14.4		<3.0
$<m_\nu>^c$		$<(2.4\text{-}4.7)$	$<(28\text{-}40)$	$<\sim27$	$<(3.3\text{-}5.0)$
$T_{1/2}(2\nu)$	$>3.6\times10^{19}$	$0.92^{+0.07}_{-0.04}\times10^{21}$	$1.1^{+0.8}_{-0.3}\times10^{20}$	$1.16^{+0.34}_{-0.06}\times10^{19}$	$>1.6\times10^{20}$

	Geochemical and milking experiments			
	^{82}Se	^{128}Te	^{130}Te	^{238}U
$T_{1/2}(0\nu,2\nu)$	$(1.3\pm0.05)\times10^{20}$	$>8\times10^{24}$	$(2.6\pm0.3)\times10^{21}$	$(2.0\pm0.6)\times10^{21}$

[a] Matrix elements of Muto et al. (89), "axial charge" = 1.0.

[b] Matrix elements of Suhonen et al. (91), "axial charge" = 1.0.

[c] Matrix elements of Engel et al. (88), α'_1 = -375±15 MeV fm^3, "axial charge" = 1/1.25.

selects the 559.1 keV gamma ray depopulating the 2+ state and electrons around 1481.6 keV (the 0+→2+ energy) are recorded in coincide with this gamma ray. At the present time no conclusive evidence for the 0+ → 2+ decay has been seen and the half-life limit is 3×10^{23} y (68% CL).

The 2ν *decay.* From an experimental point of view, the 2ν decay is difficult to observe in a high resolution spectrometer such as the Ge detector despite the expected shorter half-life of that transition. This stems from the fact that the electron intensity in 2ν decay is spread over all energies from 0 to the decay energy (2.041 MeV for ^{76}Ge) according to the four-fermion phase space distribution (see Section 6.1). At a resolution of 3 keV at 2 MeV, or about 1.5×10^{-3}, the number of electrons per energy bin is diluted accordingly. Further complications arise from the larger number of background gamma lines at low energy (see Figure 6.12). However, a recent study by Vasenko et al. (90) using an enriched ^{76}Ge detector has identified the 2ν decay in ^{76}Ge giving a half-life of $T_{1/2} = (0.9 \pm 0.1) \times 10^{21}$y (68% CL). This decay was also seen by Avignone et al. (91), with $T_{1/2} = (0.92\ ^{+0.07}_{-0.04}) \times 10^{21}$ y. These results establish the observation of the 2ν decay. A comparison with the calculated values is given in Table 6.4

We note that the Ge detector has also been used to search for double beta decay electrons from other candidate nuclei using samples that were brought in close contact with the detector, including ^{100}Mo, ^{148}Nd, and

Figure 6.13. Set-up of the Ge detector in the Gotthard experiment.

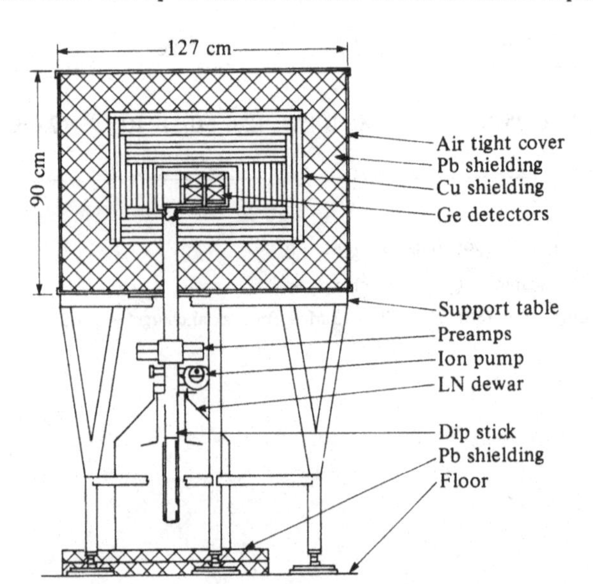

^{150}Nd (Bellotti et al. 83). Because of the poor solid angle in such a geometry, the sensitivity of these searches is not as good as for ^{76}Ge, with reported lower limits for the half-life in the neighborhood of $T_{1/2}(0\nu) > 2\times10^{18}$ y.

Experiments with semiconductor wafer stacks: ^{100}*Mo.* An interesting technique capable of significant background reduction consists of sandwiching several thin wafers of silicon semiconductor detectors (Si(Li) electron detectors) and foils of the candidate material. A double beta decay signal is expected to show up as a coincidence between two adjacent detectors. Ejiri et al. (86, 91) have studied ^{100}Mo ($E_0 = 3.034$ MeV) by preparing a sandwich of ten thin (50 mg cm^{-2}) foils of enriched ^{100}Mo interspersed with eleven 4 mm thick Si(Li) detectors. Only a few grams of ^{100}Mo (0.47×10^{23} atoms of ^{100}Mo) were available in this experiment. The electron energy resolution was 100 keV, mostly as a result of the electron energy loss in the

Figure 6.14. Set-up of the USCB-LBL Ge experiment with the NaI veto counters.

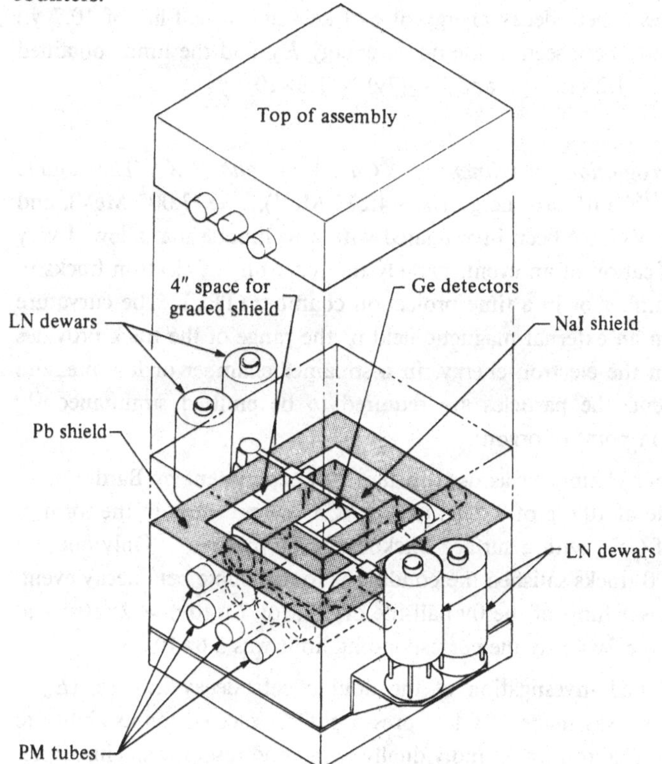

source. The experiment, installed in the Kamioka mine, has established the 2ν decay of ^{100}Mo with a half-life of $T_{1/2}(2\nu) = (1.15^{+0.30}_{-0.20}) \times 10^{19}$ y.

Low resolution ionization chambers: ^{136}Xe experiments. The isotope ^{136}Xe, which can undergo double beta decay with a decay energy of $E_0 = 2.48$ MeV, can suitably play the role of the fill gas in an ionization chamber. Here, as in the case of ^{76}Ge, the source is also the detector. In comparison to a Ge detector, an ionization chamber has a relatively poor energy resolution, and thus a less favorable signal-to-background ratio.

In an experiment at the Baksan Neutrino Laboratory in the USSR, at a depth of 850 mwe, Barabash et al. (86) have explored the double beta decay of ^{136}Xe with a 3.14 l ionization chamber pressured at 25 atm. Xenon enriched to 93% in ^{136}Xe was used, with an admixture of 0.8% H_2 to improve electron drifting. The total number of ^{136}Xe atoms was thus 20×10^{23}. Background data could be obtained conveniently by filling the chamber with natural xenon, although contamination from atmospheric ^{85}Kr presented a serious problem at low energy. (The contaminant ^{85}Kr, which is released by nuclear reactors as well as by nuclear bomb tests in the atmosphere, has a beta decay energy of 670 keV and a half-life of 10.7 y.) No excess counts were seen at the decay energy E_0, and the limits obtained were: $T_{1/2}(0\nu) > 1.2\times10^{21}$ y, and $T_{1/2}(2\nu) > 1.8\times10^{19}$ y.

Track recognition experiments: ^{48}Ca , ^{82}Se, and ^{136}Xe. The double beta decays in ^{48}Ca (decay energy $E_0 = 4.272$ MeV), ^{82}Se (3.005 MeV), and ^{136}Xe (2.48 MeV) have been investigated with a technique that allowed very specific identification of an event, namely the recording of electron tracks in a streamer chamber or in a time projection chamber (TPC). The curvature of the tracks in an external magnetic field or the range of the track provides information on the electron energy. In a streamer chamber or low pressure TPC experiment, the particles are required to be emitted simultaneously from a common point of origin.

A streamer chamber was used in the ^{48}Ca experiment by Bardin et al. (70). A sample of 10.6 g of 97% enriched ^{48}Ca was prepared in the form of a thin sheet of CaF_2 with a surface thickness of 20 mg cm^{-2}. Only one out of about 50,000 tracks satisfied the conditions for a double beta decay event, leading to a lower limit of the 0ν half-life of ^{48}Ca of $T_{1/2}(0\nu) > 2\times10^{21}$ y at 80% CL. For the 2ν decay the corresponding limit was 3.6×10^{19} y.

In a detailed investigation of the double beta decay in ^{82}Se, (E_0 = 2.995 MeV), use was made of a low pressure TPC selected for its ability to display the two electron tracks individually with good resolution. This experiment, carried out by Elliot et al. (86, 87, 91), has provided a value for the half-life for the 2ν mode, $T_{1/2}(2\nu) = (1.1^{+0.8}_{-0.3})\times 10^{20}$ y (68% CL). This result, published in 1987, represents the first laboratory observation of a double beta decay. The TPC built for this experiment is described by Moe et al. (83). A section through the chamber is depicted in Figure 6.16. Helmholz coils supply the magnetic field of 700 G and a cosmic ray veto reduces the background in this experiment carried out above ground. The source, consisting of 14 g of selenium, 97% enriched in the isotope ^{82}Se, constitutes the central electrode of the TPC. An electron pair emitted by a ^{82}Se atom in the central source foil is recognized by its projections onto a system of sense wires in the *xz* and *yz* planes. The energies of the electrons are determined from the curvature of the tracks. The sum energy spectrum of the TPC tracks is shown in Figure 6.17.

Using the same technique, Elliott et al. (91) have determined the 2ν half-life from ^{100}Mo from a 8.3 g sample of enriched $^{100}MoO_3$ and have established a 2ν half-life of $T_{1/2} = (1.16^{+0.34}_{-0.08}) \times 10^{19}$ y (68% CL).

The double beta decay of ^{136}Xe has recently been investigated by the Caltech-Neuchatel-PSI collaboration using yet another technique, a high

Figure 6.15. Portion of the Ge spectrum in the region of the decay energy of 2.041 MeV from the USCB-LBL experiment. From the absence of a peak at 2.041 MeV a lower limit of the half-life for neutrinoless double beta decay of 1.2×10^{23} y has been obtained.

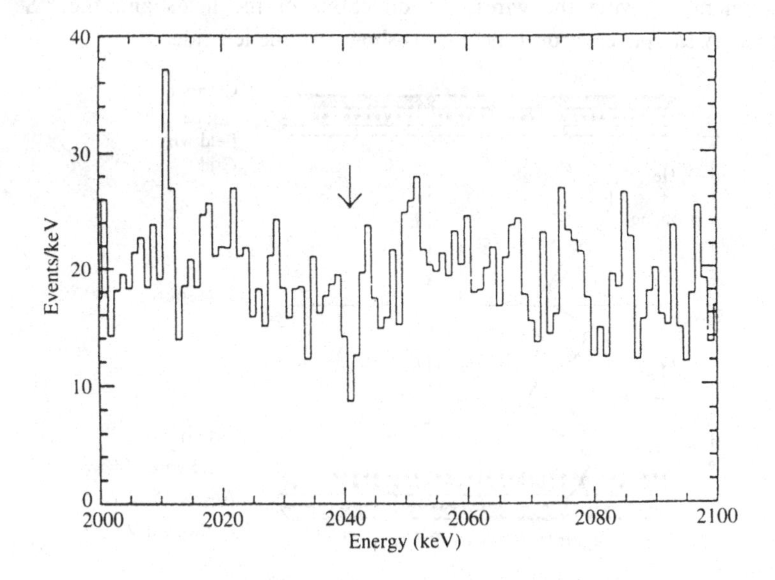

pressure TPC installed in the Gotthard tunnel (Wong et al. 91). The TPC has a fiducial volume of 207 l and was filled with enriched (62.5%) ^{136}Xe at 5 atm pressure. It thus contains 1.6×10^{25} atoms of ^{136}Xe. Secondary electrons from gas ionization by the primary double beta decay electrons drift in an electric field onto a readout plane with 168 channels, each in the x and y direction, and thus create an image of the primary electron track. The z coordinate is obtained from a measurement of the drift time. The feature of a double beta decay track is a contiguous line with a high ionization density "blob" at each end. In comparison, tracks from Comtpon electrons have only one "blob". Figure 6.18 shows two tracks, one of them a Compton electron and the other a two-electron candidate event. The measured energy resolution is 6.6% at 1.6 MeV. In the initial data, comprising 3380 hours, less than 3.5 events (90% CL) above background can be attributed to a hypothetical 0ν peak, resulting in a half-life limit of $T_{1/2} > 2.5 \times 10^{23}$ y (90% CL).

An earlier ^{136}Xe tracking experiment using a multicell proportional counter in the Gran Sasso tunnel (Bellotti et al. 91) has provided a limit of 1.2×10^{22} y for the 0ν beta decay and 1.6×10^{20} y for the 2ν decay.

Milking experiments. Two milking experiment have been performed with the goal of searching for double beta decay in ^{238}U. While ^{238}U is

Figure 6.16. Section of the TPC for the Irvine ^{82}Se double beta decay experiment, showing the wires, the directions of the fields and the ^{82}Se source. A sample electron trajectory is shown on the left side.

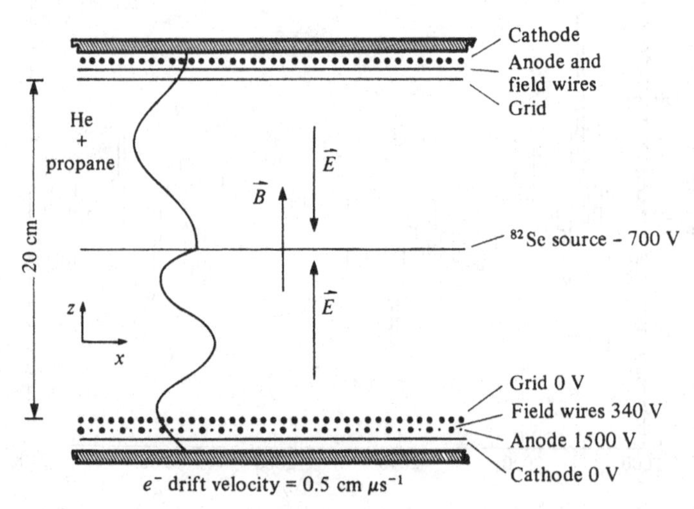

Figure 6.17. Results from the Irvine ^{82}Se experiment. The sum energy histogram of the double beta decay candidates after applying cuts for Moller scattering. The expected spectrum from 2ν is indicated by the full curve. The dashed curve is the lone electron spectrum arbitrarily scaled.

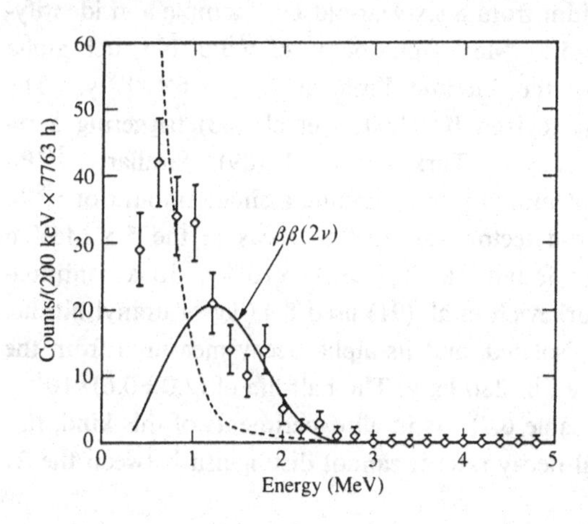

Figure 6.18 Typical tracks recorded by the TPC with 5 atm of xenon (Wong et al. 91). (a) Single electron; (b) two-electron candidate event.

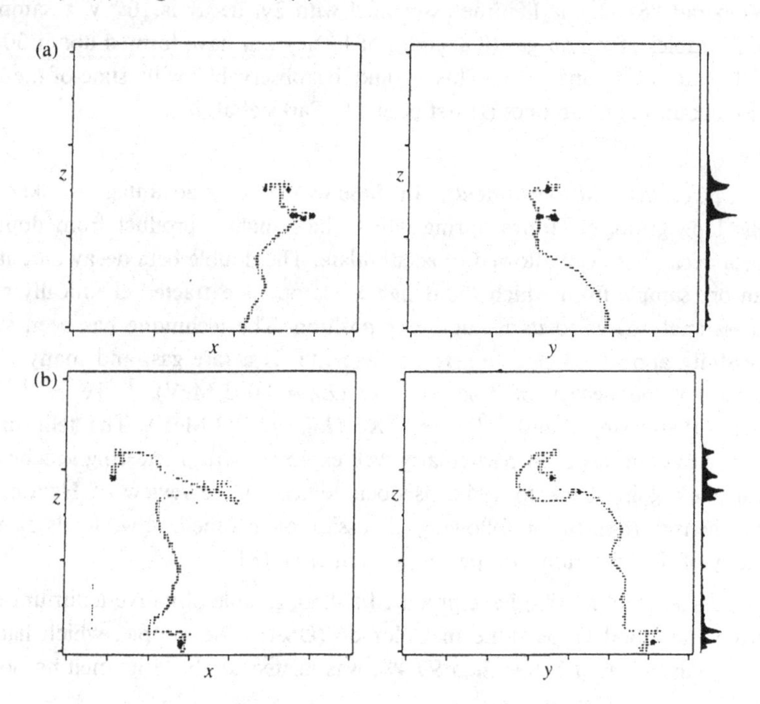

known to decay by alpha decay (with $T_{1/2} = 4.5\times10^9$ y) into the elements of the uranium series, as well as by spontaneous fission, it is energetically forbidden to beta decay into its neighbor ^{238}Np. Double beta decay leads to ^{238}Pu with $E_0 = 1.146$ MeV. Levine et al. (50) attempted to observe this decay by extracting plutonium from a six-year-old UO_3 sample and identifying it by searching for the 5.51 MeV alpha decay of ^{238}Pu. No such alpha particles were seen, fixing the lifetime limit at $T_{1/2} > 6\times10^{18}$ y. The interest in this system was revived by Haxton et al. (83) triggering a re-examination of the ^{238}U decay by Turkevich et al. (89). Similarly, ^{238}Pu was extracted from 600 g of uranyl nitrate, adding a small amount of ^{239}Pu carrier. A semiconductor α-detector was used to measure the 5.5 MeV α particles. The half-life limit found was $T_{1/2} > 4.6\times10^{19}$ y. In a continuation of this experiment, Turkevich et al. (91) used 8.47 kg of uranyl nitrate. ^{238}Pu has been chemically isolated, and its alpha decay measured from the sample that had accumulated in 280 kg y. The half-life of $(2.0\pm0.6)\times10^{21}$ y has been determined (see Table 6.7). As in all experiments of this kind, this half-life represents the total decay rate; it cannot distinguish between the 2ν and 0ν modes.

Other laboratory milking experiments are being considered. The decay of ^{136}Xe, for example, could be observed by separating the ^{136}Ba daughter atoms and counting them using laser techniques (Mitchell & Winograd 86). If the lifetime associated with 2ν decay is 10^{20} y, a sample of 10 moles of xenon gas (0.8 moles of ^{136}Xe) will have formed about 3000 ^{136}Ba atoms in one year. This should be observable with state-of-the-art atom counting techniques (Hurst et al. 79; Parks et al. 82).

Geochemical experiments. In these experiments advantage is taken of the long geological times during which the daughter product from double beta decay has been allowed to accumulate. The double beta decay parent is an ore sample from which the daughter isotope is extracted chemically and mass analyzed as to its isotopic composition. The technique has been successfully applied to cases where the daughter is a rare gas, and many data exist for the decays of ^{82}Se → ^{82}Kr ($E_0 = 3.005$ MeV), ^{128}Te → ^{128}Xe ($E_0 = 0.869$ MeV), and ^{130}Te → ^{130}Xe ($E_0 = 2.533$ MeV). The tellurium-xenon system has been particularly well explored with pioneering geochemical work going back to 1949, as documented in the review of Haxton & Stephenson (84). In the following discussion of this method we focus on the study of the tellurium isotopes by Kirsten et al. (83).

Kirsten et al. (83) have procured a 4.5 g sample of native tellurium ore from the Good Hope mine in Colorado (USA). The sample, which had a chemical purity of better than 99.4%, was heated to above its melting point in an evacuated vessel, and the escaping xenon was mass analyzed. The amount of ^{128}Xe and ^{130}Xe was found to be $(0.20 \pm 0.22)\times10^{-14}$ and $(2.11 \pm 0.08)\times10^{-11}$ cm^3 STP per gram of tellurium, respectively. (This corresponds to 1.2×10^6 and 1.27×10^{10} atoms of ^{128}Xe and ^{130}Xe, respectively.) Based on the age of the sample of $(1.31 \pm 0.14)\times10^9$ y, determined with the help of the K-Ar dating method (an age determination based on ^{40}K decay), the half-lives were found to be $T_{1/2}(128) > 8\times10^{24}$ y (95% c.l.) and $T_{1/2}(130) = (2.6 \pm 0.3)\times10^{21}$ y.

A number of corrections and possible sources of error needed to be considered carefully. The isotope abundances obtained from mass spectroscopy are always compared to the natural abundances of xenon isotopes as observed in an air sample. Among the various xenon isotopes found in the Good Hope sample, those with mass number 124, 126, 128, 134, and 136 are found to occur with the natural abundance, if normalized on 132, while those with mass number 129 and 131 show a 40% larger abundance. The isotope ^{130}Xe, on the other hand, was observed to have a 27 times larger abundance compared to an air sample, and this large enhancement was attributed to ^{130}Te double beta decay. The presence of the enhancement of isotopes ^{129}Xe and ^{131}Xe, however, signals that some other effects may contribute to altering the isotope ratios. The processes considered are fission of ^{238}U contaminants in the ore and subsequent capture of neutrons, emanating from the surrounding rock, on tellurium, as well as neutron capture on traces of ^{127}I.

As the uranium impurities are estimated to be less than 40 ppb, the fission contribution turns out to be negligible. This has been verified from the observed abundances of the fission products ^{132}Xe, ^{134}Xe, and ^{136}Xe. The isotopes ^{128}Xe and ^{130}Xe, by the way, are not fission products. As to the neutron capture on tellurium, estimates show that this process is in fact responsible for the observed excess of ^{129}Xe and ^{131}Xe. Neutron capture on traces of ^{127}I could produce ^{128}Xe, but since no excess of ^{128}Xe was observed, only a lower limit of the half-life is reported, and this process, while not ruled out, does not alter the conclusions in Kirsten's work.

Another potential error in the half-life determination may arise from incorrectly estimating the accumulation age. Incomplete gas retention in the ore may represent a possible source of error. However, the K-Ar age determination is also sensitive to possible gas losses and it may be argued that this K-Ar loss may be similar to the loss in the tellurium-xenon system.

We note that there exists another set of data obtained by a Missouri group (Hennecke et al. 75; Manuel 86; Lee et al. 91), based on tellurides from the Kalgoorlie mine, which does not agree with Kirsten's results. Hennecke finds a pronounced excess of the isotope ^{128}Xe, compared to an air sample, and thus quotes a definite half-life (rather than a limit) of $T_{1/2}(128) = (1.54 \pm 0.17)\times 10^{24}$ y. For ^{130}Te the same group reports $T_{1/2}(130) = (7 \pm 2)\times 10^{20}$ y (Manuel 86). The reason for this disagreement is presently not understood.

It has been argued that, in forming the ratio of the number of ^{128}Xe and ^{130}Xe atoms extracted from the ore, several experimental uncertainties cancel, such as those from incomplete gas retention in the ore. Kirsten's ratio is a lower limit, $T_{1/2}(128)/T_{1/2}(130) > 3000$ (95% CL), while Hennecke's ratio is a definite value, $T_{1/2}(128)/T_{1/2}(130) = 1590 \pm 50$. This ratio has been re-examined by Lee et al. (91) quoting 2380 ± 450 and 2270 ± 410, respectively, for two different Te ores. These ratios, if compared to the calculated half-life ratio for the 2ν mode of 5600 based on phase space only (see Section 6.4 and Table 6.5), lead us to the conclusion that Kirsten's results are entirely consistent with 2ν decay only and identical nuclear matrix elements for the two isotopes, while Manuel's and Lee's results would be indicative either of the presence of 0ν decay or of a difference in the nuclear matrix elements (see the discussion in Section 6.2).

Geochemical studies of ^{82}Se have provided a half-life of $T_{1/2}(82) = (1.3 \pm 0.05)\times 10^{20}$ y (Kirsten et al. 86). Again uncertainties exist in estimates of the gas retention in the selenium ore and not all groups agree on the treatment of this correction (see Haxton & Stephenson 84). However, the quoted half-life is in a reasonable agreement with the laboratory result quoted earlier.

We summarize in Table 6.7 the results for double beta decay half-lives as obtained from counting experiments (Section 6.3.1) and isotope separation experiments (Section 6.3.2). In the same table we show, for convenience, also the corresponding upper limits of the Majorana mass parameter $\langle m_\nu \rangle$ based on nuclear matrix elements of different authors (see Section 6.2).

6.4 Analysis of Neutrinoless Double Decay Data

We shall now discuss the analysis of the lifetime data for 0ν $\beta\beta$ decay. Two main issues are involved in the analysis of experimental data. The first has to do with the relation between the half-life (or its limit) and the parameters $\langle m_\nu \rangle$, $\langle \eta \rangle$, and $\langle \lambda \rangle$. The second issue involves the relation between the effective mass $\langle m_\nu \rangle$ and the true neutrino mass, i.e., the eigenvalue of the neutrino mass matrix, and, similarly, the relation between the effective lepton right-handed current parameters $\langle \eta \rangle$ and $\langle \lambda \rangle$ and the parameters η and λ in the weak Hamiltonian (6.27). Related to this are the questions, have the effective parameters $\langle \eta \rangle$ and $\langle \lambda \rangle$ a nonvanishing value for massless neutrinos, and what is the effect of neutrino mass and mixing on the relation between $<\eta>$ and η and between $<\lambda>$ and λ?

The solution of the first problem, the extraction of $<m_\nu>$, $<\eta>$, and $<\lambda>$ from the experimental data, is straightforward if one accepts the values of the coefficients C_i of Eq. (6.33) based on one particular set of nuclear matrix elements, such as one from those presented in Table 6.5. While keeping in mind the poor agreement between the calculated and geochemical 2ν $\beta\beta$ half-lives in ^{130}Te, we prefer to use a consistent set of nuclear matrix elements, rather than to resort to the arbitrary scaling procedures mentioned in Section 6.2. (The effect of scaling which amounts to dividing all nuclear matrix elements by a factor x is equivalent to multiplying $<m_\nu>$, $<\eta>$, and $<\lambda>$ by the same factor x.) It is obvious that different sets of the nuclear matrix elements (or,equivalently,of the coefficients C_i in Eq. (6.33)) lead to different limits on $<m_\nu>$, $<\eta>$, and $<\lambda>$. Perhaps, one should consider the range of the limits obtained with different sets of nuclear matrix elements, as a measure of the theoretical uncertainty. Fortunately, this range is relatively narrow for the most important parameter $<m_\nu>$.

The most stringent constraints are obtained from the ^{76}Ge decay experiments for which a half-life limit of 1.2×10^{24} years has been found (Section 6.3). From Eq. (6.33) it follows that this half-life limit confines the parameters $<m_\nu>$, $<\eta>$, and $<\lambda>$ to values inside a three-axial ellipsoid. In Figure 6.19 we show cuts through this ellipsoid by the three planes along the axes. In constructing the ellipsoid, we assumed *CP* invariance and, therefore, we used $\psi_1 = \psi_2 = 0$. From Figure 6.19 we conclude that, for our chosen set of nuclear matrix elements, the 1.2×10^{24} years half-life limit implies, for the matrix elements of Muto et al. (89) chosen as an example, that

$$
\begin{aligned}
&<m_\nu> \leqslant 1.6 \text{ eV} , \\
&|<\eta>| \leqslant 1.5\times10^{-8} , \qquad (6.36)\\
&|<\lambda>| \leqslant 2.6\times10^{-6} .
\end{aligned}
$$

These limits are obtained in a combined fit. If one assumes that $<\eta> = <\lambda> = 0$ a slightly tighter limit $|<m_\nu>| \leqslant 1.4$ eV is obtained. In either case, we see that the upper limit on $<m_\nu>$ is considerably smaller than the value of m_{ν_e} (Eq. (2.5)) obtained in the tritium decay experiments (Chapter 2). This conclusion is likely to remain true even if one were to use another more accurate set of nuclear matrix elements. We also see, as mentioned earlier, that the parameter $<\eta>$ is determined with greater sensitivity than $<\lambda>$. This is a consequence of the enhanced phase space integrals, corresponding to the recoil and p-wave effects, described in Section 6.2.

A similar analysis can be performed for other sets of nuclear matrix elements, or for other $\beta\beta$ decaying nuclei. For example the 1.2×10^{24} half-life limit in ^{76}Ge implies $<m_\nu> \leqslant$ 2.4—4.7 eV using the matrix elements of Engel et al. (88) and a range of the particle-particle interaction parameter values that result in a good agreement with the 2ν decay data. In ^{136}Xe the half-life limit 2.5×10^{23} y of Wong et al. (91) implies $<m_\nu> \leqslant$ 3.0 eV using the matrix elements of Muto et al. (89), and $<m_\nu> \leqslant$ 3.3—5.0 eV using the calculations of Engel et al. (88) and the same interval of the interaction strength as before.

The half-life ratio of ^{128}Te and ^{130}Te has been often invoked as another source of information for 0ν decay. This appears at first sight to be an excellent idea. Assuming that the addition of two neutrons to ^{128}Te does not affect the nuclear matrix elements, one concludes from Table 6.1 that the ratio of half-lives for the 2ν mode should be 5600. If the 0ν mode contributes to the decay rate, this ratio should be less than 5600. (The ratio could be as small as 24 if the decay rate is dominated by the 0ν mode, and if that mode originates entirely with the $<m_\nu>$ term.) Thus, any deviation of the ratio from 5600 would signal the presence of the 0ν mode. When discussing the experimental results (Section 6.3), we found that there is no consensus on this half-life ratio. Also, the disagreement between observed and calculated half-lives for ^{130}Te suggests that there may be important cancellations in the matrix elements $M_{GT}^{2\nu}$ (Vogel & Zirnbauer 86). The basic assumption about equality of matrix elements in ^{128}Te and ^{130}Te then becomes invalid and so does the lifetime ratio argument.

Neutrino mixing. We have discussed various consequences of neutrino mixing in Chapters 4 and 5. Here, we shall study how neutrino mixing affects the 0ν $\beta\beta$ decay parameters $<m_\nu>$, $<\eta>$, and $<\lambda>$. We shall make the plausible assumptions that a) neutrinoless $\beta\beta$ decay is caused by the virtual neutrino exchange, and b) there are no doubly-charged gauge bosons. (See the book by Kayser et al. 89 for a lucid discussion of the material in the rest of this Section.)

Let us assume that there are n lepton flavors (generations) with n left-handed Majorana neutrinos ν_{lL} and n right-handed Majorana neutrinos ν'_{lR}. A Dirac neutrino is equivalent to a pair of mass degenerate Majorana neutrinos, so our assumption accommodates Dirac neutrinos as well. Further, we shall assume that the underlying theory is of the gauge type. The neutrino mass matrix is then diagonalized by a $2n\times2n$ unitary matrix, the

Figure 6.19. Limits on $<m_\nu>$, $<\eta>$, and $<\lambda>$ derived from the experimental half-life limit of 1.2×10^{24} years for the 0ν $\beta\beta$ decay of ^{76}Ge. The curves were obtained with the values C_i of Table 6.5 based on the nuclear matrix elements of Muto et al. (89). Only the values inside the curves are allowed.

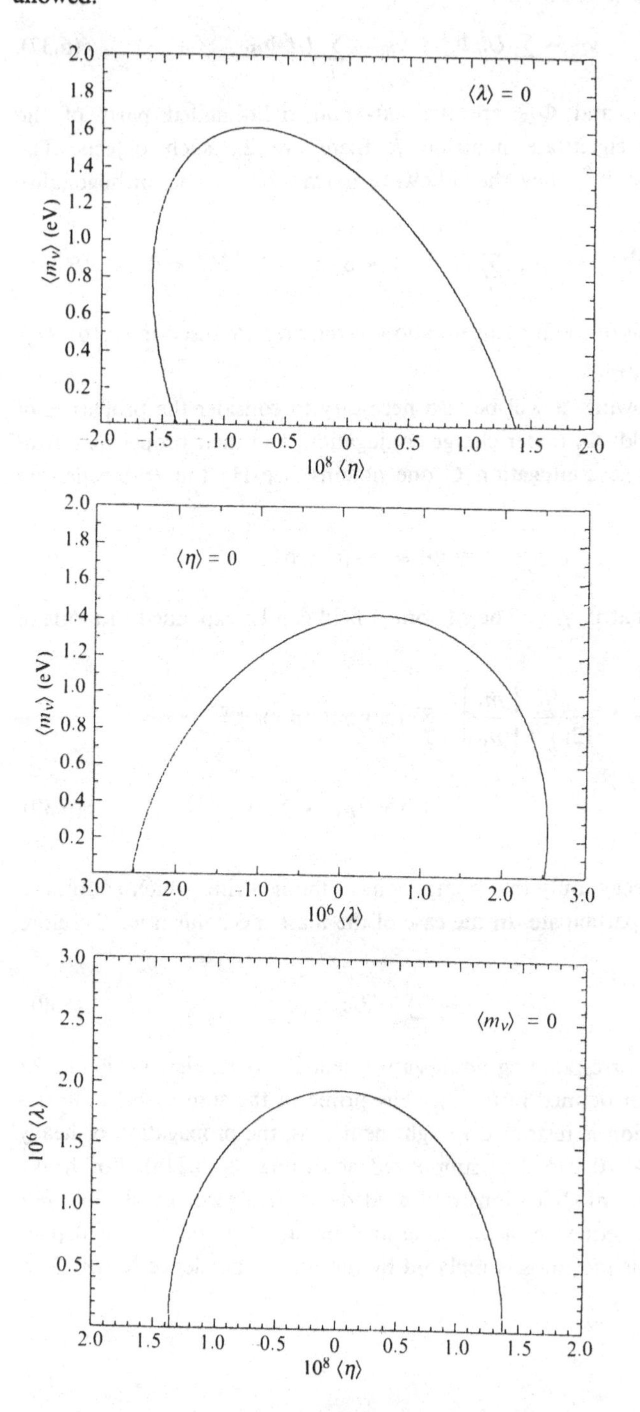

upper half of which we denote by U^L and the lower half by U^R. The neutrino flavor eigenstates are then

$$\nu_{lL} = \sum_j U_{lj}^L \Phi_{jL} \; ; \; \nu_{lR}' = \sum_j U_{lj}^R \Phi_{jR} \, . \tag{6.37}$$

The objects Φ_{jL} and Φ_{jR} are the left- and right-handed parts of the Majorana mass eigenstate neutrino j; there are $2n$ such objects. The matrices U^L and U^R obey the following normalization and orthogonality conditions

$$\sum_j U_{lj}^{L*} U_{l'j}^L = \delta_{ll'} \; ; \; \sum_j U_{lj}^{R*} U_{l'j}^R = \delta_{ll'} \; ; \; \sum_j U_{lj}^L U_{lj}^R = 0 \, . \tag{6.38}$$

(Note the relation of the present notation to the notation used in (5.16); $U_{lj}^L = V_{jl}^*$, $U_{lj}^R = V_{j(l+n)}$.)

In the following, it will be also necessary to consider the properties of the Majorana fields Φ_j under charge conjugation, and their proper quantization. Under charge conjugation C one obtains (see Haxton & Stephenson 84)

$$C\Phi_j C^{-1} \equiv \Phi_j^c = \pm \lambda_j^c C \gamma_0 \Phi_j^* \, ,$$

where C is the matrix $\gamma_4\gamma_2$. The Majorana field can be expanded in terms of Dirac spinors as

$$\Phi_j(\vec{x},0) = \int \frac{d^3p}{(2\pi)^{3/2}} \left(\frac{m_\nu}{p_0} \right)^{1/2} \sum_{\pm s} [e_j(p,s) U(p,s) e^{i\vec{p}\cdot\vec{x}} + \lambda_j^c e_j^+(p,s) V(p,s) e^{-i\vec{p}\cdot\vec{x}}] \, . \tag{6.39}$$

In 0ν $\beta\beta$ decay, only that component of the neutrino which couples to the electron can participate. In the case of the mass mechanism we therefore obtain

$$\langle m_\nu \rangle = \sum{}' \lambda_j^{c*} U_{ej}^2 m_j \, , \tag{6.40}$$

where m_j is the corresponding nonnegative neutrino mass eigenvalue and λ_j^c is the phase factor defined in (6.39). The prime in the sum (6.40) indicates that the summation is restricted to light neutrinos; the propagation of heavy neutrinos (m_ν > 10 MeV) is suppressed according to (6.21a). For heavy neutrinos, various modifications are needed (see Halperin et al. 76). For example, as a consequence of the finite nucleon size, the $(m_\nu \, e^{-m_\nu r})^2$ dependence on the neutrino mass is replaced by the m_ν^{-2} dependence for $m_\nu \gtrsim 1$ GeV (Vergados 81).

To gain a better insight into the meaning of Eq. (6.40), we must consider the constraints imposed by *CP* conservation. Under the parity transformation *P* and charge conjugation *C* the creation and annihilation operators $e^{+}(p,s)$ and $e(p,s)$ in (6.39) transform as

$$Pe_j(p,s)P^{-1} = \lambda_p^j e(-p,s)\ ;\ Ce_j(p,s)C^{-1} = \eta_c^j \lambda_j^{c*} e_j(p,s) = \pm e_j(p,s)\ .$$

(We see that the operators $e(p,s)$ transform under charge conjugation in a way that guarantees that Φ_j is a Majorana particle.) If invariance under *CP* transformation is imposed upon the Hamiltonian (6.27) and the neutrino mass term, then the mixing matrices *U* and *V* are real, and moreover, the phases η_c^j and λ_p^j are constrained by the requirement that

$$\lambda_p^j \eta_c^j = \lambda_p^1 \eta_c^1\ ,\ j=2,...,2n\ .$$

Consequently, under *CP* transformation the operators transform as

$$CPe_j(p,s)(CP)^{-1} = (\lambda_p^1 \eta_c^1)\lambda_j^{c*} e_j(-p,s)\ . \qquad (6.41)$$

The *relative CP* phases of the operators $e_j(p,s)$ and of the mass eigenstates $e_j^{+}(p,s)$ are determined, therefore, by the relative phases of λ_j^{c*}; these relative phases are ± 1. The phase factors λ_j^{c*} in (6.40) can be replaced now by the *CP* phase of the mass eigenstate *j*, a quantity which is a good quantum number of the Hamiltonian (6.27). Thus, for the case of *CP* invariance, we obtain

$$<m_\nu> = \sum_j^{2n} \lambda_j^{CP} |U_{ej}|^2 m_j\ , \qquad (6.40a)$$

where λ_j^{CP} is the *CP* relative phase of the mass eigenstate *j*. This phase is equal to ± 1 and the individual terms in (6.40a) can add or cancel each other. Wolfenstein (81) was the first to point out that cancellations in (6.40a) could also occur in *CP* conserving theories. In fact, the cancellation is complete in the limit where two mass degenerate Majorana neutrinos merge to form a Dirac neutrino.

Unless 0ν decay is seen, the upper limit for $<m_\nu>$ has no unambiguous interpretation. It might mean that there exists no neutrino with mass larger than $<m_\nu>$, but it could also mean that the individual terms in the sum (6.40) or (6.40a), possibly involving larger masses than $<m_\nu>$, cancel each other (as in the case of a Dirac neutrino, or the "pseudo-Dirac" neutrino with two almost degenerate masses). If 0ν decay is observed (and if one could verify that the mass mechanism is responsible) then one can extract from the lifetime a definite nonvanishing value of $<m_\nu>$. The following "theorem" then obviously holds true (Kayser 87): At least one Majorana neutrino is massive, and its mass m_ν is *larger* than $<m_\nu>$.

Right-handed currents. The effective parameters $<\eta>$ and $<\lambda>$ are related to the "true" parameters η and λ in the Hamiltonian (6.27) by the relations

$$<\eta> = \eta \sum' U_{ej} V_{ej} , \tag{6.42}$$

$$<\lambda> = \lambda g_V'/g_V \sum' U_{ej} V_{ej} , \tag{6.43}$$

where g_V' is the strength parameter of the hadronic right-handed current. Again, the prime indicates that summation is restricted to light neutrinos (m_j < 10 MeV). Now, it is immediately clear that $<\eta>$ and $<\lambda>$ vanish if all neutrinos are light (or massless). This is a consequence of the orthogonality condition (6.38). They also vanish for Dirac neutrinos, because the contributions of the terms j and $j+n$ cancel in that case. (See Rosen (83) and also Kayser (85) who has shown that (6.42) and (6.43) must vanish for massless neutrinos in gauge theories by considering the high energy behavior of the theory. Also, note that analogues of Eqs. (6.42) and (6.43) in the work of Haxton & Stephenson (84) contain, after the summation sign, the phase factor λ_j^{c*}, in addition to the factors $U_{ej} V_{ej}$. As explained in that work, in the case of *CP* invariant theory, which we are considering, two representations are possible; a) all masses are nonnegative and the λ_j^{c*} are ± 1, b) the masses can be negative but all λ_j^{c*} are the same. Because (6.42) and (6.43) do not contain neutrino masses explicitly, one is free to use the representation where the λ_j^{c*} do not appear.)

Concluding remarks. Actually, our statement that in the presence of right-handed currents there is no 0ν $\beta\beta$ decay for light Majorana neutrinos needs clarification. This conclusion was derived using the assumption that the neutrino mass m_j can be neglected in comparison with the momentum q of the virtual neutrino in the denominator of the neutrino propagator (see the equation following (6.20)) or, equivalently, that the mass m_j in the neutrino potential (6.21a) is negligibly small. Now, if *all* neutrinos are light, the leading terms in (6.42) and (6.43) vanish as a consequence of the orthogonality condition (6.38). The next terms in (6.42) and (6.43), obtained when the neutrino propagator (or the Yukawa potential (6.21a)) is expanded in powers of m_j, are generally nonvanishing; they are, however, small because they contain the small parameters m_j/q or $m_j r$. Therefore, even in that case the 0ν decay rate is explicitly proportional to the neutrino mass.

Thus, we see that existence of massive Majorana neutrinos is a necessary condition for a nonvanishing rate of 0ν $\beta\beta$ decay, even if weak interactions involving right-handed lepton currents exist. In contrast to the mass mechanism, the right-handed current mechanism of double beta decay requires generalization of the standard model in *two* aspects: one needs right-handed coupling and neutrino mass at the same time. It is impossible to disentangle these two aspects of the problem from the study of double beta decay alone. On the other hand, if 0ν $\beta\beta$ decay is observed, and if one can clearly recognize the right-handed mechanism (for example, by observing $0^+ \to 2^+$ transition), two significant conclusions can be drawn at once: 1) weak interactions contain coupling to the right-handed lepton currents, and, 2) there is at least one massive Majorana neutrino.

7

Massive Neutrinos in Cosmology and Astrophysics

If neutrinos have masses of several eV and if their lifetimes are comparable to, or longer than the lifetime of the universe, neutrino mass may dominate the total mass of the universe. Massive neutrinos would have a profound effect on universal expansion, primordial nucleosynthesis, and the formation and stability of structures in the universe. At the same time, cosmology and astrophysics provide important constraints on the properties of massive neutrinos. Consequently, the role of massive neutrinos in cosmology and astrophysics is an important part of the problems associated with neutrino mass. However, the various issues and techniques of cosmology and astrophysics are to a large extent beyond the scope of this book. Therefore, we present in this Chapter only a brief overview based on rather elementary considerations and refer the reader to the more rigorous derivations and discussion elsewhere, e.g., in Weinberg (72). Many problems related to the interface between cosmology and particle physics are reviewed by Steigman (79) and by Zel'dovich & Khlopov (81). The monograph by Kolb & Turner (90) treats all issues of this chapter in detail.

7.1 Cosmological Constraints on Neutrino Properties

Critical density. The simplest plausible assumption about the large scale mass and energy distribution in the universe is based on the *cosmological principle*, which states that all positions in the universe are equivalent and hence the universe is homogeneous and isotropic. This principle leads naturally to Friedmann's solution of Einstein's field equations, which describes uniform expansion of the cosmic fluid with time. The cosmological principle is consistent with observational astronomy (distribution of galaxies and radiosources, isotropy of the radio, x-ray and microwave background radiations), provided we average over distances of $\sim 10^{26}$ cm ($\sim$30 Mpc), a scale larger than typical clusters of galaxies, but significantly smaller than the radius of the visible universe ($\sim 10^{28}$ cm). At the same time, the Friedmann expansion is consistent with the observation of the cosmological red shift. This phenomenon, if interpreted as a Doppler shift of the light emitted by distant galaxies, leads to the conclusion that the velocity v of distant galaxies is to first approximation (i.e., for $r < 10^{28}$ cm) proportional to their distance from us,

$$v = H_0 r \ , \tag{7.1}$$

where H_0 is the Hubble's constant. (H_0 is actually not a constant, it is a function of time which, at present, has the value H_0. For a flat, matter dominated universe, $H_0 = 2/(3t_0)$, where t_0 is the time since the "big bang".)

In the Friedmann universe, a crucial role is played by the critical density ρ_c. If the true density ρ is equal to the critical density, the universe is "flat"; it is expanding now, but will ultimately, after an infinite time, come to rest. (If $\rho > \rho_c$, the universe is "closed" and will recontract after a finite time, while if $\rho < \rho_c$ the universe is "open" and will expand forever.)

Let us derive an expression for ρ_c, based on elementary considerations. Take a sphere of radius R and a probe particle of mass μ on the surface of this sphere. The kinetic and potential energies of this probe particle are,

$$T = \frac{1}{2}\mu\left(\frac{dR}{dt}\right)^2 \ , \quad U = -G_N \frac{4\pi}{3}\frac{R^3\rho\mu}{R} \ , \tag{7.2}$$

where G_N is Newton's constant and where we used the fact that the homogeneous matter outside the sphere does not contribute to the potential energy of the probe. Now, a steady state expansion is reached when the total energy of the probe particle vanishes ($T + U = 0$), that is for

$$\left(\frac{1}{R}\frac{dR}{dt}\right)^2 = \frac{8}{3}\pi G_N \rho_c \ ,$$

or, based on the fact that the left-hand side is nothing else than H_0^2, we obtain

$$\rho_c = \frac{3H_0^2}{8\pi G_N} = 1.9\times 10^{-29}\, h_{100}^2 \text{ g cm}^{-3} = 1.05\times 10^{4}\, h_{100}^2 \text{ eV cm}^{-3} \ . \tag{7.3}$$

In (7.3) we have used the notation h_{100} for Hubble's constant in units of 100 km s^{-1} Mpc^{-1} = $(9.8 \times 10^9 \text{ y})^{-1}$. The experimental value of h_{100} is uncertain, but an interval of 0.5—1.0 probably sandwiches the correct value.

It is customary to express the average density ρ of the universe in units of the critical density ρ_c of Eq. (7.3) using the notation $\Omega = \rho/\rho_c$. The observational value of Ω is not known reliably, but luminous parts of galaxies (i.e., baryonic matter) give a rather small value $\Omega_{GAL} \sim 0.01$ (Steigman 86). Thus, the baryonic matter in visible stars is insufficient to close the universe by approximately two orders of magnitude. On the other hand, besides counting the luminous stars in a galaxy, one can determine the gravitational mass of a galaxy in the following indirect way, based on the dynamics of the galactic motion. Galaxies are often surrounded by hydrogen gas, and from the Doppler shift of the spectral lines of this gas, one can determine its orbital velocity v_H. The total mass inside an orbit of radius r can be determined from Newton's law,

$$G_N \frac{M(r)}{r^2} = \frac{v_H^2}{r} \; ; \; \therefore \; v_H^2 = \frac{G_N M(r)}{r} , \tag{7.4}$$

where $M(r)$ is the total mass inside the radius r. Experimentally, the rotation curves are "flat", meaning that the velocity v_H does not change with distance, for distances where there is no more light (beyond the Holmberg radius). We conclude from this finding and from (7.4) that the galactic mass is still growing at these radii ($M \sim r$) and, therefore, that galaxies contain nonluminous (dark) matter which forms an extended "halo". From (7.4) and the measured velocities, one can determine the gravitational matter of galaxies including halos. A similar procedure is possible for groups and clusters of galaxies and the inferred average density is $\Omega_{DYN} \sim 0.2 \pm 0.1$, a quantity much larger than the density of luminous matter alone, but still insufficient to close the universe ($\Omega = 1$).

Flatness problem. The age of the universe, as determined by geochronology and the age of the oldest stars, or its lower limit from H_0^{-1}, is found to be $t_0 \geqslant 10^{10}$ y (the best estimate is $t_0 = 1.5 \times 10^{10}$ y). To put this number in perspective, let us note that the "natural" time scale is the Planck time t_p, a time unit obtained from the combination of the Newton's constant G_N, the speed of light c, and Planck's constant $\hbar$,

$$t_p = \left(\frac{G_N \hbar}{c^5} \right)^{\frac{1}{2}} = 5.4 \times 10^{-44} \text{ s} .$$

Thus, in "natural" units, the age of the universe is $t_0/t_p \approx 10^{61}$. Now, if $\Omega < 1$, then $\Omega(t)$ changes as t^{-1} for large cosmological times, while if $\Omega = 1$, Ω remains unity at all times. Therefore, if Ω at the present time is close to one, but not exactly one, it differed from $\Omega = 1$ by an amount of only one part in 10^{61} shortly after the Planck time. This is the so called "flatness" problem, because such a fine tuning appears "unnatural".

A very attractive solution to the flatness problem is *inflation* (Guth 81;

Linde 82; Albrecht & Steinhardt 82). In this scenario, the universe, during its early evolution, experienced an epoch of "inflation", an exponential expansion, which led to an exponentially small spatial curvature. In this case, provided the so called cosmological constant in Einstein's equations of motion is absent, the present value of Ω should be arbitrarily close to unity. (Einstein's equations contain an undetermined cosmological constant Λ which behaves as if it were the energy density of the vacuum. Most theorists believe that, for some as yet undiscovered symmetry reasons, Λ vanishes.)

Thus, we come to the conclusion that the universe contains large amounts of nonluminous matter, some (or all) of it concentrated in or around galaxies, groups, and clusters. It seems likely that the average energy density is equal to the critical density, i.e., that $\Omega = 1$. An upper limit, $\Omega \leqslant 2$, is provided by the observational estimates of the deceleration parameter. Is it possible that massive neutrinos are responsible for this "dark" matter?

Density of relic neutrinos. Let us consider early epochs of the universe, when its size was much smaller. The universe was then much hotter, and was "radiation dominated", i.e., photons and other relativistic particles contributed significantly to the energy density. It is quite easy to calculate the number densities and energy densities of the ultrarelativistic Bose and Fermi gases. These number densities are

$$n_B = \frac{g_s}{(2\pi\hbar)^3} 4\pi \int_0^\infty \frac{p^2 dp}{e^{pc/kT}-1} = \frac{g_s}{2\pi^2}\left(\frac{kT}{\hbar c}\right)^3 \times 2 \times \zeta(3) , \tag{7.5}$$

$$n_F = \frac{g_s}{(2\pi\hbar)^3} 4\pi \int_0^\infty \frac{p^2 dp}{e^{pc/kT}+1} = \frac{g_s}{2\pi^2}\left(\frac{kT}{\hbar c}\right)^3 \times \frac{3}{4} \times 2 \times \zeta(3) , \tag{7.6}$$

where g_s is the number of possible spin states, $g_s = 2$ for photons and Majorana neutrinos and $g_s = 4$ for Dirac fermions, and $\zeta(3) \approx 1.2$. Similarly, the energy densities are

$$\rho_B = \frac{g_s}{(2\pi\hbar)^3} 4\pi \int_0^\infty \frac{p^2 dp cp}{e^{pc/kT}-1} = \frac{g_s}{2} \frac{\pi^2 k^4 T^4}{\hbar^3 c^3} \times \frac{1}{15} , \tag{7.7}$$

$$\rho_F = \frac{g_s}{(2\pi\hbar)^3} 4\pi \int_0^\infty \frac{p^2 dp cp}{e^{pc/kT}+1} = \frac{g_s}{2} \frac{\pi^2 k^4 T^4}{\hbar^3 c^3} \times \frac{7}{8} \times \frac{1}{15} . \tag{7.8}$$

Let us note, in passing, that the blackbody background radiation has an experimentally determined temperature of $T = 2.7$ K at the present time; there are ~ 400 photons per cm^3, and the corresponding energy density is

$\rho_\gamma = 4 \times 10^{-34}$ g cm^{-3}, i.e., five orders of magnitude smaller than the critical value (7.3).

In the "radiation dominated" epoch, the energy density, temperature, and time are related through

$$\rho = \frac{3}{32\pi G_N} t^{-2} \; ; \; kT = \left(\frac{45\hbar^3 c^3}{32\pi^3 G_N g_s^*} \right)^{1/4} t^{-\frac{1}{2}} , \tag{7.9}$$

where g_s^* is the effective g-factor at the temperature T. It includes all ultrarelativistic particles with masses $m < kT$ (e.g., for photons, electrons, and three flavors of two-component neutrinos, $g_s^* = 1 + 7/4 + 3 \times 7/8$). Substituting the values of the constants, we find that at $t = 10^{-4}$ s the temperature was $kT = 200$ MeV and, therefore, particles with masses above 200 MeV and lifetimes shorter than 10^{-4} s ceased to exist. On the other hand, pions and nucleons were in thermal equilibrium with muons, e^+e^- pairs, photons, and also with $\nu_e\bar{\nu}_e$, $\nu_\mu\bar{\nu}_\mu$, and $\nu_\tau\bar{\nu}_\tau$ pairs. That the neutrinos remain in equilibrium can be seen from the following argument. The weak interaction cross section (in units $\hbar = c = 1$) is of the order $\sigma \sim G_F^2 E^2 \sim G_F^2 (kT)^2$, while the number density (7.6) is $\sim (kT)^3$. Thus the time between collisions is

$$t_\nu = (n_\nu \sigma v)^{-1} \sim G_F^{-2} (kT)^{-5} , \tag{7.10}$$

while the expansion time scale is determined from (7.9) as

$$t_{exp} = \frac{1}{\sqrt{G_N}(kT)^2} . \tag{7.11}$$

(Note that the muon and tau neutrinos remain in equilibrium because they interact via the neutral current weak interaction even at temperatures at which they can no longer interact via the charged current weak interaction.) Thermal equilibrium is maintained as long as $t_\nu \ll t_{exp}$. Substituting the constants into (7.10) and (7.11), we find that $t_\nu = t_{exp}$ for $kT \sim 1$ MeV; neutrinos remain in equilibrium for temperatures above this value. While in equilibrium, the number density of each Majorana neutrino flavor is proportional to the photon number density, $n_\nu / n_\gamma = 3/4$, according to (7.5) and (7.6). This ratio remains unaltered even when neutrinos "decouple", i.e., are no longer in equilibrium. However, at $t \approx 10$ s the e^+e^- pairs annihilate and contribute their energy to the photons. The entropy (proportional to ρ/T) is conserved and thus the photon number density increases by the factor $1 + 7/4 = 11/4$ (see (7.7) and (7.8)). Finally, we conclude that the present number density of each light Majorana neutrino flavor is

$$n_\nu = (4/11) \times (3/4)\, n_\gamma = (3/11) n_\gamma \approx 110 \text{ particles per cm}^3 . \tag{7.12}$$

(The corresponding number density for each light Dirac neutrino is

$\approx$ 220 particles per cm^3.) This "background neutrino sea" has not been experimentally observed so far. We discuss in Chapter 3 some of the proposals to observe it, and the difficulties associated with them.

Cosmological neutrino mass limit. The relic neutrinos, if sufficiently massive, will be nonrelativistic at the present time, and their energy density is simply their number density times their mass. Therefore, to close the universe at $\Omega = 1$ we need

$$m_\nu^M = 10^4\, h_{100}^2/110 \text{ eV}\,,\quad m_\nu^D = 10^4\, h_{100}^2/220 \text{ eV}\,, \tag{7.13}$$

for the Majorana and Dirac neutrinos, respectively. (Remembering that $0.5 < h_{100} < 1.0$, we see that the required mass is between 10 and 100 eV.) Moreover, we know from observation of the expansion rate that $\Omega \leqslant 2$ and therefore obtain from (7.3) and (7.12) the often quoted upper limit for the sum of neutrino masses

$$\sum_l \frac{g_s^l}{2} m_\nu \leqslant 200 h_{100}^2 \text{ eV}\,, \tag{7.14}$$

where the sum is over all light stable neutrino flavors with the corresponding spin degeneracy factor g_s^l.

A somewhat more stringent limit is obtained if one assumes that $\Omega = 1$, i.e., $\rho = \rho_c$. The age of the universe is then $t_0 = 2/(3H_0)$. Since, as stated above, observations require that the universe is at least $t_0 \geqslant 1.1 \times 10^{10}$ years old, we conclude that $h_{100} \leqslant 0.6$ and

$$\sum_l \frac{g_s^l}{2} m_\nu \leqslant 36 \text{ eV}\,.$$

Until now we have considered only light neutrinos, with masses m_ν less than a few MeV, and we concluded that they must obey the limit (7.14). For completeness, we should mention that cosmological considerations alone also allow the existence of relatively heavy neutrinos. Heavy neutrinos remain in equilibrium until the temperature reaches $kT \approx m_\nu/20$. Their relic number density is then (Lee & Weinberg 77)

$$n_\nu/n_\gamma \approx 10^{-7}\,(m_\nu/\text{GeV})^{-3}\,.$$

Thus, we conclude from (7.3), and using $n_\gamma = 400$, that $4 \times 10^{-5} \times m_\nu^{-2} \leqslant 2 \times 10^{-5} \times h_{100}^2$ and thus $m_\nu(\text{GeV}) \geqslant \sqrt{2}/h_{100}$. (Again, a more stringent limit $m_\nu(\text{GeV}) \geqslant 2/h_{100}$ can be obtained.) There are, therefore, two regions of allowed neutrino masses, light neutrinos below the few tens of eV range, and very heavy neutrinos above the few GeV range.

Intermediate masses between these two extremes are possible only if the corresponding neutrinos are unstable (Chapter 4) and would have decayed by now. Of course, one has to consider the corresponding decay products. For example, radiative decays would produce photons of observable fluxes. Turner (81) has analyzed the situation in detail by considering the allowed and forbidden regions in the mass vs. lifetime plane. He concludes that, from the point of view of cosmology and astrophysics, three "islands" are allowed; the above discussed "islands" of the light and heavy stable or long lived neutrinos and, in addition, a region of short lived ($T_{\frac{1}{2}} < 10^6$ s) neutrinos heavier than ~ 1 MeV.

For the unstable neutrinos one can derive a very general lifetime limit, including decays into "invisible" particles, by considering the energy density $\rho_D(t_0)$ of the decay products at the present time. Obviously, $\rho_D(t_0)/\rho_c$ must not overclose the universe. To derive the limit on the neutrino lifetime τ we follow Kolb & Turner (90). Assuming that the decay products are relativistic even now and that at $t = \tau$ the scale factor ("size of the universe") $R(t)$ varied as t^n (n=1/2 for the radiation dominated epoch and n=2/3 for the matter dominated epoch) we find that

$$\rho_D(t_0) = \Gamma(n+1)\rho_\nu(t_0)\frac{R(\tau)}{R(t_0)} , \tag{7.15}$$

where $\rho_\nu(t_0)$ is the density that neutrinos would have had, had they not decayed. Thus we see that $\rho_D < \rho_\nu$. From (7.15) it follows that, e. g., for 100 eV $\leqslant m_\nu \leqslant$ 1 MeV

$$\tau(\mathrm{s}) \leqslant \left[\frac{4\times 10^{13}}{m_\nu(\mathrm{eV})}\right]^{3/2} (\Omega h^2_{100})^{-1/2} . \tag{7.16}$$

Eq. (7.16) is valid if the neutrino lifetime τ is longer than the t_{EQ} time when the universe became matter dominated ($t_{EQ} \sim 10^3(\Omega h^{100})^{-2}$ y), and if, further, the universe was matter dominated ever since.

Primordial nucleosynthesis. The synthesis of the light elements (^{2}H, ^{3,4}He, ^{7}Li) provides a probe of the early universe and tests the standard theories of Big Bang and particle physics. (The field has been reviewed by Schramm & Wagoner 77 and by Boesgaard & Steigman 85.) The observed abundances of these elements range over nearly ten orders of magnitude. That the standard theory is able to explain these abundances is a remarkable achievement. Once the qualitative success of the standard theory has been established, one may use the detailed quantitative comparison with observations and derive limits for the nucleon abundance and for the number of flavors of the light weakly interacting particles. In particular, it is possible to show that there cannot be enough baryons to close the universe ($\Omega = 1$) and that the number of light neutrino flavors is smaller than four.

Let us show first how the helium/hydrogen mass ratio Y depends on number of neutrino flavors. As stated earlier the neutrinos "decouple" at temperatures around 10^{10} K and $t = 1$ s. At that temperature or time, the neutron to proton ratio "freezes out" because the reactions $n+e^+ \leftrightarrow p+\bar{\nu}_e$, $n+\nu_e \leftrightarrow p+e^-$, and $n \rightarrow p+e^-+\bar{\nu}_e$ are no longer in equilibrium. At the somewhat later time, and lower temperature, $t \approx 100$ s, $T \approx 10^9$ K, nuclear fusion begins with $n+p \rightarrow {}^2\mathrm{H}+\gamma$ and subsequent formation of ^{4}He and small amounts of ^{3}He and ^{7}Li. This sequence is possible because the time is short enough that neutrons have not disappeared by beta decay, the temperature is high enough that the nuclear fusion reactions are not excessively inhibited by Coulomb barriers, and at the same time the temperature is low enough that the deuterons do not photodisintegrate immediately. It turns out that in the first 20 minutes essentially all neutrons are incorporated in ^{4}He and that the primordial helium mass fraction Y is

$$Y \approx \frac{2n/p}{(n/p)+1}\,, \tag{7.17}$$

where n/p is the neutron to proton ratio at the moment of "freeze out". This ratio, according to (7.9) depends on g_s^*, and through it on the number of neutrino flavors (the predicted Y increases by about 0.01 for each additional flavor). The results of the numerical solution of the equations governing the primordial nucleosynthesis are shown in Figure 7.1 adapted from Yang et al. (84). The independent variable there is η = baryon/photon number density ratio. From observations the plotted quantities have the following values (Steigman 84): $0.22 \leqslant Y \leqslant 0.26$, $^2\mathrm{H}/^1\mathrm{H} \geqslant (1\text{—}2) \times 10^{-5}$, $(^2\mathrm{H} + {}^3\mathrm{He})/^1\mathrm{H} \leqslant (6\text{—}10) \times 10^{-5}$, and $^7\mathrm{Li}/^1\mathrm{H} \approx (1.1\pm0.4) \times 10^{-10}$. By comparing with the figure, we see that the standard model can account for the observed abundances, provided that the nucleon abundance η is within the interval $(3\text{—}7) \times 10^{-10}$ and that there are no more than four light neutrino flavors. The importance of the precise value of the neutron lifetime is also apparent.

Detailed analysis shows that the number of light neutrino flavors at equilibrium at the time of decoupling is $N_\nu \leqslant 3.4$ (Steigman 90). Among these neutrinos there could be some that are dominantly weak singlets but subdominantly coupled to the standard weakly interacting doublet neutrinos. (See Chapter 4 for the general discussion of the subdominantly coupled neutrinos.) For a relatively large interval of masses and mixing angles (Kainulainen 90), the corresponding neutrino will be in equilibrium and will, therefore, be counted in the above limit for N_ν. The nucleosynthesis limit is thus complementary to the limit for N_ν based on the study of the Z^0 width, which gives $N_\nu = 2.9 \pm 0.1$ (Dydak 90). This latter result applies only to the weak doublet neutrinos with masses $m_\nu < M_Z/2$; it is insensitive to mixing among such neutrinos and would accommodate subdominantly coupled singlet neutrinos.

Returning to the baryon density we conclude from the above η that $\Omega_{nucleon} = 0.01$—0.1, where we have taken into account the allowed interval of η and h_{100}. Nucleons, therefore, can account for the whole amount of luminous matter in the universe, and perhaps even for the Ω_{DYN} derived from the dynamics of galaxies, groups, and clusters. It appears, however,

Figure 7.1 The predicted abundances of ^{2}H, ^{3}He, and ^{7}Li (by number relative to H, denoted by A/H) and ^{4}He (by mass, denoted by Y) as a function of the nucleon to photon ratio η. The calculation used $T_{1/2}$=10.6 min for the neutron half-life. For ^{4}He the predictions are shown for three numbers of neutrino flavors, N_ν = 2, 3, 4. The "error" bar shows the range of Y which corresponds to $10.4 < T_{1/2} < 10.8$ min.

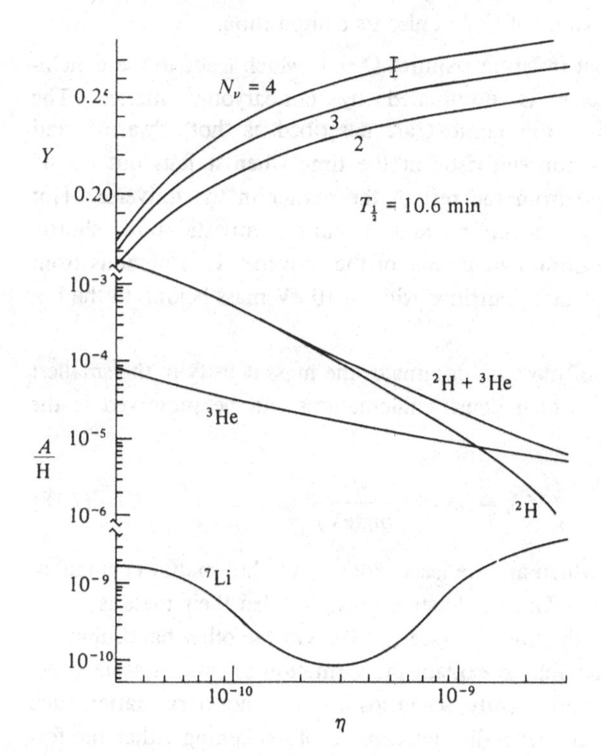

that nucleons alone cannot close the universe at $\Omega = 1$. This average energy density could be achieved only if the "dark matter" is nonnucleonic.

Neutrinos and large scale structures. Up till now we have assumed that the universe is homogeneous and isotropic. In reality, there exist structures in the density distribution, representing galaxies, groups, clusters, and superclusters ranging in mass from 10^{11} to 10^{15} solar masses $M_\odot$. (At these scales, as we well know, the universe is obviously highly anisotropic. The galactic densities are 10^5 of the average and the densities in clusters are 10^2—10^3 of the average density in the universe.) We would like to know, what effect, if any, the existence of massive (eV range) neutrinos would have on the problem of the formation of these structures in the early universe.

The blackbody background radiation is very isotropic, with $\delta T/T \leqslant 10^{-4}$ on angular scales from 1' to 90°. This implies that the universe at decoupling (or at the somewhat later time of the last photon scattering) was very smooth. The structures can begin to grow out of small density fluctuations once the universe is matter dominated. (This happened at $t_{EQ} = 1.4\times10^3(\Omega h_{100})^{-2}$ y.) In order to describe structure formation quantitatively, one has to know not only the total amount of nonrelativistic matter at t_{EQ}, i.e., the value of Ω, but also its composition.

We have seen that inflation requires $\Omega = 1$, which leads to the conclusion that the universe is dominated by nonbaryonic matter. The nonbaryonic particles in this context are described as "hot", "warm", and "cold". Cold matter is nonrelativistic at the time when it falls out of the equilibrium (decouples) from the rest of the matter in the universe. Hot matter is relativistic at decoupling and remains relativistic until shortly before it becomes the dominant matter of the universe. It is obvious from our previous discussion that neutrinos with ~ 10 eV mass belong to the hot matter.

When a particle of mass m_i dominates the mass density ρ, the smallest scale of structures on which density fluctuations can be preserved is the effective Jeans mass

$$(M_J)_i = 3\times10^{18}\frac{M_\odot}{m_i^2(\mathrm{eV})} . \tag{7.18}$$

Therefore, neutrinos, which are the least exotic of the hot matter candidates, would give only the very large scale structures provided their mass is of the order required to close the universe (see (7.13)). On the other hand, massive neutrinos would not be able to explain the formation of smaller scale structures, e.g., galaxies. Consequently, scenarios based on hot dark matter, such as massive neutrinos, are generally not capable of explaining either the formation of smaller scale structures in the universe, or the existence of dark, nonluminous matter corresponding to these scales. Moreover, the present limits on the neutrino mass (see Chapter 2) already essentially exclude electron neutrinos as dark matter candidates. (For more details and additional references we refer to the review by Klinkhamer (86), and the monograph by Kolb & Turner (90).)

7.2 Neutrinos and Supernova SN1987A

On February 23, 1987, a type II supernova was discovered in the Large Magellanic Cloud (I. Shelton, IAU Circular 4316) at the estimated distance of 50 kpc. Shortly before the arrival of the optical signal, a neutrino burst was observed in two large water Cerenkov detectors, originally built to study proton decay. (Two other detectors, one in the Mt. Blanc tunnel and the other one in the Baksan valley also reported evidence for the neutrino signal.) The supernova 1987A, the first "naked eye" supernova since Kepler's time, thus established a whole new field of neutrino astrophysics. There is a vast literature devoted to all aspects of the SN1987A discovery. Some of the review papers on the subject are: Trimble (88), Hillebrandt & Hoflich (89), Arnett et al. (89), and Schramm & Truran (90). The story of the supernova 1987A is described in Murdin (90); the collection of articles in Petschek (90) describes various aspects of the supernovae. Below we describe only the features of the SN1987A relevant to our subject, i.e., neutrino physics.

A supernova explosion is a powerful neutrino source. Generally, one expects that stars of mass larger than about eight solar masses collapse into neutron stars, or possibly black holes, when their nuclear fuel has been exhausted. During the collapse of the core of about 1.4 $M_\odot$ size a large amount of energy, of the order of the gravitational binding is released. The characteristic luminosity unit is 3×10^{53} erg, and essentially all energy is in the form of neutrino kinetic energy. The initial part of the burst, of a few ms duration, is expected to contain mostly electron neutrinos from the electron capture processes leading to the collapse, the later part of the burst contains a statistical mixture of all neutrino species. Thus, for three flavors, the electron antineutrinos will comprise $\approx 1/6$ of the total flux, with average energy of 10—15 MeV. The neutrino energy spectrum follows the Fermi thermal distribution (7.6), with $kT \approx 3$ MeV for ν_e, $kT \approx 6$ MeV for ν_μ, $\bar{\nu}_\mu$, ν_τ, $\bar{\nu}_\tau$, and kT between these two values for $\bar{\nu}_e$; all modified by neutrino absorption. The neutrino emission should decrease with time, with a time constant of the order of seconds.

All neutrinos, in particular the ν_e, could be detected via (ν,e) scattering, the recoil electrons pointing toward the source. It is not clear whether any of the detected neutrinos can be attributed to this reaction. Electron antineutrinos could also be detected via the reaction $\bar{\nu}_e p \rightarrow e^+ n$ (see Section 5.4) which has an approximately 100 times larger cross section, and the positrons are distributed essentially isotropically. All the counts in the Kamiokande and IMB detectors apparently belong to this category. (The detectors were not able to distinguish between the two reactions, except by angular distribution.) Was the observed signal really associated with the supernova? It is extremely unlikely that the signal was due to statistical fluctuation; one expects such a coincidence only once every 7×10^7 or 1×10^5 years, depending on the event signature used. Thus, keeping in mind that at least *two* detectors observed events in coincidence, it is reasonable to ascribe the signal to the burst of neutrinos accompanying the SN1987A collapse.

The signal in the Kamiokande II detector (Hirata et al. 87) consists of eleven electron events of energy 7.5 to 36 MeV, arriving within 12 s interval at 7:35:35 (± 1 min) UT on February 23; eight of these events were observed within the first two seconds.

The signal in the IMB detector (Bionta et al. 87) consists of eight events within 6 s with energy 20—40 MeV (the higher average energy is caused by the higher energy threshold of the IMB detector). The arrival time of the first neutrino has been accurately determined as 7:35:41.37 UT on February 23.

Taking into account the energy dependent efficiency and the reaction cross section, the Kamiokande and IMB signals are compatible with $kT \approx$ 4.5 MeV for the $\bar{\nu}_e$ and with the total energy output of $\approx 2 \times 10^{53}$ ergs in all six neutrino flavors. With the different thresholds of the two detectors, the fluxes of $\approx 10^{10}$ cm^{-2} s^{-1} are mutually compatible and in agreement with the expected energy output of a supernova. The duration of neutrino emission also agrees with expectations. When comparing data and models of supernovae or neutrinos (or some other aspects of the problem, such as the nuclear equation of state, etc.) one should be cautious with far reaching conclusions as one is dealing with the statistics of a small number of events. Nevertheless, even with this caveat in mind, SN1987A proved to be an extremely rich source of information. It whetted the appetite of many observers who are now eagerly waiting for the next supernova in our galaxy proper, with much larger neutrino fluxes.

In Figure 7.2 we show the energies and timing of the Kamiokande and IMB signals. An overall time shift was applied to the Kamiokande data which are uncertain to ± 1 minute.

As mentioned above two smaller liquid scintillator detectors also observed statistically compatible, but proportionally weaker signals. There were three events with 12—17 MeV in the Baksan detector (Alexeyev et al. 87) and two events with 7—9 MeV in the Mt. Blanc detector (Aglietta et al. 87). In addition, the Mt. Blanc detector reported five events almost 5 hours earlier. This cluster of events was not seen at that time by any of the other detectors. The exact nature of the earlier Mt. Blanc signal is unclear; it is likely that it was a rare statistical fluctuation of the background.

Below we discuss the constraints on the neutrino properties derived from the timing, energies, angular distribution, etc. of the observed events.

Neutrino mass. If one can determine both the velocity and energy of a particle, it is possible to deduce its mass. Indeed, from the relation

$$v = \frac{p}{E} = \frac{(E^2 - m^2)^{\frac{1}{2}}}{E},$$

it is easy to see that, for an ultrarelativistic particle, the time delay caused by a nonvanishing rest mass equals

$$\Delta t(\mathrm{s}) = 0.026(d/50\ \mathrm{kpc})(m/1\ \mathrm{eV})^2(10\ \mathrm{MeV}/E)^2\,, \tag{7.19}$$

where, in anticipation, we have chosen to measure the time in s, the mass in eV, the energy in units of 10 MeV, and the distance in units of 50 kpc =

Figure 7.2 Energies vs. time of the Kamiokande (open squares) and IMB (filled squares) neutrino events. The Kamiokande time axis has been shifted as if the first neutrinos arrived simultaneously in both detectors.

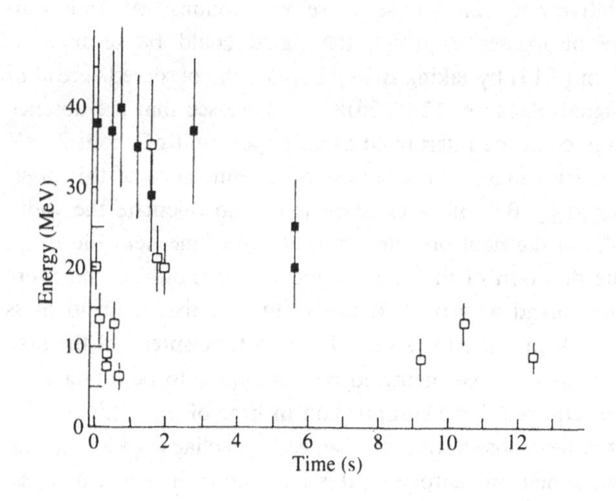

1.54×10^{21} m. If a neutrino source emits neutrinos of different energies within a time interval at the source not wider than Δt, the detector at the distance d will detect the most energetic neutrinos first, and the least energetic ones last. The neutrino mass can be calculated from the time spread of the signal. (It is obvious that useful mass information can be obtained only with extraterrestrial neutrino sources.)

The observed events, depicted in Figure 7.2, are not arranged in an obvious pattern of decreasing energy with arrival time. In addition, the duration of the signal is compatible with the expected emission time. Thus, an unambiguous determination of the neutrino mass is improbable. Also, as mentioned earlier, the $\bar{\nu}_e$ are not expected to have a well-defined temperature throughout. Since they start out with charged current interaction and end with only neutral current interaction the average energy should increase with time. An additional difficulty is the uneven number of events within the total time interval. For example, in the Kamiokande detector the first eight events are separated from the last three by seven seconds. Are the last three events "true" events, or are they (or some of them) background events?

A large number of attempts has been made to determine the neutrino mass, or obtain a significant upper limit for it, from the signal. Many of these attempts are bound to be controversial, because they are based on accepting part of the data, and rejecting other parts. It is intriguing to note, however, that the distribution in neutrino energy vs. time does not look entirely random. A random distribution would be expected if neutrinos were massless, or if the emission time is much longer than the time delay (7.19).

Here, we shall give a crude "back of the envelope" upper limit of the neutrino mass, followed by a description of a somewhat more stringent limit, based on relatively few and conservative assumptions. A crude estimate of the range of masses to which the signal could be sensitive is obtained simply from (7.19) by taking $\Delta t = 12$ s, i.e., the observed spread of the Kamiokande signal. Because $(12/0.026)^{1/2} = 21$ we see that the absence of ordering in energy could be interpreted as an upper limit of about 20 eV for the electron neutrino mass. An example of a refinement of this argument has been offered by Bahcall & Glashow (87) who calculate the width of the neutrino pulse at the neutron star as a function of the neutrino mass. They argue that the duration of the pulse at the source should be no more than twice the time spread at arrival. It turns out that the neutrino mass must be less than 11 eV in order to explain the short time spread of the first eight Kamiokande events. Larger neutrino masses appear to be implausible because they would require a large compression in time of the initially wide pulse. Obviously, the next observation of a supernova collapse in a neutrino detector would greatly help in sharpening this mass limit. It is worth stressing that even this somewhat tentative limit is comparable (or better) than the limits established by the experiments discussed in Chapter 2.

Neutrino lifetime. Since the $\bar{\nu}_e$ were observed after the trip of 50 kpc, they must have a lifetime that, after correcting for the relativistic time dilatation, assures their survival, i.e., $\gamma\,\tau \geqslant 1.6\times10^5$ y ($\gamma = E_\nu/m_\nu$). Thus, using the average energy of 15 MeV we obtain the lifetime limit of the $\bar{\nu}_e$

$$\tau/m_{\bar{\nu}_e} \geqslant 3.4\times10^5 \text{ s eV}^{-1} . \tag{7.20}$$

This limit is independent of the decay mode, but as any decay of any particle, requires a nonvanishing neutrino mass.

One can also obtain a significant constraint on the radiative decay $\nu_H \to \nu_l\gamma$ of neutrinos (see Section 4.2.1). This is possible since the "Solar Maximum Mission" (SMM) satellite (Chupp et al. 89) reported a limit for the gamma-flux of $0.3\gamma\,\text{cm}^{-2}\,\text{s}^{-1}$ in the energy interval 4.1—6.4 MeV during the 10 s interval encompassing the neutrino arrival signal. Following the procedure used in Eqs. (4.37) and (4.38) one can compare the expected photon flux with the experimental upper limit. Thus one obtains (von Feilitzsch & Oberauer 88)

$$\tau/m_{\bar{\nu}_e} \geqslant 8.3\times10^{14}\text{ s eV}^{-1} , \quad \tau/m_{\nu_{\mu,\tau}} >= 3.3\times10^{14}\text{ s eV}^{-1} . \tag{7.21}$$

Kolb & Turner (89) considered the radiative lifetime limits using the same SMM satellite data, but for different ranges of neutrino masses. Their limits are slightly more stringent than those in Eq. (7.21).

In this context it is worthwhile to consider the radiative lifetime limit based on the integrated gamma-ray flux produced by all supernovae throughout the past history of the universe. Obviously, the resulting flux of decay photons must not exceed the measured background gamma-ray flux. The limit was originally derived by Cowsik (77), and revised downwards by Kolb & Turner (89). Using the measured gamma-ray background flux of $\approx 3\times10^{-3}$ /(cm^2 sr s) for the gamma-ray energy of a few MeV, and keeping in mind the uncertainties connected with the estimates of the type II supernova rate, age of the universe, etc., one arrives at the limit

$$\tau_\nu/m_\nu \geqslant 10^{13}\text{ s eV}^{-1} , \tag{7.22}$$

less stringent, and considerably less direct, than the SN1987A based limit in Eq. (7.21).

An even more stringent limit on the radiative decay rate can be obtained when one makes use of the relation between the radiative decay and the photon-neutrino coupling effect on the neutrino-electron scattering.

If neutrinos have a nonvanishing transition magnetic μ or electric dipole moment d connecting the flavors ν_H and ν_l, the corresponding radiative decay rate is (Marciano & Sanda 77)

$$\Gamma_{\nu_H} = \frac{|\mu|^2 + |d|^2}{8\pi} \mu_H^3 \left[1 - (m_{\nu_l}/m_{\nu_H})^2\right]^3 . \tag{7.23}$$

Since the transition moments are restricted to be less than a few $\times 10^{-10}$ electron Bohr magnetons (see Section 5.7) for the electron neutrinos and about 10^{-9} for the muon neutrinos, one arrives at the lifetime limit (Raffelt 89)

$$\tau_{\nu_e} m_{\nu_e}^3 \geqslant 3.4\times 10^{18} (\mathrm{s\ eV}^3) \frac{|\mu - d|^2}{|\mu|^2 + |d|^2} \left[1 - (m_{\nu_x}/m_{\nu_e})^2\right]^{-3} , \tag{7.24}$$

for $H = e$ and $l = x$.

Neutrino magnetic moment. The time scale of the neutrino emission from supernovae is set by the fact that neutrinos cannot freely escape but must diffuse out since their mean free path is shorter than the size of the collapsing protoneutron star. Consider now what would happen if neutrinos were massive Dirac particles with magnetic moment. In the collapsing star the reactions $\nu_L e^- \rightarrow \nu_R e^-$ and $\nu_L p \rightarrow \nu_R p$ induced by the magnetic moment would produce large amounts of right-handed neutrinos, sterile with respect to the weak interaction. These neutrinos would be able to escape freely, cooling the star and modifying the corresponding time scale.

In addition, during the journey from the supernova to the Earth, the right-handed neutrinos (or some fraction of them) could flip back into the left-handed state in the residual magnetic field of the intergalactic space. Since such neutrinos would have larger energies, they would be more effectively detected by the Kamiokande and IMB detectors. This did not happen, and the SN1987A had the expected properties. Therefore, one could derive a limit on the corresponding magnetic moment. Barbieri & Mohapatra (88), Lattimer & Cooperstein (88), Notzold (88), and others, derived limits $\mu \leqslant 10^{-12} - 10^{-13}\ \mu_B$ based on such considerations.

Another possibility of neutrino-photon coupling involves a transition magnetic moment of the Majorana type that would connect, e.g., left-handed electron neutrinos to the right-handed muon antineutrinos. In that case the arguments above would not apply, since the final products of the neutrino-electron and neutrino-proton scattering will be again trapped.

Number of neutrino flavors and neutrino mixing. Unlike the neutrino properties above, the limits on the number of neutrino flavors and on neutrino mixing derived from the observation of the SN1987A are not competitive with the limits discussed elsewhere in this book. However, we mention them since they could be strengthened in future observations of the neutrino signal of a galactic supernova.

If there were more flavors of the left-handed, weakly interacting neutrinos, the total energy in each flavor would decrease. The observed flux of $\bar{\nu}_e$ agrees with expectations, provided one assumes equal partition between $N_\nu = 3$ flavors (i.e., the $\bar{\nu}_e$ carry 1/6 of the total energy). Based on such considerations, one can exclude N_ν larger than 6.

If neutrinos mix by vacuum oscillations (Chapter 4), and the oscillation length is less than 50 kpc, $\bar{\nu}_e$ will mix with the higher average energy $\bar{\nu}_{\mu,\tau}$ and thus their energy will increase as well. Obviously, this is not a very efficient mechanism of detecting neutrino oscillations. It involves, however, larger oscillation length (and thus smaller Δm^2) than anything else considered so far. On the other hand, if the Mikheyev-Smirnov-Wolfenstein matter enhanced neutrino oscillations are present, there will be an asymmetry between the resulting neutrino and antineutrino fluxes. This could result, for example, in the reduction of the initial neutronization ν_e pulse if the MSW effect mixes ν_e with ν_μ because the muon neutrinos have smaller cross section (by a factor $\approx 1/6$) for scattering on electrons.

References

Abachi, S. et al. (1986). *Phys. Rev. Lett.*, **56,** 1039.

Abazov, A. I. et al. (1991). In *Neutrino 90, Proc. 14th Int. Conf. Neutrino Physics and Astrophysics,* eds. J. Panman & K. Winter, *Nucl. Phys. B (Proc. Suppl.)* **19,** 84.

Abela, R. et al. (1981). *Phys. Lett.*, **B105,** 263.

Abela, R. et al. (1984). *Phys. Lett.*, **B146,** 431.

Adler S. L. et al. (1975). *Phys. Rev.*, **D11**, 3309.

Afonin, A.I. et al. (1983). *JETP Lett.*, **38,** 463.

Afonin, A.I. et al. (1985). *JETP Lett.*, **42,** 285.

Afonin, A.I. et al. (1987). *JETP Lett.*, **45,** 247.

Afonin, A.I. et al. (1988). *Sov. Phys. JETP,* **67,** 213.

Aglietta, M.A. et al. (1987). *Europhys. Lett.*, **3,** 1321.

Aglietta, M.A. et al. (1990). *Europhys. Lett.*, **8,** 611.

Ahrens, L.A. et al. (1985). *Phys. Rev.*, **D31,** 2732.

Ahrens, T. & Lang, T.P. (1971). *Phys. Rev.*, **C3,** 979.

Akhmedov, E. Kh. (1988). *Phys. Lett.*, **213B,** 64.

Alberico, W.M. et al. (1988). *Ann. of Phys.*, **187,** 79.

Albrecht, A. & Steinhardt. P.J. (1982). *Phy. Rev. Lett.*, **48,** 1220.

Albrecht, H. et al. (1985). *Phys Lett.*, **B163,** 404.

Albrecht, H. et al. (1988). *Phys Lett.*, **B202,** 149.

Alburger, D.E. & Cumming, J.B. (1985). *Phys. Rev.*, **C32,** 1358.

Alessandrello, A. et al. (1991). *Nucl. Instr. Meth.*, **B61,** 106.

Alexeyev, E.N. et al. (1987). *JETP Lett.*, **45,** 589.

Ali, A. & Dominguez, C.A. (1975). *Phys. Rev.*, **D12,** 3673.

Allen, R.C. et al. (1985). *Phys. Rev. Lett.*, **55,** 2401.

Altzitzoglou, T. et al. (1985). *Phys. Rev. Lett.*, **55,** 799.

Anderhub, H.B. et al. (1982). *Phys. Lett.*, **B114,** 76.

Angelini, C. et al. (1986). *Phys. Lett.*, **B179,** 307.

Apalikov, A. et al. (1985). *JETP Lett.*, **42,** 285.

Armenise, N. et al. (1990). *CERN-SPS* Proposal SPSC P254.

Arnett, W. A. et al. (1989). *Ann. Rev. Astro.*, **27,** 629.

Arnison, G. et al. (1983). *Phys. Lett.*, **B122,** 103.

Audi, G. et al. (1985). *Z. Physik,* **A321,** 533.

Avignone, F.T. et al. (1986). In *Proc. 2nd Int. Conf. Intersections of Nuclear and Particle Physics,* ed. D. F. Geesaman, New York: AIP.

Avignone, F.T. & Brodzinski, (1988). *Progr. Part. Nucl.*, **21,** 99.
Avignone, F.T. et al. (1991) *Phys. Lett.*, **B256,** 559.
Azuelos, G. et al. (1986). *Phys. Rev. Lett.*, **56,** 2241.
Bacino, W. et al. (1979). *Phys. Rev. Lett.*, **42,** 749.
Bahcall, J.N. et al. (1972). *Phys. Rev. Lett.*, **28,** 316.
Bahcall, J.N. & Primakoff, H. (1978). *Phys. Rev.*, **D18,** 3463.
Bahcall, J.N. et al. (1982). *Rev. Mod. Phys.*, **54,** 767.
Bahcall, J.N. & Glashow, S. L. (1987). *Nature,* **326,** 476.
Bahcall, J.N. & Ulrich, R.K. (1988). *Rev. Mod. Phys.*, **60,** 297.
Bahcall, J.N. (1989). *Neutrino Astrophysics.* Cambridge: Cambridge Univ. Press.
Bahcall, J.N. & Haxton, W.C. (1989). *Phys. Rev.*, **D40,** 931.
Bahcall, J.N. & Press, W.H. (1991). *Ap.J.*, **370,** 730.
Bahran, M. & G. R. Kalbfleisch (1991). Preprint OKHEP-91-005, Univ. Oklahoma.
Bagnaia, P. et al. (1983). *Phys Lett.*, **B129,** 130.
Bailin, D. (1982). *Weak Interactions.* Bristol: Hilger.
Baker, N.J. et al. (1981). *Phys. Rev.*, **D23,** 2499.
Baltz, A.J. & Weneser, J. (1987). *Phys. Rev.*, **D35,** 528.
Barabash, L.S. et al. (1986). In *Nuclear Beta Decays and Neutrino, Proc. Int. Symposium on Neutrino Mass,* Osaka, p. 153, eds. T. Kotani et al. Singapore: World Scientific.
Barbieri, R. & Mohapatra, R.N. (1988). *Phys. Rev. Lett.*, **61,** 27.
Bardin, D. Yu et al. (1971). *JETP Lett.*, **13,** 273.
Bardin, R.K. et al. (1970). *Nucl. Phys.*, **A158,** 337.
Barger, V. et al. (1980). *Phys. Rev. Lett.*, **45,** 692.
Barr, G. et al. (1989). *Phys. Rev.*, **D39,** 3532.
Belenkii, S. N. et al. (1983). *JETP Lett.*, **38,** 493.
Bellgardt, U. et al. (1988). *Nucl. Phys.*, **B299,** 1.
Bellotti, E. et al. (1983). In *Science Underground. AIP Conf. Proc.*, **96,** eds. M.M. Nieto et al. New York: Am. Institute of Physics.
Bellotti, E. et al. (1984). *Phys. Lett.*, **B146,** 450.
Bellotti, E. et al. (1986). In *86 Massive Neutrinos in Astrophysics and Particle Physics, Proc. 6th Moriond Workshop,* p. 165, eds. O. Fackler & J. Tran Thanh Van. Paris: Editions Frontières.
Bellotti, E. et al. (1991). *Phys. Lett.*, **B266,** 193
Berger, Ch. et al. (1990). *Phys. Lett.*, **B254,** 305.
Bergkvist, K.E. (1971). *Physica Scripta,* **4,** 23.
Bergkvist, K.E. (1972). *Nucl. Phys.*, **B39,** 317.
Bergsma, F. et al. (1983). *Phys. Rev. Lett.*, **128,** 361.
Bergsma, F. et al. (1984). *Phys. Lett.*, **B142,** 103.
Bergsma, F. (1986). In *Neutrino 86, Proc. 12th Int. Conf. Neutrino Physics and Astrophysics,* Sendai, p. 402, eds. T. Kitagaki & H. Yuta. Singapore: World Scientific.
Bernardi, G. et al. (1986). *Phys. Lett.*, **B166,** 479.
Bernabeu, J. et al. (1990). *Z. Phys.*, **C46,** 323.
Bethe, H. & Peierls, C.I. (1934). *Nature,* **133,** 532.
Bethe, H.A. & Longmire, C. (1950). *Phys. Rev.*, **77,** 35.

Bethe, H.A. (1986). *Phys. Rev. Lett.*, **56,** 1305.
Bieber, J.W. et al. (1990). *Nature,* **348,** 407.
Bilenky, S. M. & Pontecorvo, B. (1976). *Lett. Nuovo Cim.*, **17,** 569.
Bilenky, S.M. & Pontecorvo, B. (1978). *Phys. Rep.*, **41,** 225
Bilenky, S.M. et al. (1981). *Phys. Lett.*, **B94,** 495.
Bilenky, S.M. & Hosek, J. (1982). *Phys. Rep.*, **90,** 73.
Bilenky, S.M. & Petcov, S.T. (1987). *Rev. Mod. Phys.*, **59,** 671.
Bionta, R. et al. (1987). *Phys. Rev. Lett.*, **58,** 1494.
Blocker, C.A. et al. (1982) *Phys. Lett.*, **B109,** 119.
Blumenfeld, B. et al. (1989). *Phys. Rev. Lett.*, **19,** 2237, and JHU HEP 1289-1 to be published 1991.
Boehm, F. & Vogel, P. (1984). *Ann. Rev. Nucl. Part. Sci.*, **34,** 125.
Boehm, F. et al. (1991). *Nucl. Instr. Meth.*, **A300,** 395.
Boesgaard, A.N. & Steigman, G. (1985). *Ann. Rev. Astron. Astrophys.*, **23,** 319.
Bohr, A. & Mottelson, B.R. (1981). *Phys. Lett.*, **B100,** 10.
Bolton, R.D. et al. (1988). *Phys. Rev.*, **D38,** 2077.
Boris, S. et al. (1985). *Phys. Lett.*, **B159,** 217.
Boris, S. et al. (1987). *Phys. Rev. Lett.*, **58,** 2019.
Bouchez, J. et al. (1986). *Z.Phys.*, **C32,**499.
Bouchez, J. et al. (1988). *Phys. Lett.*, **B207,** 217.
Bouchez, J. (1989). In *Neutrino 88,* eds J. Schneps et al. p 28. Singapore: World Scientific.
Brodzinski, R.L. et al. (1985). *NIM,* **A239,** 207.
Brown, B.A. (1985). *Symposium on Nuclear Shell Model,* p. 42, eds. M.Vallieres & B.L.Wildenthal. Singapore: World Scientific.
Bryman, D.A. et al. (1983). *Phys. Rev. Lett.*, **50,** 1546.
Brucker, E.B. et al. (1986). *Phys. Rev.*, **D34,** 2183.
Bullock, F.W. & Devenish, R.C.E. (1983). *Rep. Prog. Phys.*, **46,** 1029.
Cabibbo, N. (1978). *Phys. Lett.*, **B72,** 333.
Cabibbo N. & Maiani L. (1982). *Phys. Lett.*, **B114,** 115.
Caldwell, D.O. (1986). In *Nuclear Beta Decays and Neutrino, Proc. Int. Symposium on Neutrino Mass,* Osaka, p. 103, eds. T. Kotani et al. Singapore: World Scientific.
Caldwell, D.O. et al. (1986). In *Neutrino 86, Proc. 12th Int. Conf. Neutrino Physics and Astrophysics,* Sendai, p. 77, eds. T. Kitagaki & H. Yuta. Singapore: World Scientific.
Caldwell, D.O. et al. (1987). *Phys. Rev. Lett.*, **59,** 417.
Caldwell, D.O. (1989). *Int. J. of Mod. Phys.*, **4,** 1851.
Caldwell, D.O. (1991). In *14th Europhysics Conf. on Nucl. Phys., Bratislava 1990,* ed. P Povinec, *J. Physics G Nucl. Phys.*, to be published.
Casper, D. et al. (1991). *Phys. Rev. Lett.*, 66, 2561.
Caurier, E. et al. (1990). *Phys. Lett.*, **B252,** 13.
Cavaignac, J.-F. et al. (1984). *Phys. Lett.*, **B148,** 387.
Chikashige, Y., Mohapatra, R. & Peccei, R. (1981). *Phys. Lett.*, **B98,** 265.
Ching, C.R. et al. (1989). *Phys. Rev.*, **C40,** 304.
Chupp, E.L. et al. (1989). *Phys. Rev. Lett.*, **62,** 505.
Civitarese, O. et al. (1987). *Phys. Lett.*, **B194,** 11.

Civitarese, O. et al. (1991). *Nucl. Phys.*, **A254**, 404.
Cleveland, B.T. et al. (1975). *Phys. Rev. Lett.*, **35**, 737.
Commins, E.D. & Bucksbaum, P.H. (1983). *Weak Interaction of Leptons and Quarks.* Cambridge: Cambridge University Press.
Cooper-Sarkar, A.M. et al. (1985). *Phys Lett.*, **B160**, 207.
Cowsik, R. (1977). *Phys. Rev. Lett.*, **39**, 784.
Danby, G. et al. (1962). *Phys. Rev. Lett.*, **9**, 36.
Dass, G.V. (1985). *Phys. Rev.*, **D32**, 1239.
Daum, M. et al. (1991). *Phys. Lett.*, **B265**, 425.
Davis, B.R. et al. (1979). *Phys. Rev.*, **C19**, 2259.
Davis, R. et al. (1968). *Phys. Rev. Lett.*, **21**, 1205.
Davis, R. et al. (1983). In *Science Underground, AIP Conf. Proc.* **96**, eds. M.M. Nieto et al. New York: Am. Institute of Physics.
Davis, R., Mann A.K. & Wolfenstein L. (1989). *Ann. Rev. Nucl. Part. Sci.*, **39**, 467.
De Leener-Rosier, N. et al. (1986). *Phys. Lett.*, **B177**, 228.
De Rujula, A. (1981) *Nucl. Phys.*, **B188**, 414.
Deutsch, J.P. et al. (1983). *Phys. Rev.*, **D27**, 1644.
Deutsch, J. et al. (1990). *Nucl. Phys.*, **A218**, 149.
Diemoz, M. et al. (1986). *Phys. Rep.*, **130**, 293.
Dixit, M.S. et al. (1983). *Phys. Rev.*, **D27**, 2216.
Doi, M. et al. (1981). *Phys. Lett.*, **102B**, 323.
Doi, M., et al. (1985). *Prog. of Theor. Phys. Suppl.*, **83.**
Donnelly, T.W. & Peccei, R.D. (1979). *Phys. Rep.*, **50**, 1.
Donnelly, T.W. (1981). in *Proc. of the Los Alamos Neutrino Workshop,* eds. F. Boehm & G.J. Stephenson Jr., LA-9358-C, Los Alamos.
Donoghue, J.F. et al. (1986). *Phys. Rep.*, **131**, 319.
Dorenbosch, J. et al. (1986). *Phys. Lett.*, **B166**, 473.
Durkin, L.S. et al. (1988). *Phys. Rev. Lett.*, **61**, 1811.
Drexlin, G. et al. (1991). Preprint KfK 4893, June 1991.
Dydak, F. et al. (1984). *Phys. Lett.*, **B134**, 281.
Dydak, F. (1990). *25th Int. Conf. on High Energy Physics,* CERN-PPE/91-14
Ejiri, H. et al. (1986). In *Nuclear Beta Decays and Neutrino, Proc. Int. Symposium on Neutrino Mass,* Osaka, p. 127, eds. T. Kotani et al. Singapore: World Scientific.
Ejiri, H. et al. (1987). *J. Phys. G. Nucl. Phys.*, **13**, 839.
Ejiri, H. (1991). In *Nuclear Physics in the 1990's,* eds. D.H. Feng et al., *Nucl. Phys.*, **A522**, 305c.
Ejiri, H. et al. (1991). *Phys. Lett.*, **B258**, 17.
Elliott, S.R. et al. (1986). *Phys. Rev. Lett.*, **56**, 2582.
Elliott, S.R. et al. (1987). *Phys. Rev. Lett.*, **59**, 2021.
Elliott, S.R. et al. (1991). In *14th Europhysics Conf. on Nucl. Phys.,Bratislava 1990,* ed. P. Povinec, *J. Physics G Nucl. Phys.*, to be published.
Elliott, S.R. (1991). In *Particles and Fields 91, Proc. Int. Conf. Vancouver,* Eds. D. Axen et al. Singapore: World Scientific.
Engel, J., Vogel, P. & Zirnbauer, M.R. (1988). *Phys. Rev.*, **C37**, 731.

Engel, J. et al. (1989). *Phys. Lett.*, **B225,** 5.
Engel, J., Pittel, S. & Vogel, P. (1991). *Phys. Rev. Lett.*, **67,** 426.
Enquist, K. et al. (1983). *Nucl. Phys.*, **B226,** 121.
Fackler, O. et al. (1985). *Phys. Rev. Lett.*, **55,** 1388.
Fayans, S.A. (1985). *Sov. J. Nucl. Phys.*, **42,** 540.
Feilitzsch, F. von et al. (1982). *Phys. Lett.*, **B118,** 162.
Feilitzsch, F. von & Oberauer, L. (1988). *Phys. Lett.*, **200,** 580.
Feinberg, G. & Weinberg, S. (1961). *Phys. Rev. Lett.*, **6,** 381.
Fermi, E. (1934). *Z. Physik,* **88,** 161.
Filippone, B.W. & Vogel, P. (1990). *Phys. Lett.*, **B246,** 546.
Fisher, P. (1986). In *86 Massive Neutrinos in Astrophysics and Particle Physics, Proc. 6th Moriond Workshop,* p. 615, eds. O. Fackler & J. Tran Thanh Van. Paris: Editions Frontières.
Fisher, P. et al. (1987). *Phys. Lett.*, **B218,** 257.
Frampton, P.H. & Vogel P. (1982). *Phys. Rep.*, **82,** 339.
Frank, J.S. et al. (1981). *Phys. Rev.*, **D24,** 2001.
Fritschi, M. et al. (1986). *Phys. Lett.*, **B173,** 485.
Fritschi, M. et al. (1991). In *Neutrino 90,* eds. J. Panman & K. Winter,Amsterdam: North Holland. *Nucl. Phys.* **B19**, 205.
Furry, W.H. (1939). *Phys. Rev.*, **56,** 1148.
Gaarde, C. (1983). *Nucl. Phys.*, **A396,** 127c.
Gabathuler, K. et al. (1984). *Phys Lett.*, **B138,** 449.
Gabioud, B. et al. (1979). *Phys. Rev. Lett.*, **42,** 1508.
Gabioud, B. et al. (1984). *Nucl. Phys.*, **A420,** 496.
Gaisser, T.K. & O'Connell, J.S. (1986). *Phys. Rev.*, **D34,** 822.
Gall, P.D. (1984). In *Neutrino 84, 11th Int. Conf. Neutrino Physics and Astrophysics,* p. 193, eds. E.K. Kleinknecht & E.A. Paschos. Singapore: World Scientific.
Garcia, A. (1991). *BAPS,* **36,** 2149.
Gell-Mann, M. et al. (1979). In *Supergravity.* Amsterdam: North Holland.
Gelmini, G. & Rondacelli, M. (1981). *Phys. Lett.*, **99B,** 411.
Georgi, H.M. & Glashow, S.L. (1974). *Phys. Rev. Lett.*, **32,** 438.
Georgi, H.M. et al. (1981). *Nucl. Phys.*, **B193,** 297.
Georgi H. (1984). *Weak Interactions and Modern Particle Theory.* Menlo Park: Benjamin.
Glashow, S.L. (1961). *Nucl. Phys.*, **22,** 579.
Glashow, S.L. (1991). *Phys. Lett.*, **B256,** 255.
Goeppert-Mayer, M. (1935). *Phys. Rev.*, **48,** 512.
Goldhaber, A.S. (1963). *Phys Rev.*, **140,** 761.
Grenacs, L. (1985). *Ann. Rev. Nucl. Part. Sci.*, **35,** 455.
Gribov, V. & Pontecorvo, B. (1969). *Phys. Lett.*, **B28,** 493.
Gronau, M. et al. (1984). *Phys. Rev.*, **D29,** 2539.
Grotz, K. & Klapdor, H.V. (1985). *Phys. Lett.*, **B153,** 1.
Grotz K. & Klapdor H. V. (1990). *The Weak Interactions in Nuclear, Particle and Astrophysics.* Bristol: IOP Publishing.
Guth, A.H. (1981). *Phys. Rev.*, **D23,** 347.
Hahn, A.A. et al. (1989). *Phys. Lett.*, **B218,** 365.
Halprin, A. et al. (1976). *Phys. Rev.*, **D13,** 2567.

Hampel, W. et al. (1983). In *Science Underground, AIP Conf. Proc.*, **96,** eds. M.M. Nieto et al. New York: Am. Institute of Physics.
Hampel, W. (1985). In *Solar Neutrinos and Neutrino Astronomy, AIP Conf. Proc.*, **126,** eds. M.L. Cherry et al. New York: Am. Institute of Physics.
Hasert, F.T. et al. (1973). *Phys. Lett.*, **B46,** 138.
Haxton, W.C. et al. (1981). *Phys. Rev. Lett.*, **47,** 153.
Haxton, W.C. et al. (1982). *Phys. Rev.*, **D26,** 1805.
Haxton, W.C. et al. (1983). *Phys. Rev.*, **C28,** 467.
Haxton, W.C. & Stephenson, G.J. Jr. (1984). *Prog. in Part. and Nucl. Phys.*, **12,** 409.
Haxton, W.C. (1987). *Phys.Rev.*, **D35,** 2352.
Hayano, R.S. et al. (1982). *Phys. Rev. Lett.*, **49,** 1305.
Heil, A. (1983). *Nucl. Phys.*, **B222,** 338.
Hennecke, E.W. et al. (1975). *Phys. Rev.*, **C11,** 1378.
Hetherington, D. W. et al. (1987). *Phys. Rev.*, **C36,** 1504.
Hillebrandt, W. & Hoflich, P. (1989). *Rep. Prog. Phys.*, **52,** 1421.
Hime, A. & J. J. Simpson (1989). *Phys. Rev.*, **D39,** 1837.
Hime, A. & N. A. Jelley (1991). *Phys. Lett.*, **B257,** 441.
Hirata, K. et al. (1987). *Phys. Rev. Lett.*, **58,** 1490.
Hirata, K.S. et al. (1990). *Phys. Rev. Lett.*, **65,** 1297.
Hirata, K.S. et al. (1991). *Phys. Rev. Lett.*, **66,** 9.
Hirsh, J. & Krmpotic, F. (1990). *Phys. Lett.*, **B246,** 5.
Holstein B. R. (1989). *Weak Interactions in Nuclei.* Princeton: Princeton University Press.
Huffman, A.H. (1970). *Phys. Rev.*, **C2,** 742.
Hulth, P. O. (1984). In *Neutrino 84, 11th Int. Conf. Neutrino Physics and Astrophysics,* p. 512, eds. E.K. Kleinknecht & E.A. Paschos. Singapore: World Scientific.
Hurst, G.S. et al. (1979). *Revs. Mod. Phys.*, **51,** 767.
Jeckelmann, B. et al. (1986). *Nucl. Phys.*, **A457** 709.
Jonson, B. et al. (1983). *Nucl. Phys.*, **A396,** 479.
Kainulainen, K. (1990). *Phys. Lett.*, **B244,** 191.
Kajita, T. (1991). In *Proceedings 25th ICHEP,* Singapore.
Kalyniak, P. & Ng, J.N. (1982). *Phys. Rev.*, **D25,** 1305.
Kaplan, I.G. et al. (1982). *Phys. Lett.*, **B112,** 417.
Kaplan, I.G. et al. (1983). *Sov. Phys. JETP,* **57,** 483.
Kaplan, I.G. et al. (1985). *Phys. Lett.*, **B161,** 389.
Kaplan, I.G. & Smelov, G.V. (1986). In *Nuclear Beta Decay and Neutrinos,* eds. T. Kotani et al. Singapore: World Scientific.
Kawakami, N. et al. (1987). *Phys. Lett.*, **B187,** 111.
Kawakami, N. et al. (1991). *Phys. Lett.*, **B256,** 105.
Kayser, B. (1981). *Phys. Rev.*, **D24,** 110.
Kayser, B. (1985). *Comments Nucl. Part. Phys.*, **14,** 69.
Kayser, B. (1987). In *New and Exotic Phenomena,* eds. O. Fackler and J. Tran Thanh Van, p.349. Gif-sur-Yvette: Editions Frontieres.
Kayser B., Gibrat-Debu F. & Perrier F. (1989) *The Physics of Massive Neutrinos.* Singapore: World Scientific.

Kelly, F.T. & Uberall, H. (1966). *Phys. Rev. Lett.*, **16**, 145.
Kinoshita, T. (1959). *Phys. Rev. Lett.*, **2**, 477.
Kirsten, T. (1983). In *Science Underground. AIP Conf. Proc.*, **96**, p. 396, eds. M.M. Nieto et al. New York: Am. Institute of Physics.
Kirsten, T. et al. (1983). *Phys. Rev. Lett.*, **50**, 474; and *Z. Physik*, **C16**, 189.
Kirsten, T. (1986). In *86 Massive Neutrinos in Astrophysics and Particle Physics, Proc. 6th Moriond Workshop*, p. 119, eds. O. Fackler & J. Tran Thanh Van. Paris: Editions Frontières.
Kirsten, T. et al. (1986). In *Nuclear Beta Decay and Neutrinos, Proc. Int. Symposium on Neutrino Mass*, Osaka, p. 81, eds. T. Kotani et al. Singapore: World Scientific.
Kirsten, T. (1991). In *Neutrino 90, Proc. 14th Int. Conf. Neutrino Physics and Astrophysics*, Eds. J. Panman & K. Winter, *Nuclear Physics B (Proc.Suppl.)* **19**, 77.
Klapdor, H.V. et al. (1982a). *Phys. Rev. Lett.*, **48**, 127.
Klapdor, H.V. et al. (1982b). *Phys. Lett.*, **B112**, 22.
Klapdor, H.V. & Grotz, K. (1984). *Phys. Lett.*, **B142**, 323.
Klapdor, H.V. & Grotz, K. (1985), *Phys. Lett.*, **B157**, 242.
Klapdor, H.V. et al. (1991). In *14th Europhysics Conf. on Nucl. Phys., Bratislava 1990*, ed. P. Povinec, *J. Physics G Nucl. Phys.*, to be published.
Klimenko, A.A. et al. (1984). In *Neutrino 84, 11th Int. Conf. Neutrino Physics and Astrophysics*, p. 161, eds. E.K. Kleinknecht & E.A. Paschos. Singapore: World Scientific.
Klinkhamer, F.R. (1986). In *86 Massive Neutrinos in Astrophysics and Particle Physics, Proc. 6th Moriond Workshop*, p. 33, eds. O. Fackler & J. Tran Thanh Van. Paris: Editions Frontières.
Kobzarev, I. Yu et al. (1981). *Sov. J. Nucl. Phys.*, **32**, 823.
Kolb, E.W. & Turner, M.S. (1989). *Phys. Rev. Lett.*, **62**, 509.
Kolb, E.W. & Turner, M.S. (1990). *The Early Universe* Redwood City: Addison-Wesley.
Konopinski, E.J. & Mahmoud, H.M. (1953). *Phys. Rev.*, **92**, 1045.
Konopinski, E.J. (1966). *The Theory of Beta Radioactivity.* Oxford: Clarendon Press.
Kündig, W. et al. (1986). In *Proc. Int. Nuc. Phys. Conf. : Harrogate*, eds. J.L. Durell et al. Bristol: Inst. of Physics.
Kwon, H. et al. (1981). *Phys. Rev.*, **D24**, 1097.
Kuo, T.K. & Pantaleone, J. (1987). *Phys. Rev.*, **D35**, 3437.
Langacker, P. et al. (1983). *Phys. Rev.*, **D27**, 1228.
Langacker, P. (1986). In *86 Massive Neutrinos in Astrophysics and Particle Physics, Proc. 6th Moriond Workshop*. p. 101, eds. O. Fackler & J. Tran Thanh Van. Paris: Editions Frontières.
Langacker P. (1988). In *Neutrino Physics*, eds. Klapdor H. V. & Povh B. Berlin: Springer-Verlag.
Langacker P. (1991). In *Proceedings of TASI-90*, eds. Cvetic M. & Langacker P.. Singapore: World Scientific
Lattimer, J.M. & Cooperstein, J. (1988). *Phys. Rev. Lett.*, **61**, 23.

Lee, B.W. & Weinberg, S. (1977). *Phys. Rev. Lett.*, **39,** 165.
Lee, H. & Koh, Y. (1990). *CHUPHY-1989-T1,* Preprint, Chungnam National University. Lee, J. T. et al. (1991). *Nucl. Phys.*, **A529,** 29.
Lee, T.D. & Yang, C.N. (1956). *Phys. Rev.*, **104,** 254.
Levine, C.A. et al. (1950). *Phys. Rev.*, **27,** 296.
Lewis, R.R., (1980). *Phys. Rev.*, **D21,** 663.
Li, L.F. & Wilczek, F. (1982). *Phys. Rev.*, **D25,** 143.
Lim, C.S. & Marciano, W. (1988). *Phys. Rev.*, **D37,** 1368.
Linde, A.D. (1982). *Phys. Lett.*, **B108,** 389.
Lippmaa, E.R. et al. (1985). *Phys. Rev. Lett.*, **54,** 285.
Llewellyn Smith, C.B. (1972). *Phys. Rep.*, **3,** 261.
Lu, X.Q. et al. (1989). *LAMPF Research Proposal, LA-UR-89-3764.*
Lubimov, V. et al. (1980). *Phys. Lett.*, **B94,** 266.
Lubimov, V. et al. (1981). *Sov. Phys. JETP*, **54,** 616.
Lubimov, V. (1986). In *86 Massive Neutrinos in Astrophysics and Particle Physics, Proc. 6th Moriond Workshop.* p. 441, eds. O. Fackler & J. Tran Thanh Van. Paris: Editions Frontières.
Majorana, E. (1937). *Nuovo Cimento,* **14,** 171.
Maki, Z. et al. (1962). *Prog. Theor. Phys.*, **28,** 870.
Mann, A. et al. (1991). Proposal in *Workshop on Accelerator Based Low Energy Physics,* LAMPF, Los Alamos.
Manuel, O.K. (1986). In *Nuclear Beta Decays and Neutrino, Proc. Int. Symposium on Neutrino Mass,* Osaka, p. 71, eds. T. Kotani et al. Singapore: World Scientific.
Marciano, W. & Sanda, A. I. (1977). *Phys. Lett.*, **B67,** 303.
Markey, J. & Boehm, F. (1985). *Phys. Rev.*, **C32,** 2215.
Matteuzzi, C. et al. (1985). *Phys. Rev.*, **D32,** 800.
Mikheyev, S.P. & Smirnov, A. Yu. (1985). *Sov. J. Nucl. Phys.*, **42,** 1441.
Mikheyev, S.P. & Smirnov, A. Yu. (1986). *Nuovo Cim.*, **9C,** 17.
Mikheyev, S.P. & Smirnov, A. Yu. (1987). *Sov. Phys. Usp.*, **30,** 759.
Mills, G.B. et al. (1985). *Phys. Rev. Lett.*, **54,** 624.
Minehart, R.C. et al. (1984). *Phys. Rev. Lett.*, **52,** 804.
Missimer, J. et al. (1981). *Nucl. Phys.*, **B188,** 29.
Mitchell, L.W. & N. Winograd, (1986). *Bull. Am. Phys. Soc.*, **31,** 1219.
Moe, M. et al. (1983). In *The Time Projection Chamber, AIP Conf. Proc.*, **108,** p. 37, ed. J. A. MacDonald. New York: Am. Institute of Physics.
Moe, M. et al. (1985). In *Neutrino Mass and Low Energy Weak Interactions, Proc. Telemark Conf.*, p. 37, eds. V. Barger & D. Cline. Singapore: World Scientific.
Moe, M. (1989). In *13th Int. Conf. Neutrino Physics and Astrophysics, Boston 1988,* p. 54, eds. J.Schneps et al. Singapore: World Scientific.
Moe, M. (1991). In *Neutrino 90,* eds. J. Panman & K. Winter, Amsterdam: North Holland, *Nucl. Phys. B (Proc. Suppl.)* **19,** 158.
Morales, a. (1989). In *TAUP 89,* eds. A. Bottino & P. Monacelli, p.97, Gif-sur-Yvette: Editions Frontieres.
Morales, A. (1991). In *13th Europhysics Conf. Bratislava 1990,* ed. P. Povinec, *J. Physics G Nucl.Phys.* to be published.

Moscoso, L. (1991). In *Neutrino 90,* eds. J. Panman & K. Winter, Amsterdam: North Holland, *Nucl. Phys. B (Proc. Suppl.)* **19,** 147.
Murdin, P. (1990). *End in Fire,* Cambridge: Cambridge Univ. Press.
Muto, K. & Klapdor, H.V. (1988). *Phys. Lett.,* **B201,** 420.
Muto, K. et al. (1989). *Z. Phys.,* **A334,** 187.
Nemethy, P. et al. (1981). *Phys. Rev.,* **D23,** 262.
Nieves, J.F. (1983). *Phys. Rev.,* **D28,** 1664.
Notzold, D. (1988). *Phys. Rev.,* **D38,** 1658.
Oberauer, L. et al. (1987). *Phys. Lett.* **B198,** 113.
O'Connell, T.S. (1972). In *Few Particle Problems,* eds. I. Slaus et al. Amsterdam: North Holland.
Ogawa, K. & Horie, H. (1989). In *Nuclear Weak Process and Nuclear Structure,* p.308, eds. M. Morita et al., Singapore: World Scientific.
Ohi, T. et al. (1985). *Phys. Lett.,* **B160.,** 322.
Ohi, T. et al. (1986). In *Neutrino 86, Proc. Int. Conf. Neutrino Physics and Astrophysics, Sendai, 1986,* p. 69, eds. T Kitagaki & H. Yuta, Singapore: World Scientific.
Okun, L.B. (1982). *Leptons and Quarks.* Amsterdam: North Holland.
Pal, P.B. & Wolfenstein, L. (1982). *Phys. Rev.,* **D25,** 766.
Parke, S.J. & Walker, T.P. (1986). *Phys. Rev. Lett.,* **57,** 2322.
Parks, J.E. et al. (1982). *Thin Solid Films,* **108,** 69.
Pasierb, E. et al. (1979). *Phys. Rev. Lett.,* **43,** 96.
Pauli, W. (1930). *Letter to the Physical Society of Tubingen,* unpublished; the letter is reproduced in Brown, L.M. (1978), *Physics Today,* **31,** No. 9, 23.
Perl, M.L. et al. (1975). *Phys. Rev. Lett.,* **35,** 1489.
Perl, M.L. (1980). *Ann. Rev. Nucl. Part. Sci.,* **30,** 299.
Petcov, S. T. (1977). *Sov. J. Nucl. Phys.,* **25,** 340; *ibid.* **25,** 698.
Petcov, S. T. (1988). *Phys. Lett.,* **200B,** 373.
Petschek, A. G., ed. (1990). *Supernovae,* New York: Springer Verlag.
Pietschmann, H. (1983). *Weak Interactions Formulae, Results, and Derivations.* Vienna: Springer.
Pizzochero, P. (1987). *Phys. Rev.,* **D36,** 2293.
Pontecorvo, B. (1946). *Chalk River Lab. Report # PD-205,* (unpublished).
Pontecorvo, B. (1958) *Sov. Phys. JETP* **6,** 429.
Pontecorvo, B. (1967) *Zh. Eksp. Theor. Fiz.,* **53,** 1717.
Primakoff, H. & Rosen, S.P. (1959). *Rep. Prog. Phys.,* **22,** 121.
Racah, G. (1937). *Nuovo Cimento,* **14,** 327.
Raffelt, G. (1985). *Phys. Rev.,* **D31,** 3002.
Raffelt, G. (1989). *Phys. Rev.,* **D39,** 2066.
Reeder, D.D. (1984). In *Neutrino 84, 11th Int. Conf. Neutrino Physics and Astrophysics,* p. 504, eds. E.K. Kleinknecht & E.A. Paschos. Singapore: World Scientific.
Reines, F. & Cowan, C.L. (1953). *Phys. Rev.,* **90,** 492.
Reines, F. et al. (1974). *Phys. Rev. Lett.,* **32,** 180.
Reines, F. et al. (1980). *Phys. Rev. Lett.,* **45,** 1307.
Reines, F. (1983). *Nucl. Phys.,* **A396,** 469c.

Reusser, D. et al. (1991a). *Phys. Lett.*, **B255,** 143.

Reusser D. et al. (1991b). *Phys. Rev.*, to be published.

Review of Particle Properties (1990), eds. Part. Data Group, *Phys. Lett.*, **B239,** 1.

Robertson, R.G.H. et al. (1986). In *Neutrino 86, Proc. 12th Int. Conf. Neutrino Physics and Astrophysics,* Sendai, p. 49, eds. T. Kitagaki & H. Yuta. Singapore: World Scientific.

Robertson R.G.H. & Knapp D.A. (1988). *Ann. Rev. Nucl. Part. Sci.*, **38,** 185.

Robertson, R.G.H. et al. (1991). *Phys. Rev. Lett.*, **67,** 957.

Rosen, S.P. (1982). *Phys. Rev. Lett.*, **48,** 842.

Rosen, S.P. (1983). *Lecture Notes on Mass Matrices,* LASL preprint, unpublished.

Rosen, S.P. & Gelb, J.M. (1986). *Phys. Rev.*, **D34,** 969.

Rowley, J.K. et al. (1985). In *Solar Neutrinos* and *Neutrino Astronomy, AIP Conf. Proc.* **126,** eds. M.L. Cherry et al. New York: Am. Institute of Physics.

Sakurai, J.J. (1964). *Invariance Principles and Elementary Particles,* Princeton: Princeton University Press.

Salam, A. & Ward, J. C. (1964). *Phys. Lett.*, **13,** 168.

Salam, A. (1968). In *Nobel Symposium,* **No. 8,** ed. N. Swartholm. Stockholm: Almquist & Wiksell.

Schramm, D.N. & Wagoner, R.V. (1977). *Ann. Rev. Nucl. Part. Sci.*, **27,** 37.

Schramm, D.N. & Truran, J.W. (1990). *Physics Rep.* **189,** 89.

Schreckenbach, K. et al. (1981). *Phys. Lett.*, **B99,** 251.

Schreckenbach, K. et al. (1983). *Phys. Lett.*, **B129,** 265.

Schreckenbach, K. (1984). *Technical Report 84SC26T,* Institute Laue-Langevin, Grenoble.

Schreckenbach, K. et al. (1985). *Phys. Lett.*, **B160,** 325.

Shaevitz, M.H. (1983). In *Proc. Int. Conf. Leptons Photons,* p. 132, eds. D.G. Cassel & D.L. Kreinich. Ithaca: Cornell University Press.

Shoji, Y. et al. (1986). In *Nuclear Beta Decays and Neutrino, Proc. Int. Symposium on Neutrino Mass,* Osaka, p. 294, eds. T. Kotani et al. Singapore: World Scientific.

Shrock, R.E. (1980). *Phys. Lett.*, **B96,** 159.

Shrock, R.E. (1981). *Phys. Rev.*, **D24,** 1232, 1275.

Shrock, R.E. (1982). *Nucl. Phys.*, **B206,** 359.

Simpson, J.J. (1981). *Phys. Rev.*, **D24,** 2971.

Simpson, J.J. (1983). *Phys. Rev.*, **D30,** 1110.

Simpson, J.J. et al. (1984). *Phys. Rev. Lett.*, **53,** 141.

Simpson, J.J. (1985). *Phys. Rev. Lett.*, **54,** 1891.

Simpson, J.J. & A. Hime (1989). *Phys. Rev.*, **D39,** 1825.

Skouras, L.D. & Vergados, J.D. (1983). *Phys. Rev.*, **C28,** 2122.

Smith, L.G. & Wapstra, A.H. (1975). *Phys. Rev.*, **C11,** 1392.

Smith, L.G. et al. (1981). *Phys. Lett.*, **B102,** 114.

Sobel, H. (1986). In *Neutrino 86, Proc. 12th Int. Conf. Neutrino Physics and Astrophysics,* Sendai, p. 148, eds. T. Kitagaki & H. Yuta. Singapore: World Scientific.

Staggs, S. T. et al. (1989). *Phys. Rev.*, **C39,** 1503.

Staudt, A. et al. (1990). *Europhysics Lett.*, **13,** 31.
Stech, B. (1980). *Unification of the Fundamental Particle Interactions,* p. 23, eds. S. Ferrara et al.. New York: Plenum Press.
Steigman, G. (1979). *Ann. Rev. Nucl. Part. Sci.*, **29,** 313.
Steigman, G. (1984). In *Inner Space/Outer Space, Proc. Workshop,* to be published.
Steigman, G. (1986). In *86 Massive Neutrinos in Astrophysics and Particle Physics, Proc. 6th Moriond Workshop,* p. 51, eds. O. Fackler & J. Tran Thanh Van. Paris: Editions Frontières.
Steigman, G. (1990). In *Fundamental Symmetries in Nuclei & Particles,* p. 208, eds. H. Henrikson & P. Vogel. Singapore: World Scientific.
Stockdale, I.E. et al. (1985). *Phys. Rev. Lett.*, **52,** 1384.
Stodolsky, L. (1975). *Phys. Rev. Lett.*, **34,** 110.
Suhonen, J., et al. (1990). *Phys. Lett.*, **B237,** 8.
Suhonen, J., et al. (1991). *Nucl. Phys.*, submitted for publication.
Sur, B. et al. (1991). *Phys. Rev. Lett.*, */fB66,* 2444.
Taylor, G. N. et al. (1983). *Phys. Rev.*, **D28,** 2705.
Taylor, J.C. (1978). *Gauge Theories of Weak Interactions,* Cambridge: Cambridge University Press.
Tomoda, T. & Faessler, A. (1987). *Phys. Lett.*, **B199,** 475.
Tomoda, T. (1991). *Rep. Progr. Phys.*, **54,** 53.
Toussaint, D. & Wilczek, F. (1981). *Nature,* **289,** 777.
Tretyakov, E.F. et al. (1975). *Bull. USSR Acad. Sci.*, *Phys. Ser.*, **39,** 102.
Trimble, W. (1988). *Rev. Mod. Phys.*, **60,** 859.
Tsuboi, T. et al. (1984). *Phys. Lett.*, **B143,** 293.
Turck-Chieze, S. et al. (1988). *Ap.J.*, **335,** 415.
Turkevich, A. et al. (1989). In Los Alamos Workshop on *The Breaking of Fundamental Symmetries in Nuclei*, unpublished.
Turkevich, A. et al. (1991). *Phys. Rev. Lett.*, submitted for publication.
Turner, M. K. (1981). In *Neutrino 81,* ed. R. Cence et al. Honolulu: University of Hawaii.
Ushida, N. et al. (1986). *Phys. Rev. Lett.*, **57,** 2897.
Vasenko, A.A. et al. (1990). *Mod. Phys. Lett.*, **A5,** 1299.
Vasenko, A.A. et al. (1991). *Neutrino 90, Proc. 14th Int. Conf. Neutrino Physics and Astrophysics,* eds. J. Panman & K. Winter, *Nucl. Phys. B (Proc. Suppl.),* **19,** contributed paper quoted by Moe (91).
Vidyakin, G.S. et al. (1990a). *Sov. Phys. JETP,* **66,** 243, and **71,** 424.
Vidyakin, G.S. et al. (1990b). *JETP Lett.*, **51,** 245.
Vergados, J.D. (1981). *Phys. Rev.*, **C24,** 640.
Vergados, J.D. (1986). *Phys. Rep.*, **133,** 1.
Vogel, P. et al. (1981). *Phys. Rev.*, **C24,** 1543.
Vogel, P. (1984a). *Phys. Rev.*, **D30,** 1505.
Vogel, P. (1984b). *Phys. Rev.*, **D29,** 1918.
Vogel, P. et al. (1984). *Phys. Lett.*, **139B** 227.
Vogel P. & Fisher P. (1985). *Phys. Rev.*, **C32,** 1362.
Vogel P. & Zirnbauer, M.R. (1986). *Phys. Rev. Lett.*, **57,** 3148.
Voloshin, M. B. et al. (1986). *Sov. Phys. JETP,* **64,** 446; *Sov. J. Nucl. Phys.*, **44,** 440.
Vuilleumier, J.-L. et al. (1982) *Phys. Lett.*, **B114,** 298.

Vuilleumier, J.-L. (1986). *Rep. Prog. Phys.,* **40,** 1293.
Walecka, J.D. (1975). In *Muon Physics II,* eds. V.W. Hughes & C.S. Wu, New York: Academic Press.
Walter, H.K. (1984). *Nucl. Phys.,* **A434,** 409c.
Wang, K.C. et al. (1984). *Neutrino 84, 11th Int. Conf. Neutrino Physics and Astrophysics,* p. 177, eds. E.K. Kleinknecht & E.A. Paschos. Singapore: World Scientific.
Wark, D. & F. Boehm (1986). In *Nuclear Beta Decays and Neutrino, Proc. Osaka Symposium 1986,* p. 391, eds. T. Kotani et al. Singapore: World Scientific.
Weinberg, S. (1967). *Phys. Rev. Lett.,* **19,** 1264.
Weinberg, S. (1972). *Gravitation and Cosmology.* New York: Wiley.
Wilkerson, J. F. et al. (1987). *Phys. Rev. Lett.,* **58,** 2023.
Wilkerson, J. F. et al. (1991). In *Neutrino 90, Proc. 14th Int. Conf. Neutrino Physics and Astrophysics,* eds. J. Panman & K. Winter, *Nucl. Phys. B (Proc. Suppl.)* **19,** 215.
Wilkinson, D. H. (1982). *Nucl. Phys.,* **A377,** 474.
Willis, S.E. et al. (1980). *Phys. Rev. Lett.,* **44,** 522.
Witten, E. (1980). *Phys. Lett.,* **B91,** 81.
Wolfenstein, L. (1978). *Phys. Rev.,* **D17,** 2369.
Wolfenstein, L. (1981). *Phys. Lett.,* **B107,** 77.
Wong, H.T. et al. (1991). *Phys. Rev. Lett.,* **67,** 1218.
Wotschack, J. (1984). In *Neutrino 84, 11th Int. Conf. Neutrino Physics and Astrophysics,* p. 117, eds. E.K. Kleinknecht & E.A. Paschos. Singapore: World Scientific.
Yamazaki, T. et al. (1984). *Neutrino 84, 11th Int. Conf. Neutrino Physics and Astrophysics,* p. 183, eds. E.K. Kleinknecht & E.A. Paschos. Singapore: World Scientific.
Yanagida, T. (1979). *Workshop on Unified Theory and Bayon Number of Universe,* unpublished.
Yang, J. et al. (1984). *Astrophys. J.,* **281,** 493.
Yasumi, S. et al. (1986). *86 Massive Neutrinos in Astrophysics and Particle Physics, Proc. 6th Moriond Workshop,* p. 579, eds. O. Fackler & J. Tran Thanh Van. Paris: Editions Frontières.
Zacek, G. et al. (1986). *Phys Rev.,* **D34,** 2621.
Zacek, V. et al. (1985). *Phys Lett.,* **B164,** 193.
Zamick L. & Auerbach, N. (1982). *Phys. Rev.,* **C26,** 2185.
Zdesenko, Yu.G. (1980). *Sov. J. Part. Nucl.,* **11,** 542.
Zdesenko, Yu.G. et al. (1982). In *Neutrino 82, Proc. Int. Conf.,* p. 209, ed. A. Frenkel. Budapest: Central Research Institute for Physics of the Hungarian Academy of Sciences.
Zel'dovich Ya. B. (1952). *DAN SSSR,* **86,** 505.
Zel'dovich, Ya. B. & Khlopov, M. Yu. (1981). *Sov. Phys. Uspekhi,* **24,** 755.
Zhao, L.et al. (1990). *Phys. Rev.,* **C42,** 1120.
Zlimen, I. et al. (1991) *Phys. Rev. Lett.,* **67,** 560.

Index

Printed in the United States
By Bookmasters